Chemistry of Taste

ACS SYMPOSIUM SERIES **825**

Chemistry of Taste

Mechanisms, Behaviors, and Mimics

Peter Given, Editor
Pepsi-Cola Company

Dulce Paredes, Editor
Kraft Foods

American Chemical Society, Washington, DC

Library of Congress Cataloging-in-Publication Data

American Chemical Society. Meeting (219th : 2000 : San Francisco, Calif.)

Chemistry of Taste : mechanisms, behavior, and Mimics / Peter Given, Dulce Paredes, editor[s].

p. cm.—(ACS symposium series ; 825)

"Developed from a symposium sponsored by the Division of Agricultural and Food Chemistry at the 219th National Meeting of the American Chemical Society, San Francisco, California, March 26–30, 2000."

Includes bibliographical references and index.

ISBN 0–8412–3734–4

1. Taste—Congresses. 2. Chemoreceptors—Congresses.

I. Given, Peter, 1952- II. Paredes, Dulce, 1958- III. American Chemical Society. Division of Agricultural and Food Chemistry. IV. Title. V. Series.

QP456 A447 2000
612.8′7—dc21 2002018336

The paper used in this publication meets the minimum requirements of American National Standard for Information Sciences—Permanence of Paper for Printed Library Materials, ANSI Z39.48–1984.

Distributed by Oxford University Press

PRINTED IN THE UNITED STATES OF AMERICA

Foreword

The ACS Symposium Series was first published in 1974 to provide a mechanism for publishing symposia quickly in book form. The purpose of the series is to publish timely, comprehensive books developed from ACS sponsored symposia based on current scientific research. Occasion-ally, books are developed from symposia sponsored by other organizations when the topic is of keen interest to the chemistry audience.

Before agreeing to publish a book, the proposed table of contents is reviewed for appropriate and comprehensive coverage and for interest to the audience. Some papers may be excluded to better focus the book; others may be added to provide comprehensiveness. When appropriate, overview or introductory chapters are added. Drafts of chapters are peer-reviewed prior to final acceptance or rejection, and manuscripts are prepared in camera-ready format.

As a rule, only original research papers and original review papers are included in the volumes. Verbatim reproductions of previously published papers are not accepted.

ACS Books Department

Contents

Preface xi

Mechanisms of Taste

1. **Hypothesis of Receptor-Dependent and Receptor-Independent Mechanisms for Bitter and Sweet Taste Transduction: Implications for Slow Taste Onset and Lingering Aftertaste** 2
   M. Naim, S. Nir, A. I. Spielman, A. C. Noble, I. Peri, S. Rodin, and M. Samuelov-Zubare

2. **Modulation of Two Second Messengers in Bitter Taste Transduction of Agriculturally Relevant Compounds** 18
   A. I. Spielman, W. Yan, V. Krizhanovsky, S. Rosenzweig, T. Yamamoto, and M. Naim

3. **Molecular Mixture Models: Connecting Molecular Events to Perception** 32
   Daniel M. Ennis

Genetics and Physiology

4. **Genetics of Sweet Taste** 40
   Gary K. Beauchamp, Danielle R. Reed, Michael G. Tordoff, and Alexander A. Bachmanov

5. **Genetic Markers, Taste Responses, and Food Preferences** 52
   Adam Drewnowski

6. **Clustering Bitter Compounds via Individual Sensitivity Differences: Evidence Supporting Multiple Receptor–Transduction Mechanisms** 65
   Jeannine F. Delwiche, Zivjena Buletic, and Paul A. S. Breslin

7. **Neural Representation of Sweet Taste in the Cortex of the Monkey**......78
Thomas R. Scott and Carlos R. Plata-Salamán

8. **Age-Related Chemosensory Losses: Effect of Medications**......94
Susan S. Schiffman, Jennifer Zervakis, Brevick G. Graham, and Holly L. Westall

Olfactory Mechanisms and Flavor Release

9. **Olfactory Receptor Neuron Signaling**......110
Stuart Firestein

10. **Olfactory Psychophysics**......123
Richard L. Doty

11. **Dose–Response Curves of Odor and Taste Stimuli: Influence of Sweetening Agents**......140
Christine Vuilleumier, Isabelle Cayeux, and Maria Inés Velazco

12. **Communicating with Chemicals**......158
Anthony Blake

13. **Taste Release and Its Effect on Overall Flavor Perception**......166
Tracey A. Hollowood, Jim M. Davidson, Lucy DeGroot, Rob S. T. Linforth, and Andrew J. Taylor

Taste, Smell, and Trigeminal Interactions

14. **Chemoreception of Fat**......180
Timothy A. Gilbertson and Insook Kim

15. **Astringency and Bitterness of Flavonoid Phenols**......192
A. C. Noble

16. **Pungency and Tingling: Sensations and Mechanisms of Trigeminal Chemical Sensitivity**......202
Bruce Bryant and Igor Mezine

Taste Preference and Consumer Models

17. **Sensory Drivers of Liking and Sensory Preference Segmentation** 214
Howard R. Moskowitz

18. **Internal and External Preference Mapping: Understanding Market Segmentation and Identifying Drivers of Liking** 227
Jean-Xavier Guinard

19. **Measurement of Emotion in Olfactory Research** 243
Stephen Warrenburg

Analytical Approaches

20. **In Vitro Taste Sensors: Technology and Applications** 262
Glenn Roy and John T. McDevitt

21. **Mimicking the Mammalian Sense of Taste Through Single-Component and Multicomponent Analyte Sensors** 276
Sheryl L. Wiskur, Axel Metzger, John J. Lavigne, Stephen E. Schneider, Eric V. Anslyn, John T. McDevitt, Dean Neikirk, and Jason B. Shear

22. **Electronic and Computational Olfaction** 289
Alan Gelperin and J. J. Hopfield

23. **Cross-Reactive Optical Sensing Arrays** 318
Caroline L. Schauer, Shannon E. Stitzel, and David R. Walt

Indexes

Author Index 333

Subject Index 335

Preface

This book is the culmination of a symposium sponsored by the ACS Division of Agricultural and Food Chemistry on the chemistry of taste, held in San Francisco, March 28–30, 2000. We invited key researchers from various scientific disciplines, who all share the common goal of understanding the complexities of taste and how it can be applied in medical, academic, and industrial research.

The objectives of the symposium and this book are twofold. First is to gain a broader understanding of human taste perception by linking different taste research disciplines, such as flavor chemistry, chemoreception, physiology, genetics, neurobiology, and psychophysics. Second is to explore advanced analytical and taste perception methodologies for potential consumer products and sensory and behavior research applications.

We have expanded the definition of taste to that of the "taste experience." Before, during, and after taste stimuli affect the tongue and nasal passages, a myriad of sensory signals are produced by chemicals released upon stimulation. These signals cover basic taste modalities, smell, and chemosensory irritation. The signals are collectively integrated by the central nervous system as sensory and cognitive perception. The authors cover the various aspects of the taste experience from sensory perception to cognition.

This book is unique because it integrates a wide spectrum of taste research disciplines. We begin with understanding taste mechanisms and interactions using animal and human models, linking them to larger consumer models of perception, cognition, and behavior, and finally, exploring new and advanced analytical and sensory measures. We hope that by understanding how taste is perceived and measured, we can then explore innovative ways to modify taste perceptions and predictably map consumer responses.

The topics will be grouped as they were presented during the symposium:

- *Mechanisms of Taste*—Authors look at various chemoreception mechanisms of basic taste qualities (sweet, sour, bitter, and salty), where taste stimuli bind to receptor cells on the tongue and produce sensory (transduction) effects. These studies provide an understanding of taste receptor mechanisms and related structure–activity relationships.
- *Genetics and Physiology*—We probe the impact of nature and nurture in how people perceive taste. The authors discuss recent findings attributing individual differences in taste perception to differences in genetic composition, biochemical taste mechanisms, and physiological states.
- *Olfactory Mechanisms and Flavor Release*—Similar to research in taste receptors, the latest advances in molecular biology and genetics have started to unlock how the olfactory system detects and discriminates different odorants. The authors discuss the current consensus model of olfactory signaling and the latest tools used in assessing olfactory function. The perceived combination of taste and olfactory stimuli has been defined as flavor. We discuss recent developments in quantitative measurement of flavor release in the human nose and its application in flavor perception and the enjoyment of food.
- *Taste, Smell, and Trigeminal Interactions*—To further expand the taste experience, we explore the interaction of trigeminal chemoreception with taste and smell. In addition to taste and olfaction, the trigeminal nerve is the third sensory pathway that is sensitive to chemical stimuli. The authors discuss how a variety of chemical compounds trigger trigeminal sensations, such as pain, temperature, textural, and other tactile cues.
- *Taste Preference and Consumer Models*—Sensory and product attributes by themselves do not determine whether a consumer will consume a particular food. The individual's liking of various taste attributes, his or her physiological needs, and the context in which products are tasted are key determinants of consumption over a sustained period of time. The authors will explore how we can study consumer taste preference, consumption, and purchasing behavior. This knowledge can provide a competitive advantage to companies by identifying great-tasting products that meet consumers' wants and needs.
- *Analytical Approaches*—Finally, the underlying biochemical taste mechanisms have spurred advanced analytical approaches that can predict human taste responses based on innovative chemical analyses that mimic human senses. The authors share their research on next-

generation taste sensors that could expand their use beyond quality-control applications to broader and more sensitive research applications, such as high-throughput sensory screening.

We hope that the chapters in this book will enable us to gain a deeper understanding of human taste perception. We can then use and build on the information in modifying taste perception, predicting sensory responses, and increasing consumer acceptability of food and pharmaceutical and oral products.

The symposium was co-sponsored by 21 corporate sponsors, including Alex Fries, Bell Flavors, Coca-Cola Company, Danisco, Dragoco, Firmenich, Fortitech, Givaudan Roure, International Flavors and Fragrances, Keebler, Kraft Foods, M&M Mars, Mead Johnson, Monsanto, Nutrinova, Pepsi-Cola Company, Procter & Gamble, Takasago, Tate and Lyle, Unilever, and Virginia Dare.

We thank the ACS Division of Agricultural and Food Chemistry and the ACS Books Department for encouraging this publication and our corporate sponsors for their generous support of the symposium. Our final thanks go to the authors of the symposium for sharing their breakthrough research and co-editing the chapters in this book.

Peter Given
Pepsi-Cola Company
100 Stevens Avenue
Valhalla, NY 10595
peter.given@pepsi.com
Phone:: 914-742-4563

Dulce Paredes
Kraft Foods
555 South Broadway
Tarrytown, NY 10591
dparedes@kraft.com
Phone: 914-335-6209

Mechanisms of Taste

Chapter 1

Hypothesis of Receptor-Dependent and Receptor-Independent Mechanisms for Bitter and Sweet Taste Transduction: Implications for Slow Taste Onset and Lingering Aftertaste

M. Naim[1], S. Nir[2], A. I. Spielman[3], A. C. Noble[4], I. Peri[1], S. Rodin[1], and M. Samuelov-Zubare[1]

[1]Institute of Biochemistry, Food Science and Nutrition and [2]Seagram Center, Department of Soil and Water Sciences, Hebrew University of Jerusalem, Rehovot 76–100, Israel
[3]New York University College of Dentistry, New York, NY 10010
[4]Department of Viticulture and Enology, University of California, One Shields Avenue, Davis, CA 95616

Signal messengers such as cAMP, IP_3 and cGMP in taste cells following bitter and sweet taste stimulation can be monitored in real time, in the subsecond range. However, many amphipathic sweeteners and bitter tastants are slow in taste onset and linger, and the molecular basis for these temporal properties is ill-defined. The bitter tastants quinine and cyclo(Leu-Trp), and the non-sugar sweetener saccharin, permeate rapidly through liposomes and taste cells. Furthermore, amphipathic bitter tastants appear to interact with and/or permeate phospholipid-based bitter taste inhibitors. Thus, bitter taste is masked by preventing access of these tastants to taste cells. Such tastants can stimulate responses in cells that are not related to taste cells, and are therefore unlikely to contain taste receptors. It is hypothesized that due to their rapid permeation into taste cells, these tastants may activate the downstream transduction components directly, in addition to their action on G-protein-coupled receptors. The delayed temporal properties produced by many bitter tastants and non-sugar sweeteners may be related to this phenomenon.

Signal transduction of bitter and sweet tastants, as well as that of umami taste, appears to involve G-protein-coupled receptors (GPCRs) (*1-4*). Activation of effector enzymes such as adenylyl cyclase, phospholipase C and phosphodiesterases have been proposed, and changes in the levels of signal molecules such as cAMP, cGMP and IP_3 following sweet and bitter taste stimulation (*5-11*) may, as in olfaction (*12*), be in the 25- to 500-ms time range (*5, 9*). However, the taste intensity of many non-sugar sweeteners (though not sugars) and bitter tastants may be slow in onset and may linger, sometimes up to a few minutes (*13, 14*). The slow rate of onset and the lingering aftertaste seem to be taste-peripheral phenomena as shown in electrophysiological and behavioral studies with experimental animals (*15, 16*), but the molecular basis for these temporal properties is ill-defined. In contrast to sugar sweeteners, both non-sugar sweeteners (sulfamates, oximes, flavonoids, amino acids, dipeptides, proteins, guanidines) and bitter tastants (peptides, terpenes, alkaloids, xanthines, flavonoids) represent molecules with extremely diverse chemical structures. It has been recently proposed that multiple bitter receptors may serve as the target for multiple bitter tastants (*1*). Nevertheless, there is one feature that many non-sugar sweeteners and bitter tastants share: most are amphipathic, i.e., contain both hydrophobic and hydrophilic domains. Surprisingly, little attention has been given to the possibility that such tastants may, as already mentioned for other amphipathic compounds (*17-20*), permeate the plasma membrane and interact with downstream signal-transduction components directly, by means of receptor-independent mechanisms, in addition to their action on GPCRs. In this paper, we discuss recent results of studies designed to explore such a hypothesis.

Temporal Properties of Sweet and Bitter Tastants

High sugar intake has been linked to metabolic disorders and dental caries, suggesting that it should be limited. The ability to synthesize potent non-caloric sweeteners has increased significantly in recent years (*21, 22*). Most non-carbohydrate sweeteners, however, have very different temporal properties than saccharides. Typically, they exhibit slow rates of onset so that it takes longer for maximum of intensity to be reached. And they are characterized by very persistence aftertaste (*14, 23, 24*). As shown in Fig. 1 (top), the sweetness of the dipeptide aspartame, lasts longer than that of an equally sweet sucrose solution. The lingering aftertaste of some other non-sugar sweeteners (*25*) may last much longer. Although there is enormous commercial potential, modification of the temporal properties of low-caloric sweeteners to resemble the sweetness of sucrose or products such as high-fructose corn syrup has been limited by the lack of understanding of the molecular basis of sweet taste sensation.

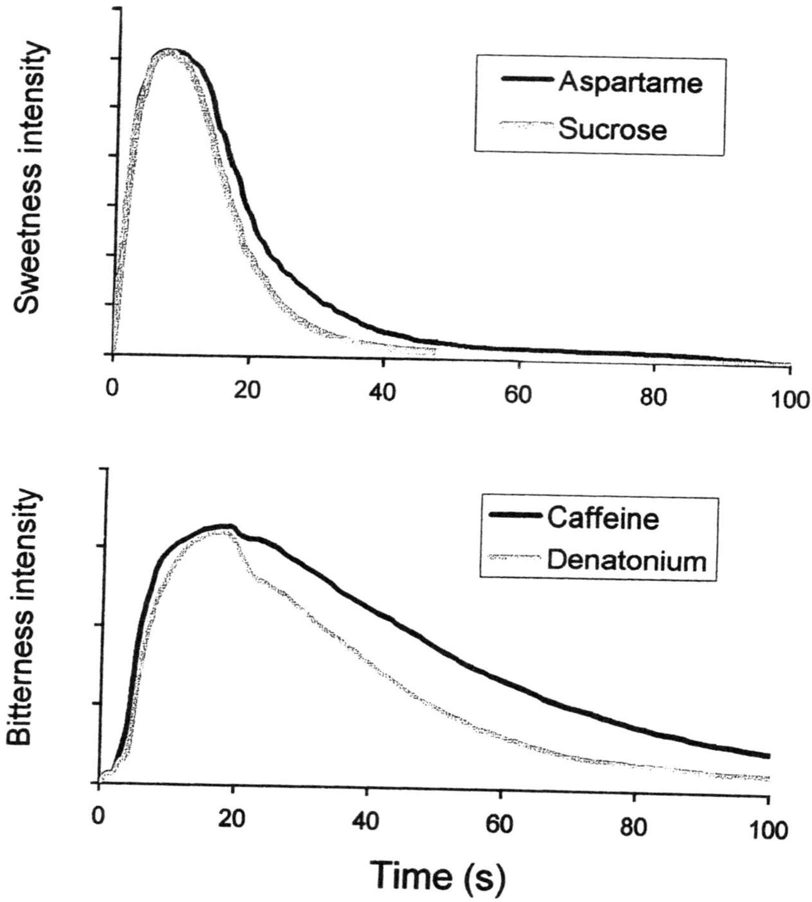

Figure 1. Average time intensity curves for sweetness of aspartame and sucrose (Noble, A.C., unpublished) (top) and for bitterness of caffeine and denatonium (Cubero-Castillo, E. and Noble, A.C., unpublished) (bottom). Values are derived from 18 subjects for sweetness and 12 for bitterness.

From an evolutionary perspective, the aversive response of humans and other mammals to bitter-taste substances has been useful for survival, since many toxic constituents are bitter. However, today, the range of foods available is more diverse (*26, 27*). Many bitter foods are not only safe for consumption but contain bitter constituents (e.g. phenolic compounds) that provide nutritional benefits (*28*). Despite this, these foods are often eliminated from our diets because of their unacceptable and lingering bitterness. As illustared in Fig. 1 (bottom), caffeine and denatonium benzoate persists much longer than the sweetness elicited by aspartame or sucrose (Fig. 1 top). The rate of decay of bitterness differs between caffeine and denatonium. Although the bitterness was equi-bitter, caffeine had a slower rate of decay of bitterness over time. Previously, caffeine was shown to decay slower than quinine (*29*). The development of universally applicable bitter masking agents or bitterness-lingering inhibitors requires an understanding of molecular events at the taste-cell level.

Both electrophysiological and behavioral studies have indicated that the lingering aftertaste of some non-sugar sweeteners can be determined in experimental animals (*15, 16*). The fact that the temporal properties can be monitored during electrophysiological recordings of peripheral taste nerves suggests that these properties are due to events occurring at the taste cell level, probably related to biochemical responses at the taste receptor and/or along the signal transduction pathway. Interestingly, the water sweet aftertaste that reemerged with water after tasting a sweetener which possesses a lingering aftertaste was used as a behavioral parameter (*16*) to quantify the sweet persistence of some sweeteners.

Little attention has been paid to basic aspects of the slow taste onset and lingering aftertaste of bitter tastants and non-sugar sweeteners. For sweet molecules, an "orderly queue" of stimulus molecules forming near the sweet taste receptor site was hypothesized; sucrose molecules pass through queues quickly while sweeteners with lingering aftertaste do so slowly (*14*). According to DuBois and Lee (*25*), a sensation of sugar taste follows the specific and rapid diffusion of molecules to the receptor site, followed by binding. For non-sugar sweeteners, diffusion may occur predominantly to non-receptor sites on the receptor protein. Secondary diffusion to a sweet receptor site results in a delayed taste response.

Amphipathic Bitter and Sweet Tastants May Interact with Liposomal Membranes and Translocate into Liposomes

For the amphipathic tastants to interact directly with membrane transduction components located downstream of the GPCRs in a taste cell, they need to be able to permeate deeply into the plasma membrane and/or translocate to its cytosolic side. Amphipathic tastants have been reported to interact with lipid bilayers and liposomes to induce changes in membrane potential (*30, 31*). In a recent study (*32*), we employed fluorometric detection of small membrane pores (diffusion potential) (*33*) to monitor changes in diffusion potential in small unilamellar vesicles (SUV) made up of either phosphatidyl choline (PC) with cholesterol or azolectin, based on the method described by Kumazawa et al. (*31*). Upon addition of the bitter casein-derived peptide, cyclo(Leu-Trp), the bitter tastant quinine and the sweetener saccharin, each induced an immediate, concentration-dependent increase in fluorescence intensity, indicating marked interactions of tastant molecules with membrane liposomes. Subsequently, translocation experiments were conducted to explore whether amphipathic tastants can translocate into liposomes. For this purpose, multilamellar vesicles (MLVs) were prepared from PC and cholesterol. We used a translocation assay based on auto-fluorescence of the aforementioned tastants and fluorescence-quenching by KI (*34*), whose penetration through liposomal membranes is extremely slow. The idea was that the fluorescence of tastant molecules that translocate, and either remain bound to the internal side of the membrane or disperse in the internal medium would be protected from quenching by KI. KI was added externally to a suspension containing relatively large concentrations of liposomes and tastants in order to maximize the effect of tastant binding and translocation. A remarkable proportion of the bitter tastants cyclo(Leu-Trp) and quinine and the non-sugar sweetener saccharin translocated within seconds (Fig. 2). After 20 s, the fraction of translocated cyclo(Leu-Trp) reached about 9%, that of quinine about 28% and that of saccharin about 3%. Translocation continued throughout the 30-min measurement period. The extent of tastant translocation through the liposomes followed the order: quinine > cyclo(Leu-Trp) > saccharin.

Via what mechanism do amphipathic tastants permeate or translocate through liposomal membranes? The permeabilities of most non-electrolytes through lipid bilayer membranes can be adequately explained by the solubility-diffusion model (*35*). The ionic charges of the three fluorescent tastants under the experimental conditions are shown in Table I. The cyclo(Leu-Trp) peptide is uncharged under these conditions (*36*). Quinine is slightly cationic due to its pKa (8.52) (*37*). Hence, significant fractions of quinine remain in their uncharged form which, indeed, correlates with the solubility-diffusion model and the ability of quinine to intercalate into the liposomes. Saccharin (pKa 1.8) (*38*) is an anion under these conditions. However, saccharin appears to be more lipophilic than would be inferred from its dissociation constant (*39*). This may explain its moderate permeation of the liposomes in the present experiments.

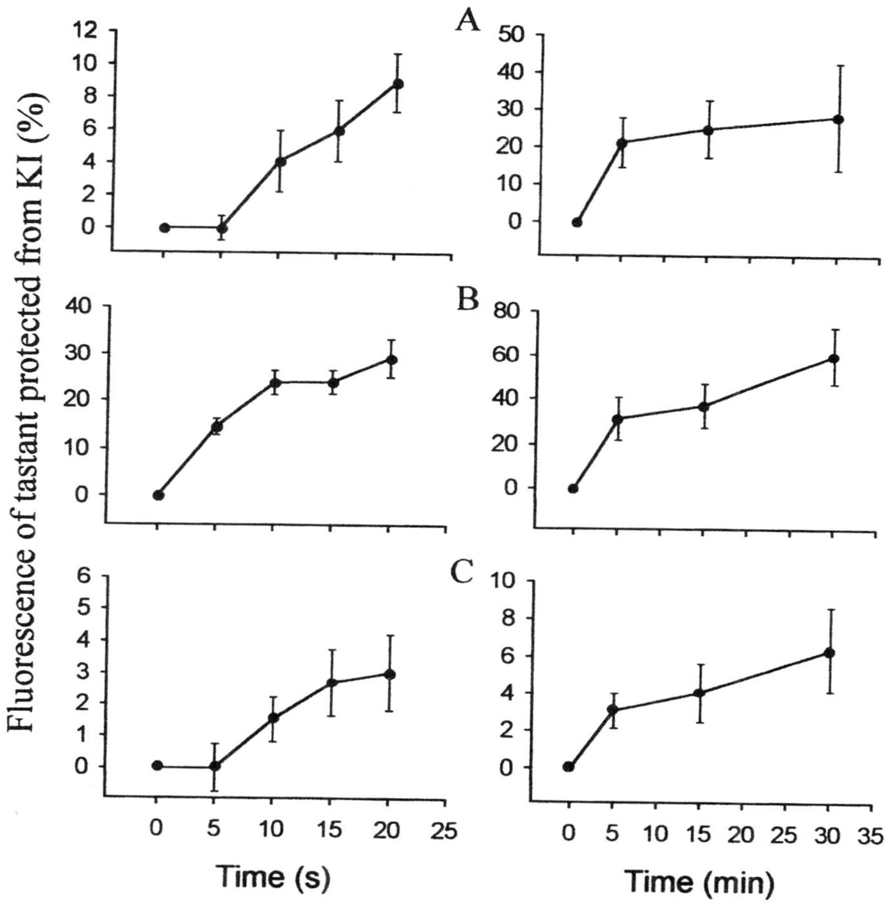

Figure 2. Fluorescence of cyclo(Leu-Trp) (A), quinine (B) and saccharin (C) protected from KI. Incubation times were up to 20 s (left graphs) and up to 30 min (right graphs). Concentrations were 0.5 mM for cyclo(Leu-Trp) and quinine, and 1.5 M for saccharin. Lipid (MLV) concentration was 18 mM. Reproduced with permission from ref. (32). Copyright 2000 American Physiological Society.

Table I. Ionic charge of amphipathic tastants during experiments

Tastant	pH of solutions	Relevant pK values	Electric charge
Cyclo(Leu-Trp)	7.0	No mobility (pH 6-8)	Uncharged
Quinine	7.25	8.52	Slightly cationic
Saccharin	6.8	1.8	Anion

Amphipathic Bitter and Sweet Tastants May Permeate Taste Cells

Following the outcome of the liposome experiments and to further explore the hypothesis of amphipathic tastant access to downstream transduction components, the aim of the next experiments (*32*) was to use the auto-fluorescence of the aforementioned tastants to investigate whether they can permeate taste cells. Circumvallate (CV) taste-bud sheets were prepared from rat tongues by means of sub-epithelial collagenase treatment. Separated CV sheets were incubated in a Tyrode's solution with either quinine (2 mM), cyclo(Leu-Trp) (1 mM), or saccharin (20 mM) at 30°C for 30 s or 5 min. These tastant concentrations are compatible with those that elicit taste sensation in rats. After incubation, using a fine forceps, taste-bud sheets were transferred five times to different test tubes containing Tyrode's solution to wash all tastant molecules from the extracellular medium. Cells were then permeabilized by freeze-thawing (-70°C) in deionized water, and membranes were removed by centrifugation (30 min, 28,000g, 4°C). The intracellular contents were collected and lyophilized, and tastant levels in CV and epithelial were determined with an L-7100 Merck-Hitachi HPLC equipped with F-1050 fluorescence and L-7455 UV-visible diode-array detectors set up in a sequence for concomitant determinations, using a LiChrospher® 100 RP-18 column (5 µm, 250 mm, 4 mm, Merck) with an RP-18 precolumn. This set-up allowed simultaneous tastant determinations by the fluorescence and UV-visible detectors. Peaks of the three tastants were identified and quantitated with known markers injected alone or co-injected with unknown samples using the fluorescence detector. The UV-visible spectrum and the correlation of the unknown peaks with those of the saccharin and cyclo(Leu-Trp) standards were monitored for further identification using the diode-array detector.

Table II. Estimated accumulation of amphipathic tastants in circumvallate (CV) taste cells

Tastant	Extracellular conc. (mM) during incubation	Intracellular conc. (mM) after 30 s	after 300 s
Saccharin	20	140 ± 1	NA
Quinine	2	15.4 ± 5.2	77 ± 31
Cyclo(Leu-Trp)	1	3.5 ± 0.5	8.0 ± 1.5

Values are the means ± SEM of 2-3 samples injected twice through the HPLC. Each sample was derived from 2-6 rats. NA – not available. Adapted with permission from reference (32). Copyright 2000, American Physiological Society.

The results (Table II) demonstrate that these amphipathic tastants permeate rapidly and accumulate inside living taste cells. Direct identification using the fluorescence and UV-visible diode-array HPLC detectors, including UV spectral analysis, clearly indicated the presence of peaks corresponding to saccharin, quinine, and cyclo(Leu-Trp) in the CV tissue incubated for 30 s with these tastants. It should be noted that the structural organization of the taste buds was kept intact in these taste-bud sheets, rather than proceeding to the next step of obtaining single taste buds. We thus assume that we minimized tastant contact with the basal side of the taste cell and maximized tastant entry through the taste pore. Even a relatively weakly fluorescing molecule, such as saccharin, could be detected inside the taste cells within 30 s of incubation of isolated taste buds, using a fluorescent HPLC detector. If this initial accumulation is assumed to increase linearly with time, then the average rate of saccharin accumulation in the CV cells during the first 30 s is 4.7 mM/s. Under the same conditions, the average accumulation rates for quinine and the cyclo(Leu-Trp) peptide are about 500 μM/s and 117 μM/s, respectively. This means that within 1 to 2 s, the levels of these tastants inside the taste cells are compatible with those needed to elicit taste sensation (*40-42*). These rapid rates of tastant accumulation in taste cells occurred even though tastant elimination from inside to the extracellular medium, as in the case of MDR1, had probably been activated (*43*). The dynamic accumulation of amphipathic tastants was also monitored *in situ* by confocal microscopy using single CV taste buds (*32*). Tastants were clearly localized to the cell cytosol.

Therefore, the hypothesis of receptor-independent mechanisms is supported as access of these tastants to transduction components located downstream of GPCRs within the time-course of taste sensation is very likely. This hypothesis is further supported by the fact that amphipathic tastants such as the sweeteners saccharin, cyclamate, and neohesperidin dihydrochalcone (NHD) and the bitter tastant quinine were found to be potent direct activators of G-proteins *in vitro* (*44*). Possible receptor-independent mechanisms may indeed explain the ability of some amphipathic tastants to affect transduction pathways and cellular responses in cells not related to taste. These include depolarization of neuroblastoma cells by some bitter tastants (*45*), activation of adenylyl cyclase cascades by saccharin in liver and muscle membranes (*46*) and in fat-cell membranes (*47*), and stimulation of insulin release from isolated pancreas islets by the non-sugar sweetener acesulfame K (*48*). We found that the bitter tastants cyclo(Leu-Trp) and limonin, and the sweeteners NHD and saccharin stimulate pigment movement in frog melanophores, similar to the native hormone melatonin. The presence of taste receptors in all of these cells is very unlikely.

Amphipathic Bitter Tastants Interact with Phospholipid-Based Bitter Taste Inhibitors

Kurihara, Katsuragi and co-workers (*49-51*) have proposed that a lipoprotein, PA-LG, made up of phosphatidic acid (PA) and β-lactoglobulin (LG), is a specific bitter-taste inhibitor that does not affect the responses to other stimuli such as salts, acids and sugars. Subsequently, it was found that lecithin and some other phospholipids may also inhibit bitter taste sensation (*52*) and Kao Corporation (Tokyo, Japan) developed a masking additive (BMI-60) composed of phosphatidyl inositol (40%), PA (15-20%), phosphatidylethanolamine (10-15%) and PC (5-6%). The mechanism by which these phospholipids inhibit bitterness was proposed to be related to their strong affinity to the bitter receptor sites, hence masking bitterness (*52*). It was hypothesized that receptor sites for bitter substances on the taste cell membrane are hydrophobic and those for other stimuli such as salts, acids and sugars are hydrophilic, thus the binding of PA-LG to hydrophobic sites explains its selective inhibition of bitterness (*51*). However, the ability of the phospholipid-based bitter taste inhibitors to inhibit/mask bitterness induced by different bitter tastants may vary considerably. The following experiments propose an alternative/additional mechanism via which the phospholipid-based inhibitors mask bitterness. Due to the hydrophobicity of the bitter tastants, they may either interact with the external surface of the micelles (or liposomes) formed by the phospholipid-based inhibitors, or they may permeate such vesicles. In either case, these effects may diminish the bitter tastant's access to the taste receptor cells. MLVs were prepared from PC and cholesterol and the aforementioned translocation assay based on fluorescence-quenching by KI (*34*) was conducted. MLVs (18 mM) were incubated with either quinine (0.5 mM) or cyclo(Leu-Trp) (0.5 mM) for 20 s, with and without the presence of 0.25, 0.5 and 1% BMI-60. The percentage of tastant fluorescence protected from KI-quenching was then determined. Subsequently, we tested the effect of BMI-60 on quinine and cyclo(Leu-Trp) permeation of CV taste-bud cells during 5- and 2-min incubations, respectively. The content of cyclo(Leu-Trp) and quinine in the taste cells was then determined by HPLC (*32*).

About 4% translocation of quinine into MLVs was found after a 20-s incubation (Fig. 3A). Interestingly, protection of quinine fluorescence was increased to 6, 8.5 and 11.5% when 0.25, 0.5 and 1% BMI-60 was present, respectively. Most surprisingly, when quinine was incubated with 1% BMI-60, with no MLV present, fluorescence protection was 11%, almost threefold than that resulting from incubation of quinine with MLVs alone. There was a concentration-dependent effect of this fluorescence protection by BMI-60. Thus BMI-60, being a phospholipid, formed micelles, into which quinine was able to permeate; most probably, the affinity of quinine to the BMI-60 micelles was higher than that to MLVs. Protection of cyclo(Leu-Trp) from KI-quenching was 15% when incubated alone with MLVs (Fig. 3B). The addition of BMI-60 (0.5% and 1%) resulted in reduced protection, rather than the increased protection found for quinine. Some slight fluorescence protection of cyclo(Leu-Trp) occurred when incubated with BMI-60 micelles without MLVs. It seems that cyclo(Leu-Trp) interacted with BMI-60 (probably with its external membrane) without being translocated into the BMI-60 vesicles, such that less cyclo(Leu-Trp) molecules are available for translocation into MLVs.

In line with the liposome results, 1% BMI-60 significantly inhibited (more than 80%, $p < 0.01$) the permeation of quinine into rat CV taste cells (Fig. 4A). Evidence that BMI-60 actually inhibits quinine's access to taste cells is thus provided. Concentrations of 0.25 and 0.5% BMI-60 were much less effective, but within the experimental uncertainty the inhibition was proportional to BMI-60 concentration. The magnitude of inhibition of the casein-derived cyclo(Leu-Trp) peptide's permeation into taste cells by 1% BMI-60 (Fig. 4B) was much lower ($p < 0.05$) than that of quinine. Interestingly, these results are in line with those of Katsuragi et al. (*52*) indicating that the phospholipid-based bitter inhibitors are less effective in the inhibition of bitterness induced by whey peptide, a product of casein hydrolysate. In conclusion, these experiments suggest that BMI-60 inhibits bitter taste either by interaction with bitter tastants (e.g., binding of amphipathic tastants to the BMI-60 micelle surface), or by the fact that some amphipathic tastants

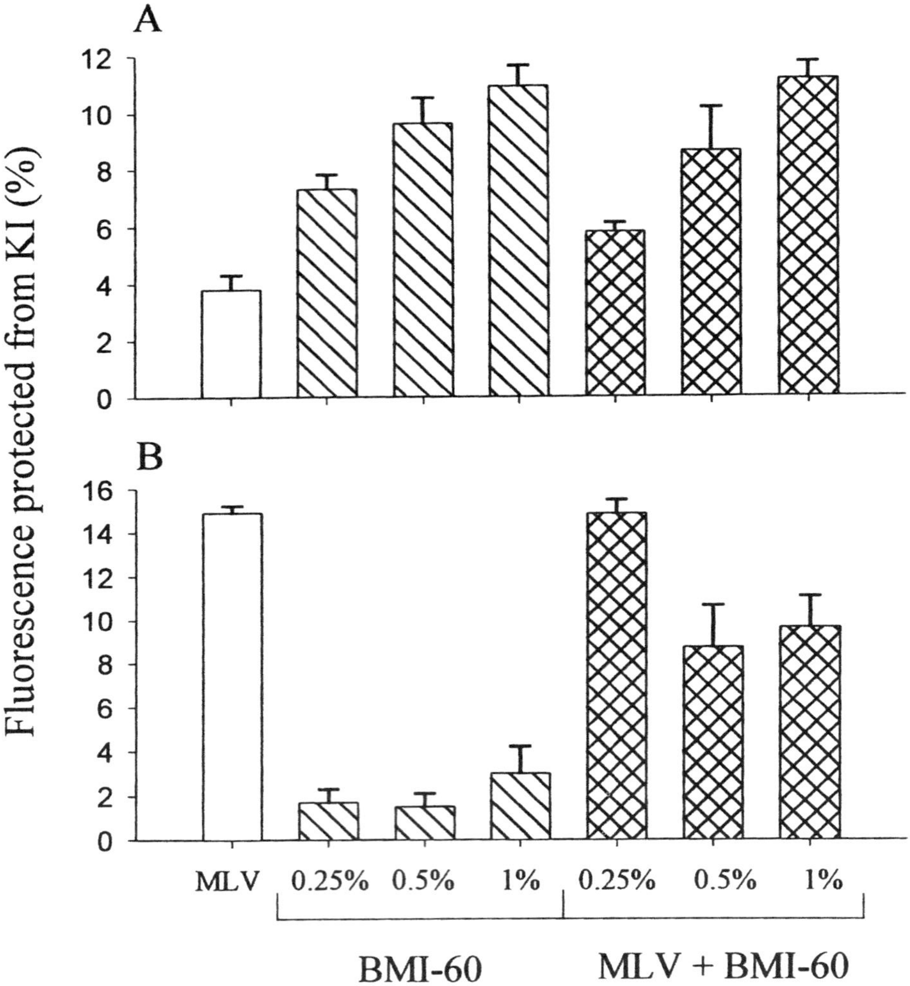

Figure 3. Translocation (expressed as fluorescence protected from KI-quenching) of 0.5 mM quinine (A) and 0.5 mM cyclo(Leu-Trp) (B) into multilamellar vesicles (MLVs) and the bitter-taste inhibitor BMI-60. MLVs concentration was 18 mM lipid. Concentrations of BMI-60 are indicated on the X-axis. Values are means and SEM of triplicates for each data point.

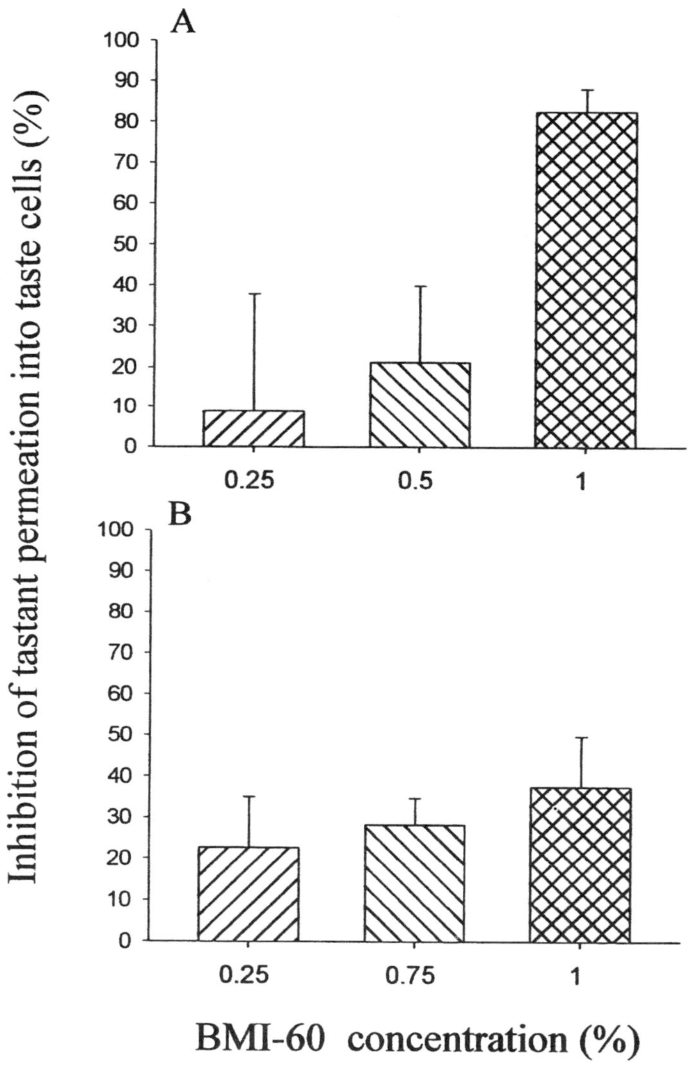

Figure 4. Inhibition of permeation of quinine (A) and cyclo(Leu-Trp) (B) into rat CV taste-bud cells in the presence of various concentrations of BMI-60. Values are the mean and SEM of three rat samples for each data point, each subjected twice to HPLC.

penetrate the BMI-60, to its interior micelles. In either case, these effects decrease the bitter tastant's access to the taste receptor cells.

Concluding Remarks and Research Needs

Understanding the molecular basis of slow rate of taste onset and lingering aftertaste induced by bitter tastants and non-sugar sweeteners has great economic potential with respect to various food and drink products. Although these temporal properties are related to peripheral events at the taste cell level, the nature of such events is unknown. Because multiple signal-transduction pathways appear to occur at the taste cell level for both bitter and sweet taste sensation, the delineation of signal mechanisms responsible for the delayed temporal properties is further complicated. Here we propose that the concomitant action of amphipathic tastants on GPCRs and direct activation of transduction components downstream of the GPCRs (e.g., G-proteins and/or effector enzymes) after permeation into taste cells may be occurring and that this combined action may be related to slow rate of taste onset and lingering. The hypothesis of amphipathic-tastant-induced receptor-dependent and receptor-independent mechanisms is supported by the following: (a) some bitter tastants and non-sugar sweeteners, due to their amphipathic properties, permeate taste cells rapidly, (b) some are direct activators of G-proteins *in vitro*, and (c) they may, under some circumstances, induce depolarization in the membranes of living cells, and stimulate transduction pathways and physiological responses in a variety of cells and tissues that are not related to taste cells and are therefore unlikely to contain taste receptors. Further research is needed to explore the mechanisms by which amphipathic tastants permeate taste cells and to verify whether they also permeate cells that are not taste cells. Finally, to explore the link between receptor-independent activation of signal transduction pathways and the temporal delay properties of sweet and bitter taste sensation, time-course measurements of the release of signal molecules in controlled reconstitution systems are required.

Acknowledgments

This research was supported by grant I-2518 from BARD, The United States-Israel Binational Agricultural Research & Development Fund. We thank Dr. Y. Katsuragi of Kao Corporation, Tokyo, Japan, for kindly supplying BMI-60 samples. We acknowledge the excellent technical assitance of Ms. O. Agamy and Mr. A. Maarabi.

Literature Cited

1. Adler, E.; Hoon, M. A.; Mueller, K. L.; Chandrasheker, J.; Ryba, J. P.; Zuker, C. S. *Cell* **2000,** *100*, 693-702.
2. Chaudhari, N.; Landin, A. M.; Roper, S. D. *Nature Neurosci.* **2000,** *3*, 113-119.
3. Lindemann, B. *Physiol. Rev.* **1996,** *76*, 719-766.
4. Chandrasheker, J.; Mueller, K. L.; Hoon, M. A.; Adler, E.; Feng, L.; Guo, W.; Zuker, C. S.; Ryba, J. P. *Cell* **2000,** *100*, 703-711.
5. Krizhanovsky, V.; Agamy, O.; Naim, M. *Am. J. Physiol. Cell Physiol.* **2000,** *in press*.
6. Bernhardt, S. J.; Naim, M.; Zehavi, U.; Lindemann, B. *J. Physiol. (Lond.)* **1996,** *490*, 325-336.
7. Cummings, T. A.; Powell, J.; Kinnamon, S. C. *J. Neurophysiol.* **1993,** *70*, 2326-2336.
8. Hwang, P. M.; Verma, A.; Bredt, D. S.; Snyder, S. H. *Proc. Natl. Acad. Sci. USA* **1990,** *87*, 7395-7399.
9. Spielman, A. I.; Nagai, H.; Sunavala, G.; Dasso, M.; Breer, H.; Boekhoff, I.; Huque, T.; Whitney, G.; Brand, J. G. *Am. J. Physiol. Cell Physiol.* **1996,** *270*, C926-C931.
10. Striem, B. J.; Pace, U.; Zehavi, U.; Naim, M.; Lancet, D. *Biochem. J.* **1989,** *260*, 121-126.
11. Wong, G. T.; Gannon, K. S.; Margolskee, R. F. *Nature* **1996,** *381*, 796-800.
12. Breer, H.; Boekhoff, I.; Tareilus, E. *Nature* **1990,** *345*, 65-68.
13. Noble, A. C. *Physiol. Behav.* **1994,** *56*, 1251-1255.
14. Birch, G. G.; Latymer, Z.; Hollaway, M. *Chem. Senses* **1980,** *5*, 63-78.
15. Hellekant, G. In *Thaumatin*; Witty, M., Higginbotham, J. D., Eds.; CRC Press: London, 1994, pp 99-113.
16. Rogatka, H.; Naim, M.; Zehavi, U. *Chem. Senses* **1985,** *10*, 269-277.
17. Aridor, M.; Rajmilevich, G.; Beaven, M. A.; Sagi-Eisenberg, R. *Science* **1993,** *262*, 1569-1572.
18. Higashijima, T.; Burnier, J.; Ross, E. M. *J. Biol. Chem.* **1990,** *265*, 14176-14186.
19. Mousli, M.; Bueb, J.-L.; Bronner, C.; Rouot, B.; Landry, Y. *Trends Pharmacol. Sci.* **1990,** *11*, 358-362.
20. Takesono, A.; Cismowski, M. J.; Ribas, C.; Bernard, M.; Chung, P.; Hazard III, S.; Duzic, E.; Lanier, S. M. *J. Biol. Chem.* **1999,** *274*, 33202-33205.

21. DuBois, G. E.; Walters, D. E.; Kellog, M. S. In *Sweet-Taste Chemoreception*; Mathlouthi, M., Kanters, J. A., Birch, G. G., Eds.; Parthenon Press: London, 1993, pp 237-267.
22. Nofre, C.; Tinti, J.-M. *Food Chem.* **1996,** *56*, 263-274.
23. Matysiak, N. L.; Noble, A. C. *J. Food Sci.* **1991,** *56*, 823-826.
24. Pfeiffer, J. F.; Boulton, B. B.; Noble, A. C. *Food Qual. Pref.* **2000,** *11*, 129-138.
25. DuBois, G. E.; Lee, J. F. *Chem. Senses* **1983,** *7*, 237-247.
26. *Bitterness in Foods and Beverages*; Rouseff, R. L.; Elsevier: Amsterdam, 1990, pp. 356.
27. Beauchamp, G. K.; Kurihara, K.; Ninomiya, Y.; Sato, T.; Smith, D.; Yamamoto, T. *Physiol. Behav.* **1994,** *56*, 1121-1266.
28. Frankel, E. N.; Waterhouse, A. L.; Teissedre, P. L. *J. Agric. Food Chem.* **1995,** *43*, 890-894.
29. Leach, E. J.; Noble, A. C. *Chem. Senses* **1986,** *11*, 339-345.
30. Brand, J. G.; Feigin, A. M. *Food Chem.* **1996,** *56*, 199-207.
31. Kumazawa, T.; Nomura, T.; Kurihara, K. *Biochemistry* **1988,** *27*, 1239-1244.
32. Peri, I.; Mamrud-Brains, I.; Rodin, S.; Krizhanovsky, V.; Shai, Y.; Nir, S.; Naim, M. *Am. J. Physiol. Cell Physiol.* **2000,** *278*, C17-C25.
33. Shai, Y.; Bach, D.; Yanovsky, A. *J. Biol. Chem.* **1990,** *265*, 20202-20209.
34. Lange, R.; Anzenbacher, P.; Muller, S.; Maurin, L.; Balny, C. *Eur. J. Biochem.* **1994,** *226*, 963-970.
35. Stein, W. D. *Physiol. Rev.* **1997,** *77*, 545-590.
36. Minamiura, N.; Matsumura, Y.; Yamamoto, T. *J. Biochem.* **1972,** *72*, 841-848.
37. *Handbook of Chemistry and Physics;* Lide, D. R.; Frederikse, H. P. R.; CRC Press: Boca Raton, 1994.
38. Williamson, D. S.; Nagel, D. L.; Markin, R. S.; Cohen, S. M. *Food Chem. Toxicol.* **1987,** *25*, 211-218.
39. Ziesenitz, S. C.; Siebert, G. *Z. Ernahrungswiss.* **1988,** *27*, 155-169.
40. Cagan, R. H.; Maller, O. *J. Comp. Physiol. Psychol.* **1974,** *87*, 47-55.
41. Grill, H. J.; Norgren, R. *Brain Res.* **1978,** *143*, 263-279.
42. Shiba, T.; Nunami, K.-I. *Tetrahedron Lett.* **1974,** *6*, 509-512.
43. Jakob, I.; Hauser, I. A.; Thevenod, F.; Lindemann, B. *Am. J. Physiol. Cell Physiol.* **1998,** *274*, C182-C191.
44. Naim, M.; Seifert, R.; Nurnberg, B.; Grunbaum, L.; Schultz, G. *Biochem. J.* **1994,** *297*, 451-454.
45. Kumazawa, T.; Kashiwayanagi, M.; Kurihara, K. *Brain Res.* **1985,** *333*, 27 - 33.

46. Striem, B. J.; Naim, M.; Zehavi, U.; Ronen, T. *Life Sci.* **1990,** *46*, 803-810.
47. Dib, K.; Oget, I.; Wrisez, F.; El Jamali, A.; Aguie-Aguie, G.; Correz, C.; Lambert, B. *Int. J. Obes.* **1996,** *20*, 15-20.
48. Liang, Y.; Maier, V.; Steinbach, G.; Lalic, L.; Pfeiffer, E. F. *Horm. Metabol. Res.* **1987,** *19*, 285-289.
49. Katsuragi, Y.; Kurihara, K. *Nature* **1993,** *365*, 272-273.
50. Katsuragi, Y.; Yasumasu, T.; Kurihara, K. *Brain Res.* **1996,** *713*, 240-245.
51. Katsuragi, Y.; Sugiura, Y.; Otsuji, K.; Kurhara, K. *Biochim. Biophys. Acta* **1996,** *1289*, 322-328.
52. Katsuragi, Y.; Mitsui, Y.; Umeda, T.; Otsuji, K.; Yamasawa, S.; Kurihara, K. *Pharmaceut. Res.* **1997,** *14*, 720-724.

Chapter 2

Modulation of Two Second Messengers in Bitter Taste Transduction of Agriculturally Relevant Compounds

A. I. Spielman[1], W. Yan[1], V. Krizhanovsky[2], S. Rosenzweig[1], T. Yamamoto[3], and M. Naim[2]

[1]Basic Science Division, New York University College of Dentistry, New York, NY 10010
[2]Institute of Biochemistry, Food Science and Nutrition, Hebrew University of Jerusalem, Rehovot 76–100, Israel
[3]Department of Behavioral Physiology, Faculty of Human Sciences, Osaka University, Osaka 565–0871, Japan

The formation of chemical cellular signals in response to taste stimulation may be initiated in sub-second time range. We are using a Quench Flow Module (QFM) and a Fast Pippeting System (FPS) to monitor cellular signals in rat taste tissue homogenates and intact cells in response to bitter taste stimulants that are important to the acceptance of agricultural food products. Stimulation by grape-derived catechin and the citrus-derived naringin increased the formation of the IP_3 signal with parallel decline of cAMP. Under similar experimental conditions, the citrus-derived limonin reduced cAMP whereas casein-derived cyclo(Leu-Trp) bitter dipeptide produced a rapid and transient increase of cAMP. cGMP was not affected by any of the four stimuli. These results further emphasize the presence of multiple transduction pathways and modulation of two parallel second messengers in bitter taste sensation of some bitter compounds.

Introduction

Bitter taste may have evolved as an aversive response to potentially toxic compounds. Today, the range of available foods is diverse including many bitter products (e.g. phenolic compounds) that are not only safe but also highly desirable to provide nutritional benefits (*1*). In some cases, bitterness is desirable (e.g. grapefruit, aperitifs, beer, tonic water, seasoning) and its lingering aftertaste may dominate the overall flavor (*2*). Despite their usefulness, often these foods are eliminated from our diets because of their unacceptable bitterness. While consumption of green vegetables and fruits has increased due to nutritional recommendations, bitterness reduces acceptance of lettuce, yams, asparagus, cabbage, cauliflower, broccoli, cucumbers, squash, olives and olive oils, persimmon, grapes, strawberry, apples and many other fruits and vegetables (*1,3,4*). In many instances, the consumer or producer uses sugars and dressings to mitigate bitterness.

Very often, bitterness does not occur naturally, but it is produced during food processing. Naringin (flavanone glycoside) and especially limonin (a triterpene) are the main bitter tastants in citrus products (*1*). Limonin, which is present as a tasteless precursor in citrus fruit, contributes bitterness when the open D ring of its precursor closes upon juice extraction to form limonin. Thermal processing of avocado cannot be used as a preservation method because bitter compounds are formed upon heating (*5*). Tomatine (a glycoalkaloid) and naringin induce bitterness in tomato purees and pastes (*4*). Increased solubility of soy proteins by enzyme hydrolysis can produce bitter peptides (*6*) and hydrophobic amino acids, and peptides which accumulate during proteolysis of milk proteins (e.g. due to bacteria) may detract from the flavor of cheese (*7*).

With only limited success, extensive technology has been applied to reduce (by removing bitter compounds or using specific enzymes to break down bitterants) or mask (adding sugar in most cases) bitterness in processed agricultural food products. The power of consumers demanding less bitter foods is enormous. The market value of bitter food products made palatable by the addition of sugar alone has been estimated at $150 billion per year in the US (*8*). In addition to the high cost of sucrose, public health professionals globally advise reducing sugar intake. The development of universally applicable bitter masking agents or bitterness inhibitors requires an understanding of molecular events at the taste-cell level. The current investigation continues previous biochemical studies on bitter taste transduction that measured signaling molecules in the subsecond timeframe (*9-13*). We have monitored such cellular events for agriculturally relevant bitter compounds limonin, naringin (citrus-derived), catechin (grape-derived) and the cyclic leucine-tryptophane peptide derived from caseine. We have used intially behavioral and electrophysiological recordings in rats to identify the gustatory thresholds and levels of aversiveness for each compound. This was followed by measurement of second messengers using the quench flow module (QFM) and the fast pippeting system (FPS).

Materials and Methods:

1. Behavioral experiments

Limonin, naringin, catechin, and quinine were purchased from Sigma Chemical Co. (St Louis, MO). Cyclo (Leu-Trp), a casein-hydrolysate-derived cyclo dipeptide which is about two to threefold more bitter than caffeine was synthesized (>99% purity) by Bachem Co. (Bubendorf, Switzerland). Naive male Sprague-Dawley rats (Anilab, Tal Shahar, Israel) weighing 170-200 g were acclimated to our laboratory for 5 days before testing. Animals were individually housed in cages (40X30X20 cm) at a temperature of 23°C ± 2°C with a cycle of 12 hr light and 12 hr darkness. Each cage was fitted at the front with two drinking valves (Edstrom Industrial, Inc. WI) connected through 8 cm plastic tubes to two 40 mL syringes. The two drinking valves were used during testing with an additional valve fitted at the back of the cage to supply tap water in between tests. Rats were fed on lab chow (Anilab) and water ad lib in their home cages. Rats were then trained during a 5 day period (1 hr daily) to approach a preferred solution of 10 mM sodium saccharin vs. water in a two-choice situation (*14*). At the end of the training period, more than 90% of the liquid intake was from the saccharin solution. Animals were then subjected to a brief-exposure (10 min) preference test between a single concentration of a bitter tastants in water vs. water conducted between 6-7 P.M. in the dark cycle using a red lamp for illumination. Positional effects were avoided by randomizing the location. Paired t-tests were performed on the preference data to reveal significant difference from the 50% percent preference. Percent preference was defined as: intake of tastant (mL)/total volume intake (mL) X 100.

2. Electrophysiological experiments:

Sprague Dawley male rats were deprived of water for 5-6 hr before the experiment. The animals were deeply anesthetized by intraperitoneal injection of pentobarbital (50 mg/Kg) and urethane (0.7 g/Kg). The trachea was canulated and the glossopharyngeal nerves were exposed and cut centrally by the lateral approach. Recording of nerve responses was conducted by placing a small flow-chamber on the circumvallate (CV) taste papilla through which a test solution was continuously passed. The magnitude of the nerve responses is relative to that of 2mM quinine solution used as a reference of 100. Analysis of variance and post hoc LSD tests were used to evaluate the statistical significance of the data.

3. Signal transduction experiments

QFM experiments.

Foliate and CV papillae, which contain the largest concentration of taste buds, were punched out and trimmed to remove all muscle, connective and salivary tissue (*15*). Tissue were then homogenized centrifuged (1000g) and the supernatant (50-100 μg protein/ml) was used for quench flow experiments. We used a special quench-flow instrument (QFM-5, BioLogic, France and Molecular Kinetics, US) which allows us to monitor responses in any time range between 10 msec and 1 sec. Details of the experimental set up were reported previously (*11-12*).

Fast release of signal molecules in response to bitter taste stimulation using intact taste cells (FPS experiments).

Cyclic nucleotide (cAMP and cGMP) levels in taste cells in response to bitter taste stimulation were monitored. Rats were anesthetized with ether and then decapitated. CV taste papilla and nonsensory EP sheets were prepared from rat tongue by means of sub-epithelial collagenase treatment (4 mg/mL, 0.28 U/mg collagenase; and trypsin inhibitor, 4 mg/mL) for 30 min in Tyrode's solution. The dissected sheet of CV tissue was cut into symmetrical halves. Each half was transferred to an Eppendorf tube containing 10 μl ice-cold Tyrode's solution. EP sheets were treated identically and used as control tissue. Then, 50 μl of a single concentration of a bitter tastant ("start" solution) and 18 μl of 60% (v/v) perchloric acid ("stop" solution) were injected consecutively via the FPS (see further on). Samples were then frozen at -70°C for at least 30 min, thawed on ice and centrifuged (4°C, 10,000g, 10 min). Supernatants were removed, neutralized to pH 6.2 with 2 M potassium carbonate solution, and recentrifuged as before. Supernatants were removed, kept on ice, acetylated and used for second messenger determination by RIA and protein determination by the Bradford assay (*16*). Analysis of variance was performed on the data using the JMP statistics program (SAS Institute Inc., USA) and paired t-test was used to determine the difference between the basal and stimulated levels.

To monitor of the release of signal molecules in intact taste cells at time range below 500 ms an improved FPS was built (Gertron Co. Ltd., Petach-Tikva, Israel). This instrument delivers "start" and "stop" solutions consecutively into an Eppendorf test tube at predetermined, short time intervals. The system contains: 1) a controller, a PC (ACER laptop model Acernote 350PC) with application software; 2) pump-induced air pressure with an electronic timer which transfers signals to two miniature three-way valves (LFAA1201610H or equiv.); 3) valves which are driven with a "spike and hold" driver, reducing the turn-on time to 1.3 ms and turn off-time to 3.2 ms; 4) two Gilson micropipettes (P-200) with sustaining accuracy of ± 1% down to 50 μL. No loss in accuracy was found after connecting the air tubes to the micropipette tip holder. The application software was a timing generator for: a) start valve opening time, b) stop valve opening time, and c) delay (between end of start to beginning of stop). The function of the time (20 to 300 ms) that it takes for the "start/stop" solutions to depart from the pipette tips, depending on a given volume (10 to 150 μL), was calculated based on measurements using a Kodak motioncorder analyzer (model 1000) at a rate of 400 frames/s. To record the actual time needed for the "start/stop" solutions to fully leave the pipette tip and reach the Eppendorf tube (injection time), air pressure was set to 13 mm Hg above atmospheric pressure. These calculations resulted in the formula: $T(\text{ms}) = 2.15 * \text{volume}\ (\mu L) + 5.3$ for nonviscous solutions. The delay between "start" and "stop" (interval between start valve closing and stop valve opening) was also verified by the aforementioned technique. Regression analysis of the volume of the solutions injected vs. injection time resulted in $R^2 > 0.95$.

The two micropipettes were positioned in the Eppendorf tube to achieve a strong vortex with fast homogeneous mixing, as evidenced by filming (with the above Kodak analyzer) the mixing process with a colored 50 μL (methylene blue) start solution injected into an Eppendorf tube containing 20 μL Tyrode solution (the volume of the incubation mixture used in this study). Mixing time was defined from the time at which the colored solution started to enter the tube to the time at which, upon termination of injection, the liquid inside the

Eppendorf tube completely settled, i.e., there was no detected turbulence. A very fast mixing value, essentially overlapping the injection time, was observed, and therefore, this value was ignored for the above calculations. Thus, when setting different incubation times (delay between start and stop solutions) of 220 and 350 ms (e.g., Fig. 7), we defined these time periods as the length of time the taste cells were exposed to a predetermined concentration of taste stimulus.

Results

1. Behavioral and electrophysiological experiments

Rats show aversive responses to compounds that taste bitter to humans and were selected as the animal experimental model to study bitter taste transduction in this project. In light of paucity of sensory data related to the response of rats to bitter tastants occurring in agricultural food products, we first tested the rat's taste responses to selected bitter stimuli such as naringin and limonin (both bitter tastants in citrus products, (*1*), cyclo(Leu-Trp) which is a casein-derived bitter tastant in cheese (*17*) and catechin (in wine, *18*).

Brief exposure preference tests, where the taste stimuli are available for only a few minutes, are considered more valid measurements of taste with little confounding by postingestional factors (*19*). The results presented in Fig. 1 clearly indicate that the preference of rats to solutions containing either cyclo(Leu-Trp), limonin and naringin were below the 50% preference level (compared to a significant preference for saccharin solution used as a control) indicating that these tastant solutions are aversive during the brief exposure preference tests. Similar behavioral responses were found in for Sprague Dawley and Long Evans rats strains. Preference tests with catechin (30 mM) resulted in only a trend towards aversion (mean percent preference was 39 ± 8, $p < 0.11$, $n = 11$). The electrophysiological data were in line with the behavioral data indicating significant responses of the glossopharyngeal taste nerve following stimulation of the area of the CV taste papilla by these bitter tastants (Fig. 2). One may therefore conclude that the bitter compounds tested is rejected by rats possibly because of their bitterness.

2. Signal transduction experiments

The results of time-course for the release of IP_3 and cAMP in response to bitter taste stimulation in taste tissue homogenates (*in vitro* experiments) by catechin, naringin, cyclo(Leu-Trp) and limonin are shown in Figs 3 through 6 using the QFM. There was no stimulation or inhibition in cGMP release in response to stimulation by these tastants. Catechin stimulation produced rapid and transient significant elevation of IP_3 within 200 ms (Fig. 3, top). There was also a drop in cAMP at 50 ms which was concentration dependent. A similar pattern was observed after stimulation by naringin (Fig. 4) though the reduction of cAMP at 50 ms is much more modest. Therefore, catechin and possibly naringin may initiate a similar pattern of bitter taste transduction. Stimulation by cyclo(Leu-Trp) shows a significant elevation in cAMP at time range above 100

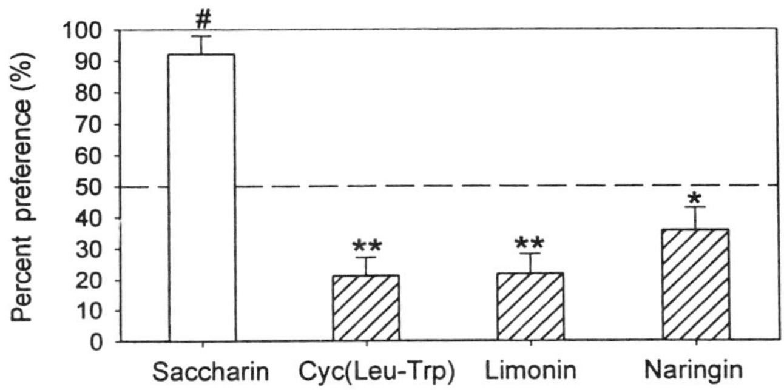

Figure 1. Two-choice short-term (10 min) preference tests between water and solution containing bitter taste stimulus. Saccharin sweet solution (10 mM) was used as a control. Percent preference is the intake (mL) of a tastant solution divided by total intake from both choices and multiply by 100. 50% preference line indicates neither preference nor aversion. Concentrations of bitter stimuli were 1 mM, 33.3 μM and 1.5 mM for cyclo(Leu-Trp), limonin, and naringin, respectively. Values are the means and SEM of 15 rats for each experiment. * Indicates significant ($p < 0.05$) aversion, ** significant ($p < 0.01$) aversion, # indicates significant ($p < 0,01$) preference.

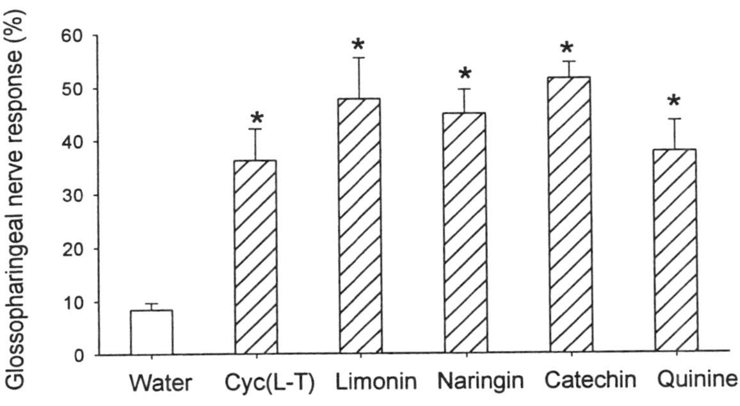

Figure 2. Electrophysiological taste responses of the glossopharyngeal nerve of rats to taste stimulation by bitter stimuli. Tastant concentrations were 1 mM, 40 μM, 3 mM, 10 mM and 0.2 mM for cyclo(Leu-Trp), limonin, naringin, catechin, and quinine, respectively. Values are means and SEM of 5 rats for each data point and are relative to the response by 2 mM quinine taken as standard (100).
* significant ($p < 0.05$) nerve response.

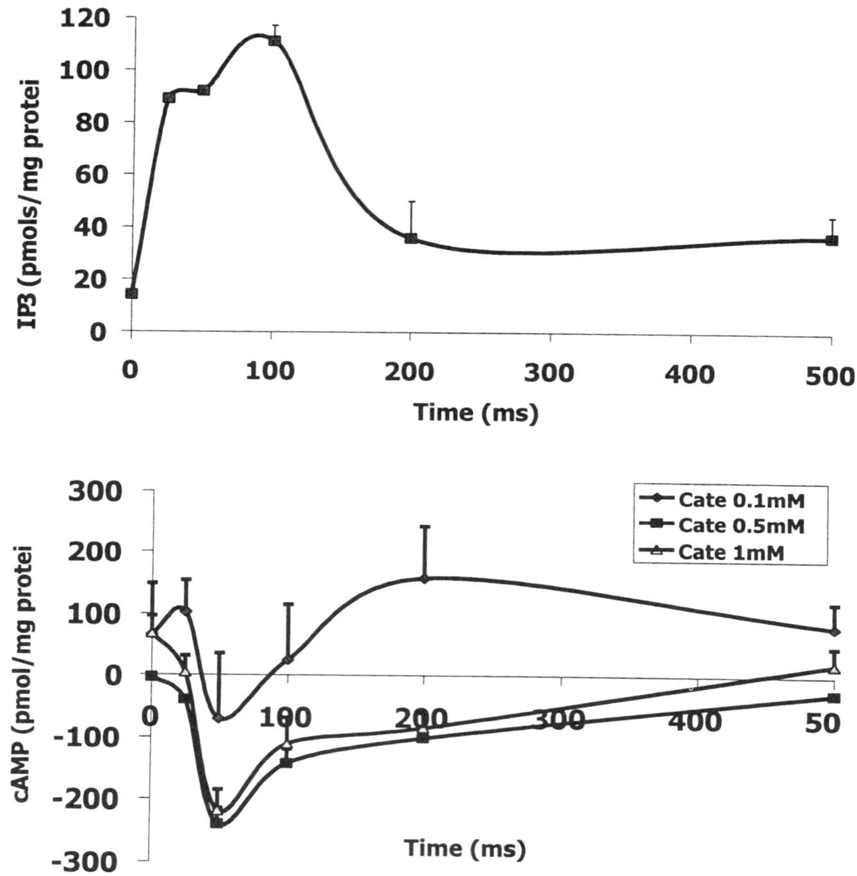

Figure 3. Time-course of catechin-stimulated IP_3 (top) and cAMP (bottom) in rat taste tissue. Data are the mean of three experiments done in triplicates. Data are expressed as net IP_3 and cAMP (pmols/mg of protein) after subtracting the basal levels from stimulated and nongustatory levels from gustatory levels. For IP_3 experiment the concentration of catechin was 1 mM. Note a rapid increase in IP_3 at 50-100 ms and a drop in cAMP at 50 ms that is concentration dependent.

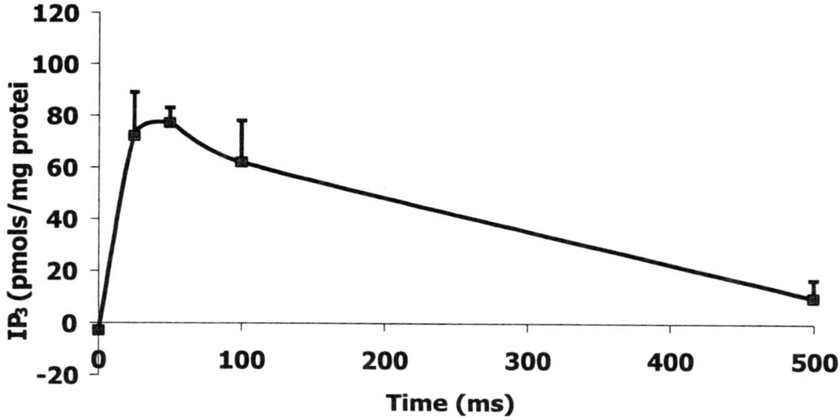

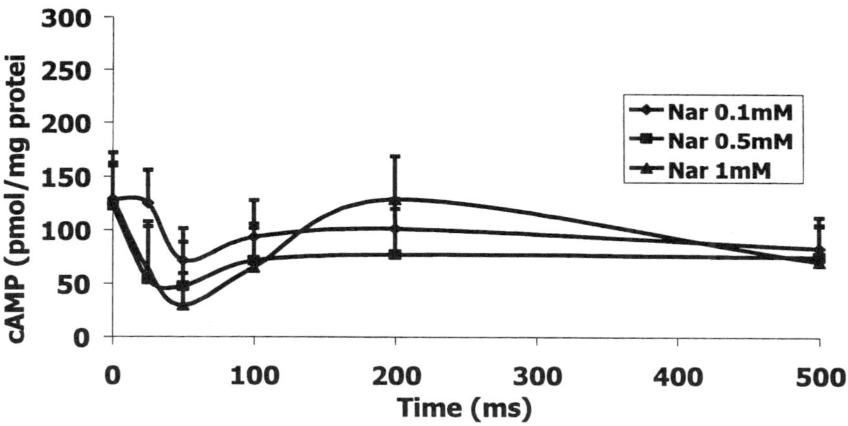

Figure 4. Time-course of naringin-stimulated IP_3 (top) and cAMP (bottom) in rat taste tissue. Data are the mean of three experiments done in triplicates. Data are expressed as net IP_3 and cAMP (pmols/mg of protein) after subtracting the basal levels of IP_3 and cAMP from stimulated, and nongustatory levels from gustatory levels. For IP_3 experiment the concentration of naringin was 1 mM. Note a rapid increase in IP_3 at 50-100 ms and a drop in cAMP at 50 ms that is concentration dependent.

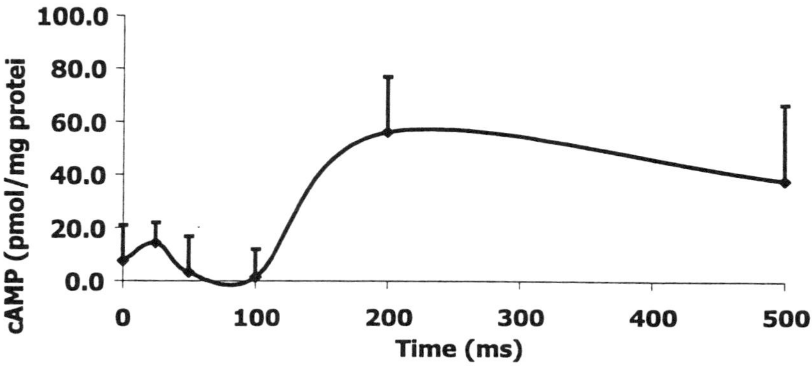

Figure 5. Time-course of cyclo(Leu-Trp)-induced cAMP in rat taste tissue. Data are the mean of three experiments done in triplicates. Data are expressed as net IP_3 and cAMP (pmols/mg of protein) after subtracting the basal levels. For IP_3 experiment the concentration of cyclo(Leu-Trp) was 0.5 mM. Note a drop in IP_3 production at 50 ms and a modest increase in cAMP at 200 ms.

ms. Limonin stimulation produced a reduction in cAMP at about 100 ms and a significant elevation in IP_3 at a time above 100 ms (Fig. 6).

The results for the release of cyclic nucleotides in intact taste cells (CV) and intact non-sensory epithelial cells (EP) (representing *in situ* experiments) using the FPS are shown in Fig. 7. Experiments were performed with and without the presence of the non-specific PDE inhibitor, IBMX. No IP_3 determinations were performed. Stimulation (220 ms) by cyclo(Leu-Trp), in the presence of IBMX, produced significant elevation in cAMP in the CV cells but not in EP (Fig. 7A), suggesting a tissue specific effect. There was no such effect when IBMX was not present. Limonin stimulation (350 ms) reduced cAMP in the CV but not in EP, but such a reduction was dependent on the absence of IBMX (Fig. 7B). Stimulation by naringin (220 ms) did not affect cAMP level (Fig. 7C). Stimulation of CV and EP cells by the three bitter tastants did not affect cGMP level at these time ranges of stimulation.

A summary of all signaling cascades is shown in Table I.

Table I: Summary of taste transduction pathways stimulated by selected bitter tastants that occur in agricultural food products.

Tastant	Food Product	Signals		
		cAMP	cGMP	IP_3
Limonin	Citrus	↓	-	↑
Naringin	Citrus	↓?	-	↑
Cyclo(Leu-Trp)	Cheese	↑	-	NA
Catechin	Wine	↓	-	↑

Arrows: ↑ - elevation; ↓ - reduction; - neither reduction nor elevation; NA not available

Discussion

Signal transduction experiments using QFM and FPS provided the early time-course (below 200 ms) for the fast and transient release of signal molecules such as cAMP and IP_3 in response to stimulation by the selected bitter tastants. These *in-vitro* experiments were conducted with taste tissue homogenates, *which* allow biochemical control of events at the cellular level. The fast pipetting system (FPS) allows us to determine the fast release of the above signal molecules at time periods of 75 ms and up from intact cells, conditions that are more physiological but require the use of many animals.

A major advantage of the biochemical experiments with QFM is the possibility of using the same tissue across different time periods so that

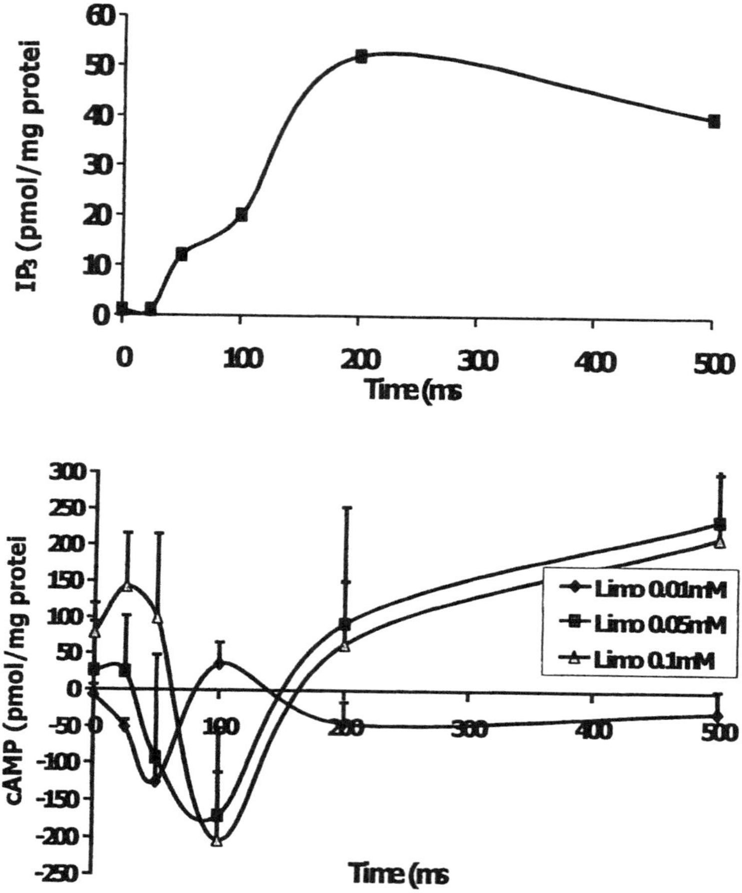

Figure 6. Time-course of limonin-stimulated IP_3 (top) and cAMP (bottom) production in rat taste tissue. Data are the mean of two experiments done in triplicates. Data are expressed as net IP_3 and cAMP (pmols/mg of protein) after subtracting the basal levels. For IP_3 experiment the concentration of limonin was 0.1 mM. Note a modest increase in IP_3 production at 200 ms and a drop in cAMP at 100 ms.

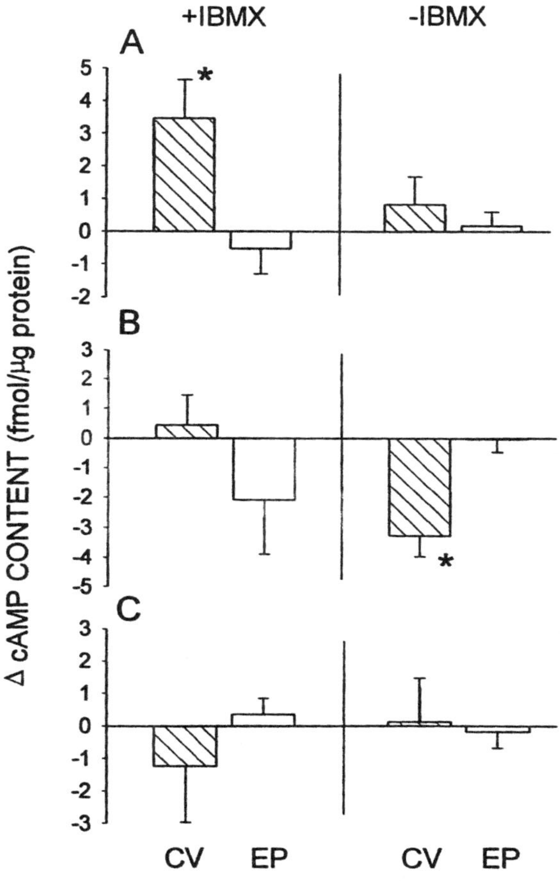

Figure 7. Changes in cAMP level in intact CV taste cells (hatched bars) and in intact non-sensory epithelial cells (EP)(open bars) following 220 ms stimulation by 1 mM cyclo(Leu-Trp) (A), 350 ms stimulation by 35 µM limonin (B) and 220 ms stimulation by 1.7 mM naringin (C). Values are above basal level and are the means and SEM of 20 rats for each data point. * Indicates significant ($p < 0.0005$) difference from the basal level.

collecting time course data is reliable and easier. Therefore, signal transduction experiments with QFM and FPS are complementary to each other. Interestingly, although these two experimental systems were different in their technology and the experimental concept, the signal transduction results of the QFM and FPS have led to similar conclusions.

The reduction in cAMP due to stimulation by limonin and catechin fits well with the α-gustducin or α-transducin subunits activating PDE transduction pathway (*20*). Such a conclusion is supported by the fact that IBMX (a PDE inhibitor) abolished the reduction of cAMP in intact CV cells in response to stimulation by limonin (Fig. 7B). The increase in IP_3 in response to stimulation by the catechin and limonin (Fig. 4 and 6, top) may result from activation by the βγ-subunits of the same G-proteins, similar to denatonium (*13*). The transduction results for the cheese-derived bitter peptide cyclo(Leu-Trp) (Fig 5 and Fig. 7A) indicated an elevation in cAMP, a pathway which was previously suggested (*12, 21*). In the case of cyclo(Leu-Trp)-stimulated cAMP increase, IBMX had to be present, at least when intact cells were used during 220 ms period. Thus, an inhibition of cAMP-PDE was needed to show the cAMP increase. This may suggest that this bitter dipeptide activated the adenylyl cyclase cascade.

The results (see summary Table 1) in both experimental systems suggest multiple transduction mechanisms for bitter taste and are in line with recent studies in other laboratories (*22-23*) Limonin stimulation in both systems induced a transient drop in cAMP level within the initial 200 ms. The bitter dipeptide cyclo(Leu-Trp) induce an elevation in cAMP at around 200 ms. A clear transient IP_3 increase was found following stimulation by naringin and catechin. None of the tested bitter tastants affected the level of cGMP within the first few hundred ms period. A lack of effect on cGMP was evident although we found that xanthine bitter tastants such as caffeine and theophylline stimulated a transient cGMP increase in taste tissue (*12*).

Acknowledgments

This research was supported by grant I-2518 from BARD, The United States-Israel Binational Agricultural Research & Development Fund. We thank the excellent technical assistance by Ms. O. Agamy and Mr. E. Maaravi.

References

1. *Bitterness in Food and Beverages*; Rouseff, R.L.; Elsevier: Amsterdam, 1990, pp. 356.
2. Ginnard, J. X.; Hong, D.Y.; Zoumas-Morse, C.; Budwig, C.; Russell, G. *Physiol. Behav.* **1974,** *56,* 1256-63.
3. Haslam, H; Lilley, T.H. *Crit. Rev. Fd. Sci. Nutr.,* **1988,** *27,* 1-40.
4. Fenwick, G.R.; Curl, C.L.; Griffiths, N.M.; Heaney, R.K.; Price, K.R. in *Bitterness in Food and Beverages*; Rouseff, R.L.; Ed.; Elsevier: Amsterdam, 1990, pp. 205-250.

5. Maga, J.A. in *Bitterness in Food and Beverages*; Rouseff, R.L.; Ed.; Elsevier: Amsterdam, 1990, pp. 81-101.
6. Roy, G.M. *Trends Food Sci. Technol.,* **1992,** *3,* 85-91.
7. Sullivan, J.J.; Jago, G.R. *Aust. J. Diary Technol.,* **1972,** *27,* 98.
8. Roy, G.M. *Crit Rev. Food Sci. Nutr.,* **1990,** *29,* 59-71.
9. Bernhardt, S. J.; Naim, M.; Zehavi, U.; Lindemann, B. *J. Physiol.* **1996,** *490,* 325-336.
10. Spielman, A. I.; Nagai, H.; Sunavala, G.; Dasso, M.; Breer, H.; Boekhoff, I.; Huque, T.; Whitney, G.; Brand, J. G. *Am. J. Physiol.* **1996**, *270*, C926-C931.
11. Rosenzweig, S., Yan, W., Dasso, M. Spielman, A.I. *J. Neurophysiol.,* **1999,** *81,* 1661-65.
12. Huang, L.; Shanker, Y.G.; Dubaiskaite, J.; Zheng, J.Z.; Yan, W.; Rosenzweig, S.; Spielman, A.I.; Max, M.; Margolskee, R.F. *Nature (Neurosci.)* **1999,** *2,* 1055-1062.
13. Lindemann, B. *Physiol. Rev.,* **1996,** *76,* 719-766.
14. Naim, M.; Rogatka, H.; Yamamoto, T.; Zehavi, U. *Physiol. Behav.,* **1982,** *28,* 979-986.
15. Spielman, A.I.; Brand, J.G.; Wysocki, L. *Chem. Senses,* **1989,** *14,* 841-846.
16. Bradford, M.A. *Anal. Biochem.,* **1976,** *72,* 248-254.
17. Shiba, T.; Nunami, K.-I. *Tetrahedron Lett,.* **1974**, *6*, 509-512.
18. Noble, A. C. *Physiol. Behav.* **1994**, *56*, 1251-1255.
19. Cagan, R. H.; Maller, O. *J. Comp. Physiol. Psychol.,* **1974**, *87*, 47-55.
20. Ruiz-Avila, L.; McLaughlin, S.K.; Wildman, D.; McKinnon, P.J.; Robichon, A.; Spickofsky, N.; Margolskee, R.F. *Nature,* **1995,** *376,* 80-85.
21. Price, S. *Nature,* **1973,** *241,* 54-55.
22. Ogura, T; Kinnamon, S.C. *J. Neurophysiol.,* **1999,** *82,* 2657-2666.
23. Tsunenari, T; Kurahashi, T.; Kaneko, A. *J. Physiol. (Lond),* **1999,** *519,* 397-404.

Chapter 3

Molecular Mixture Models: Connecting Molecular Events to Perception

Daniel M. Ennis

Institute for Perception, 7629 Hull Street, Suite 200, Richmond, VA 23235

Molecular mixture models are discussed that connect events at the periphery to taste perception. Application of the models to mixtures of glucose and fructose shows evidence for the existence of a transducer in human sweet taste and provides an hypothesis to explain the superior sweetening power of fructose to glucose. These mathematical models may be a useful alternative to the use of animal models to describe human chemosensory mechanisms.

Humans are not rats or spiny lobsters. This obvious fact about humans becomes important when non-human animals are used as models of human biology. Although there are many practical reasons to study chemical sensing in non-human animals, human chemosensory mechanisms are not identical to those of the many animal species used to model them. In order to connect peripheral events in humans to chemosensory perception, it would seem prudent to develop models of peripheral events based on data obtained from humans. Since this data is usually in the form of psychophysical measurement, there is a need to build mathematical models that relate molecular events involving human interactions with chemicals in the environment to the perception of those interactions. To be reliable, these models necessarily require weak assumptions (those with minimal demand on credibility) about the connection between events at the periphery and those centrally.

Mixtures occur ubiquitously in the environment and we are continually exposed to them. The effects of mixtures on the chemical senses are complex and puzzling. One of these effects is called "synergy." This term is used loosely to describe effects that cannot be explained by additivity of the components alone, and is used even more loosely to describe effects greater than "expected." There is a lack of clarity concerning the meaning of additivity and synergy in mixture research. Can we define synergy satisfactorily and develop models to predict it? More generally, can we use data from human psychophysical studies on mixtures to draw conclusions about molecular mechanisms that occur on the tongue or at the olfactory epithelium without direct biochemical information?

Mixtures

Our senses are constantly exposed to chemical mixtures. Even when single substances are presented as odorants or tastants, they may undergo chemical or biological changes leading to mixtures at receptor sites. Since foods and beverages are mixtures of potential chemosensory stimulants, mixtures also occur in commercial applications. One of the most extensive commercial applications of mixtures is the use of high fructose corn syrups. These mixtures predominantly contain different levels of glucose and fructose. These sweeteners along with others containing sucrose, glucose and high intensity sweeteners are used very broadly in foods and beverages throughout the food industry.

Synergy and Isoboles

When two substances produce the same type of effect; such as cardiac stimulation, sweet taste or a particular odor quality; the concentrations of each compound in a mixture that produce the same level of response can be determined. An isobole is a plot of the concentrations of the substances that produce the same level of effect. In a taste or olfactory experiment, the points on one of these plots are referred to as *points of subjective equality*.

Two isoboles are shown in Figure 1. The line marked "addition" shows a linear relationship between the concentrations of A and B that produce equal effects. From the curve marked "synergy", it can be seen that as the concentration of A increases, the corresponding amount of B decreases but not in a linear fashion. In fact, the total concentration of both substances together is less than that predicted from a linear isobole.

Points of subjective equality for sweetness have been determined in humans for various concentrations of glucose and fructose. The isoboles are convex. This suggests a synergistic relationship between these sweeteners. It would be useful to provide a molecular explanation for the occurrence of synergy in humans. The challenge is to explore this hypothesis without biochemical evidence and to test molecular models using psychophysical data.

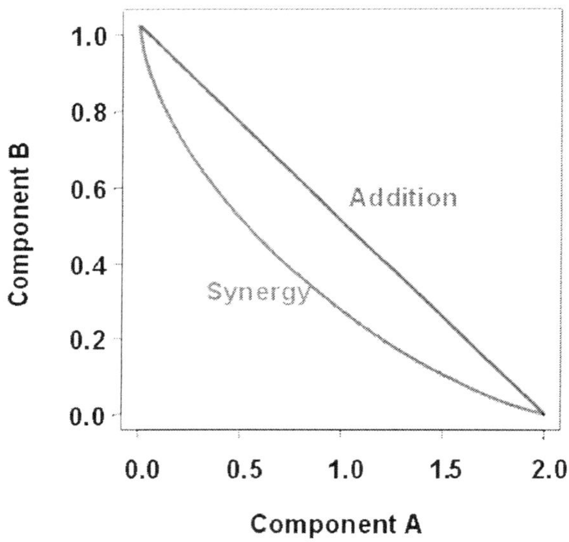

Figure 1. Linear and non-linear isoboles

Models Based on the Law of Mass Action

When a substance (A), an agonist, binds a receptor (R) to produce an agonist-receptor complex (AR), it is assumed that this binding process is reversible, $A + R \leftrightarrow AR$.

From the Law of Mass Action

$$[AR]/[R_t] = K_a[A]/\{1+K_a[A]\}, \qquad (1)$$

where [A] represents the concentration of A, K_a is the association constant for A binding to R, and $[R_t]$ is the total receptor concentration. Under the assumption that the effect produced, such as sweet taste intensity or response to a drug, is a linear function of activated receptors, equation 1 can be used to model the fraction of the total response associated with A as a function of [A]. When modeling isobole data this linearity assumption is strong and unnecessary and more robust and credible mixture models can be developed without it.

These new models involve three improvements:

1. Replace linearity with a monotonicity assumption,
2. Allow for the possibility of a transducer entity, and
3. Include mixture effects.

The monotonicity assumption states that as [AR] increases, the measured response always increases or always decreases. This assumption allows for a broad range of dose-response functions, including the linear function, but does not demand any form of proportionality. An important feature of a monotonic function is that it is *one-to-one.* A function is *one-to-one* if and only if $f(x_1) = f(x_x)$ implies $x_1 = x_2$. For the purpose of this discussion, $f(x_1)$ and $f(x_2)$ are percepts (such as sweetness intensity) and x_1 and x_2 are chemical entities in the neural chain from stimulation to perception. This idea of a *one-to-one* mapping is used extensively in the development of molecular mixture models (1). The chemical entitiles x_1 and x_2 may result from receptor binding, transducer binding or both.

The need to introduce a transducer entity arises from research on the transduction mechanism. The transduction mechanism is a process by which a stimulus induces a potential change in a receptor cell. Compound A binds receptor R to produce complex AR. This complex interacts with transducer T resulting in the formation of ART. It is now assumed that the intensity of the percept is monotonically related to [ART]. These reactions can be represented as A + R ↔ AR and AR + T ↔ ART.

The association constant, K_a, for the initial binding to a receptor is a measure called *affinity* and the association constant, K_{ar}, for the second reaction is a measure called *efficacy*. A compound could have a high affinity and no efficacy such as a β-blocker or taste inhibitor. In general, a compound's effect depends on its various levels of affinity and efficacy. These ideas have been used to develop a general model (*1, 2*) for the mixture effects of any number of compounds binding any number of receptors and transducers. This model explains the synergistic effects of glucose and fructose in mixtures.

Application to Glucose/Fructose Mixtures

De Graaf and Frijters (*3*) found the total concentration of mixtures of fructose and glucose that are as sweet as glucose alone at various target concentrations (0.125, 0.25, 0.5, 1.0 and 2.0 molar). The mixtures contained fructose at 25%, 50%, 75% and 100%. From these data it is possible to construct isoboles for glucose and fructose showing the points of subjective equality. De Graaf and Frijters used the method of constant stimuli to find the points of subjective equality. This method involves paired tests of a target concentration of glucose and particular glucose/fructose mixtures to determine the sweeter stimulus. From these data the points of subjective equality were derived.

Figure 2 shows two special cases of the general mixture model. In both cases independent systems are assumed. In one case a transducer is included and in the other there is no transducer. The transducer model fits the data (the marked points) significantly better than the alternative. Although more complex special cases of the general mixture model also fit the glucose/fructose data, these cases have been discussed elsewhere (*4*) and will not be discussed here.

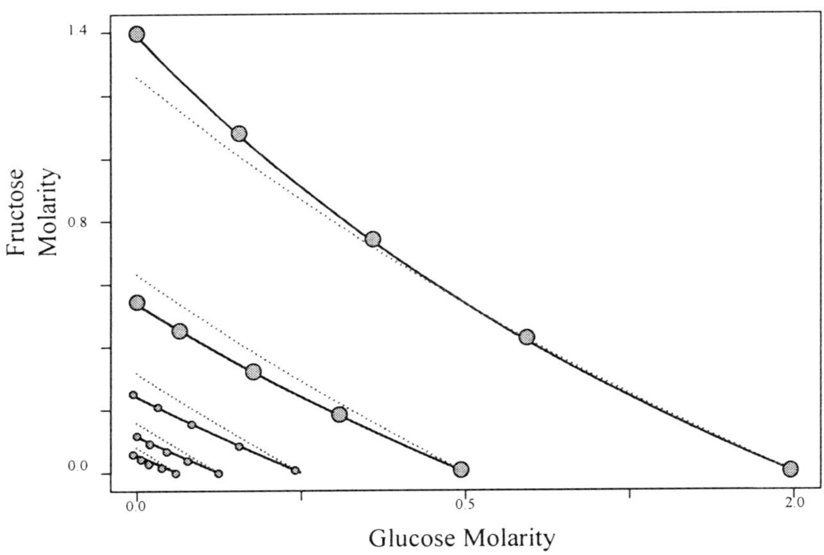

Figure 2. Isobole data and two model fits. The model with the transducer is the solid line, the model without the transducer is the dotted line.

The isobole equation for the mixture model without a transducer and assuming independent receptors is

$$[B_m] = k\ \{[A]-[A_m]\}/\{1+2[A_m]K_a + [A][A_m]\ K_a^2\},$$

where $k = K_a/K_b$, the ratio of affinities.

The isobole equation for the independent receptor-transducer model (*1*) is

$$[B_m] = \alpha\ \{[A]-[A_m]\}/\theta \qquad (2)$$

where $[A_m]$ and $[B_m]$ are the concentrations of glucose and fructose in the mixture, $[A]$ is the target concentration of glucose, α and θ are parameters that depend on the affinities and efficacies of A and B, and θ additionally depends on [A] (*1*).

Applications to Food and Beverage Products

In this article, the discussion focused on models for mixture effects in simple systems. For more complex systems, such as a carbonated beverage, mixture effects of product components for one sensory attribute may depend on other attributes. The ideas discussed are a useful starting point for these applications, but may require modification to take other taste qualities into account. Central suppression of sweetness by sourness is one possible effect that may need to be considered. Mixture research on food and beverage products should determine what extensions of the mixture models need to be made.

Implications

Linear isoboles cannot explain synergy. Simple binding to common receptors and transducers produce linear isoboles. Independent receptor and transducer systems with simple binding will predict synergy and it can be seen from Figure 2 that this model accounts for the synergistic effect of glucose/fructose mixtures in humans.

Psychophysical data, such as that of De Graaf and Frijters (*3*), can be used to make inferences about molecular mechanisms at the periphery. This is possible because the type of intensity matching data used only requires a weak (monotonic) assumption about the connection between events at the periphery and mental percepts. Models using the Law of Mass Action were based on peripheral binding parameters. From these parameters, inferences can be drawn about the presence of a transducer, the type of binding and the relative affinity and efficacy of different chemosensory stimulants. Examination of the fit to the mixture data shows that fructose may be sweeter than glucose because of its greater affinity (K_b is greater than K_a) though weaker efficacy (K_{br} is less than K_{ar}). Antagonists, or blockers, have high affinity (they occupy receptors) but no efficacy. This leads to the unexpected conclusion that although fructose is sweeter than glucose in humans it may be more similar to an inhibitor of sweet taste since its sweetness is driven by its affinity. It is not an inhibitor because it has a non zero efficacy.

References

1. Ennis, D. M. *Food Chem.* **1996**, 56, 329-335.
2. Ennis, D. M. *Chem. Senses* **1991**, 16, 1-17.
3. De Graaf, C.; Frijters, J.E.R. *Chem. Senses* **1986**, *11*, 295-314.
4. Ennis, D.M. *Food Tech.* **1991**, *45*, 140, 142, 145.

Genetics and Physiology

Chapter 4

Genetics of Sweet Taste

Gary K. Beauchamp, Danielle R. Reed, Michael G. Tordoff, and Alexander A. Bachmanov

Monell Chemical Senses Center, 3500 Market Street, Philadelphia, PA 19104

Individual differences in the perception of sweetness can be due to effects of environment, genes and their interaction. The study of individual differences due to genes can shed light on the mechanisms of sweet perception. Human studies have demonstrated substantial individual differences in perception and preference for sweet substances but there is little evidence as to the basis for these differences. In contrast, animal model studies with the mouse, the organism of choice for molecular genetics studies, have clearly demonstrated that differences in sensitivity to, and preferences for, sweetness among strains of inbred mice are due to genetic factors. Genetic mapping studies in our laboratories and others have shown that one locus, *Sac* (located on distal chromosome 4) plays a prominent role in sweet perception in mice and, most likely, in humans. Recent studies have provided strong evidence that *Sac* is a sweet taste receptor (called *Tas1r3*). This work sets the stage for a much fuller understanding of the initial events in sweet taste perception and may provide new insights into possible genetic contributions to human individual differences in perception and preference for sweetened foods and beverages.

Introduction

Genes code for proteins that act throughout the pathway from the stimulus-receptor interactions in the mouth to conscious perception of sweetness. All of these steps are under genetic control during development in all individuals of a species. A major goal is to map such complex pathways and to study the interplay of developmental and environmental factors in gene expression.

One approach to this overall goal of understanding the genetic factors involved in sweet perception is to focus on individual differences within a species in sweetness perception. This approach searches for or creates (e.g., by inducing mutations) polymorphisms in shared genes in a population. It attempts to identify and explain differences such as why one person finds a given level of sweetener weak when another finds it sickeningly strong. This difference in the perception of sweet taste intensity could be due to non-genetic difference between the two individuals, for example different nutritional states. But it could also be due to genetically-determined individual differences in the binding characteristics of a sweet receptor.

In the current essay, we concentrate mainly on the evidence relevant to the genetics of individual differences in sweetness perception. We have liberally borrowed from and updated where appropriate, our review that dealt with this topic among others (1).

Individual variation in sweet taste perception and preference in people

A prerequisite for a genetic influence on individual differences in sweet taste perception and preference is for reliable individual differences to exist. Several studies suggest that there are substantial individual differences in perceived intensity of, and liking for, sweetness. For example, Thompson et al (2) categorized subjects into one of two types, based on their reaction to a series of sucrose concentrations. Type I subjects had a rise, and then a decline in liking for sucrose as the concentration increased; Type II subjects had a rise in the liking for increasing concentrations, and as the concentration rose higher, there was a plateau rather than a decrease in pleasantness ratings.

Both Pangborn (3) and Meiselman (4) noted the extreme degree of individual variation in sweet preferences. Likewise, Witherly and Pangborn (5) observed that large within-group variation overshadowed between-group differences when obese and lean adults gave hedonic ratings for sweet stimuli. Faurion et al (6) tested human subjects for their sweetness recognition thresholds or their perceived intensity ratings of sweet-tasting stimuli. They concluded that variation within individuals is small compared with variation between individuals. In summary, there is diversity in the hedonic and sensory experience of humans to sweet tastes and it is reasonable to believe that some of this diversity may be genetic in origin. However, as indicated below, most studies designed to investigate heritable variation in sweet preference in humans have not indicated a strong genetic component.

Genetic studies of sweetness

Studies designed to test the contributions of genetic and experimental variability on sweet perception and preference in humans fall into two categories: family members are compared to see if the degree of biological relatedness co-varies with similarities in behavior (a family study), or twin pairs are compared to see if monozygotic twins are more behaviorally similar compared with dizygotic twins (a twin study). Studies employing these techniques have several limitations due to methodological constraints. Preferences for individual food items are generally assessed by pencil and paper questionnaires, and may not reflect a person's actual eating behavior with regard to those foods. Some foods, particularly sweet ones, are preferred much more strongly by all subjects than are other foods. If all people under study (e.g., all twin pairs) like a food very much, then ceiling effects may reduce individual variation, which precludes genetic analysis. Food diaries may provide more accurate estimates of the foods subjects eat, but because food intake is usually only measured for a brief period (3-7 days), it may not be representative of the subjects' food selection habits. Diary data may also be inaccurate due to reporting biases of subjects.

With these limitations in mind, we review studies of preferences for individual food items generally regarded as sweet followed by studies of sweet carbohydrate selection patterns. More detail on individual food items not considered sweet and on fat selection patterns can be found in Reed et al (1).

Family and twin studies of preferences for individual sweet food items.

Studies of food preferences of families and twins have indicated that genetic differences cannot account for individual differences (7). Studies of family resemblance for individual food items, both sweet and non-sweet, have demonstrated few or no correlations between first degree relatives (8-13, 7). For example, Ritchey & Olson (13) found that pre-school children's degree of liking for sweet foods was uncorrelated with the degree of parental liking for sweet foods. However, a limitation of these studies is that taste preferences change over the life span (14-16), and studies of parent-child correlations may be at some disadvantage in detecting any genetic differences. Similarly, twin studies provide little support for the hypothesis that genetic differences are involved in individual differences in sweet food preferences (Table 1).

Table 1. Twin studies of preferences for individual sweet substances (taken from Reed et al (1))

Food item	N (MZ/DZ)[a]	Significance[b]	Reference
Candy	232/223	ns	(17)
Cereal, sweetened	14/21	+	(18)
Doughnut	13/10	ns	(19)
Honey	13/10	ns	(19)
Ice cream	13/10	ns	(19)
Jam	13/10	ns	(19)
Jam, jelly, syrup	232/223	ns	(17)
Peppermint	38/34	ns	(20)
Snack cake	14/21	ns	(18)
Soda/cola	14/21	ns	(18)
Sucrose	146/165	ns	(21)
Sugar	38/34	ns	(20)
Sugar in coffee	232/223	+	(17)
Sugar in tea	232/223	+	(17)

[a] number of monozygotic and dizygotic twin pairs

[b] ns is not significant; + is significant

Family and twin studies of sweet taste sensitivity and preference

Cavalli-Sforza (22) reviewed family and twin studies of food preferences and concluded, "food preferences are largely determined by cultural transmission and individual experience. It is hard, however, to exclude genetics entirely". Likewise, Pérusse & Bouchard (23) concluded that 20% of the variance associated with fat and carbohydrate preference is genetic, and they further concluded that, "the literature indicates a rather moderate role of heredity in energy intake and food preferences" but "a low heritability level does not mean that genes have nothing to do with nutritional habits."

Based on the evidence currently available, we also conclude that for individual sweet food items and for sucrose solutions, genetic heritability is low. Greene et al (20) reported no heritable component for sucrose preference, although they did find a difference in the liking for sweet substances between Afro-American and Caucasian-Americans (see also Bacon et al (24)). Krondl et al (19), studying sweet taste recognition thresholds in monozygotic and dizygotic twin pairs, found that the heritability for sweet sensitivity was 0.52, which approached but did not achieve statistical significance.

A bitter taste polymorphism and sweet food preferences.

Some people are able to taste very low concentrations of the bitter compound phenylthiocarbamide (PTC) and its chemical relative 6-n-propylthiouracil (PROP), while other people are much less sensitive to its bitter taste. This taste polymorphism is genetically determined and is thought to depend upon a single gene although the gene has not yet been identified (25-26). If this genetically determined bitter taste polymorphism influences sweet taste perception and preference, and there is some indication that it does, this is evidence that individual differences in sweet taste have a genetic basis. For example, people who perceive PROP as extremely bitter may also perceive sucrose to be sweeter and rate it as hedonically less pleasing than do medium PROP tasters or non-tasters (27), perhaps because the extreme intensity of the sweet taste is offensive (28). However, this observation has not been replicated in some reports. Kang et al (29) grouped Korean subjects by whether or not they liked sugar; among the group who disliked sugar, there was a trend toward a higher frequency of PROP nontasters. Likewise, Drewnowski et al (30) failed to detect a relationship between the hedonic appreciation of sucrose and the perception of the bitter taste of PROP. Bartoshuk (31) have argued that the failures to find a relationship between PROP status and sweet taste perception is due to methodological and scaling problems. More research is needed to resolve this issue.

Summary of human work

There appears to be substantial individual variation among people in sweet food preferences and sweet taste perception. In spite of this, few studies using techniques designed to detect genetic influences on these differences have found them. Does this mean that genes are not involved in individual variation? Probably not since relatively few studies have been conducted and the methods used have often not been very powerful. Indeed, animal model studies with rodents, as reviewed below, provide convincing evidence that genetic differences within a species have profound effects on sweet taste perception and preference.

Animal model studies

Sweet perception and preference in mice.

Many studies have demonstrated that inbred strains of rodents differ in their preference for the sweetener saccharin and that the liking for saccharin can breed true in outbred strains of rodents (see (1) for references). Fuller (32) investigated the inheritance pattern of sweet intake in the mouse by studying two inbred strains of mice (C57BL/6J and DBA/2J) that differed in their saccharin preference. Based on saccharin intake and preference of the two strains and their F1, F2 and backcross generations, he concluded that an allele of a gene existed in the C57BL/6J strain (Sac^b) that conferred an increased "incentive value" to ingested saccharin. The results of his breeding experiments suggested a dominant mode of inheritance for the Sac^b allele.

Stockton and Whitney (33) also investigated sucrose and glucose preferences of several inbred strains of mice and their F_1 generations. Inbred strains of mice showed differences in their preference and intake of both sucrose and glucose solutions; for some concentrations of sucrose, the amount of variance accounted for by genes was over 50%. The sucrose preferences of the F_1 mice were between the high- and low-preferring strains for some concentrations, and resembled the high-sucrose preferring strain for other concentrations, suggesting an additive or dominant mode of inheritance. Glucose preference yielded lower levels of heritability compared with sucrose preference; glucose is less sweet than sucrose to humans at similar concentrations and the lower heritability values of glucose may be a result of this attenuated sweet taste.

After Lush (34) confirmed the existence of the *Sac* locus, he crossed mice from the high and low preferring strains and the F_1 progeny of this cross were backcrossed to the low-saccharin preferring strain. All of the resulting mice fell into two groups (likers or non-likers) based on their intakes of acesulfame (a high intensity, non-nutritive sweetener) and saccharin, suggesting a single locus was responsible for much of the strain difference.

The *Sac* locus was first mapped to the distal portion of mouse chromosome 4 by measuring the saccharin preference of 19 strains of BXD recombinent inbred (RI) mice (35); this location was confirmed by BXD RI mapping, and the genotyping of backcross progeny from strains of mice high and low in sweet preference (36). Subsequently, as described in more detail below, Bachmanov et al (37) mapped the *Sac* locus to a small region of distal chromosome 4 using the F_2 generation from a high-saccharin preferring strain (C57BL/6ByJ) and a low-saccharin preferring strain (129/J). A subsequent report from Blizard et al (38) also suggested that the *Sac* locus maps to distal chromosome 4.

Some inbred strains of mice find the amino acid D-phenylalanine (D-phe) to be sweet (i.e., taste similar to sucrose) while other strains apparently do not (reviewed in (39)). Breeding experiments using tasters and non-tasters of D-phe and their F_1 and F_2 hybrid generations suggested that a single genetic locus largely determined this phenotype (*dpa*). Linkage tests using several markers suggested that the *dpa* locus maps close to the *b* (brown) and *Mup-1* loci on proximal chromosome 4 (40).

Because the *Sac* locus and the *dpa* locus both were reported to influence the preference for sweet substances, it was unclear whether they represented alleles of the same gene, or whether *Sac* and *dpa* were distinct genes, each with two or more alleles. Capeless and Whitney (41) attempted to resolve this issue by testing five inbred strains of mice for their preferences for D-phe and saccharin. They concluded that if *Sac* and *dpa* loci are allelic, then at least three alleles exist, but they did not rule out the possibility that two independent loci, each with two alleles, could also account for their observations. Subsequent linkage studies by Bachmanov et al (37) have concluded that the *dpa* locus and the *Sac* locus are separate based on their chromosomal positions.

Other investigators have measured the volume of sweet solutions drunk by F_2 mice and mice from RI strains, and have mapped several other chromosomal regions which co-segregate with sweet preference (see Table 3 in Reed et al. (1) for a detailed summary).

The *Sac* locus and the T1R family of putative taste receptors.

One possibility is that the *Sac* locus codes for a sweet receptor. In an independent line of investigation, there has been a considerable effort to identify sweet taste receptors from taste bud tissue (42-47). Hoon et al (48) identified a putative sweet taste receptor called TR1 (Taste Receptor 1 now renamed T1R1). This protein has a deduced amino acid sequence representative of a seven-transmembrane domain G protein-coupled receptor. Its role in sweet taste reception was suggested based in part on its expression in the apical ends of a subset of taste receptor cells. The other evidence implicating T1R1 in sweetness detection involves possible co-localization of the T1R1 gene with *Sac* (48). Studies on *Sac* demonstrate that in addition to its role in controlling strain differences in behavioral acceptance of sweeteners as reviewed above, it also affects afferent responses of gustatory nerves to sweeteners (37). This strongly suggests that *Sac* determines peripheral sweet taste perception and thus may code for a sweet taste receptor. Therefore, allelic variation in the T1R1 gene (*Tas1r1*) may determine the phenotypes associated with *Sac*.

To assess *Tas1r1* as a candidate gene for *Sac*, we (49) compared TR1 cDNA sequences expressed in the tongue tissues of C57BL/6ByJ (B6) and 129/J (129) mouse strains. These two strains have different alleles of *Sac*, with the B6 allele determining higher gustatory neural and ingestive behavioral responses to sweeteners (see Figure 1 in 37). Next, using *Tas1r1* sequence variation between the B6 and 129 strains and our genetic mapping panels originating from these two strains, we conducted a high-resolution linkage analysis of the chromosomal locations of *Tas1r1* and *Sac*. This linkage analysis has confirmed that the *Tas1r1* gene is located on distal chromosome 4, as was initially reported by Hoon et al (48). However, we have found that the *Tas1r1* gene maps proximal to *Sac*. Thus, *Tas1r1* and the saccharin preference locus, *Sac*, are coded by linked but different genes. This finding led us to proceed with positional cloning of the *Sac* locus. We narrowed the location to a small 194 kb portion of chromosome 4 and identified the 12 genes in this region (49-51). Only one of these genes, now called *Tas1r3*, encodes a protein that resembles the known structure of a taste receptor. Furthermore, we have found that strains of mice with low and high preference for sweeteners have potentially important differences in DNA sequences for this gene. Finally, we conducted a breeding experiment in which the *Tas1r3* gene from the high preferring B6 strain was bred onto the low preferring 129 strain. The resulting congenic strain showed a high preference for sweeteners. From these data we concluded that *Tas1r3* was very likely a sweet taste receptor.

Simultaneously with our work, several other groups of scientists were conducting complimentary studies aimed at identifying potential sweet receptors

(51-56). These investigators also identified *Tas1r3* as a putative sweet taste receptor. Moreover, it was suggested in two of these publications (53-54) that *Tas1r3* may dimerize with the other two members of this receptor family, *Tas1r1* and *Tas1r2* and that these dimers may be the actual sweet receptor. More recently, Nelson et al (55) reported that when *Tas1r3* and *Tas1r2* are simultaneously expressed in a non-taste cell system certain sweeteners activate the system. This, in combination with the demonstration by this group that, by using direct gene transfer (transgenic) techniques, low preferring mice can be transformed into high preferring ones, provides very strong confirmation that the *Tas1r3* receptor plays a major role in sweet taste reception.

Summary and prospectus

We now have the beginnings of a new understanding of how sweet taste perception is initiated in the periphery, at the receptor level. Many questions still need to be addressed, however. How many receptor combinations (e.g., combinations of *Tas1r3* with *Tas1r2* and/or *Tas1r1*) are actually involved in detecting sweeteners? Are there additional receptors, for example for sweet amino acids? How does the structure of these receptors vary among species that respond differently to sweet compounds? Exactly how do these receptors bind sweet molecules? Can human differences in preference and perception of sweeteners be explained, as they apparently are in inbred mice, by differences in receptor sequences?

Clearly, many fundamental questions remain. Nevertheless, the progress made in understanding sweet taste reception during the last two to three years has been remarkable. We can anticipate many more exciting discoveries to be made, and many of the questions listed above to be answered, in the coming years. This is a very exciting time for taste research.

Acknowledgement

This research was supported by NIDCD 00882.

References

1. Reed, D.R.; Bachmanov, A.A.; Beauchamp, G.K.; Tordoff, M.G. Price, R.A.; *Behav. Genet.* **1997**, *27*(4), 373-387.

2. Thompson, D.A.; Moskowitz, H.R.; Campbell, R.G. *Physiol. Behav.* **1977** *19*, 335-337.
3. Pangborn, R.M. In *Carbohydrate sweeteners Foods and Nutrition*; Koivistoinen., P.; Hyvonen, Eds; Academic press: London, 1980; pp. 87-110.
4. Meiselman, H.L. In *Sweetness*; Dobbing, J., Ed.; Springer-Verlag: Great Britain, 1987; p 261-276
5. Witherly and Pangborn 1980
6. Faurion A. In *Progress in Sensory Physiology*; Autrum, H., Ottoson, D., Perl, E.R., Schmidt, R.F., Shimazu, H., Willis, W.D., Eds.; Springer-Verlag: Heidelberg, Germany, 1987; pp. 129-201.
7. Rozin, P. *Appetite* **1991**, *16*, 93-102.
8. Birch L.L. *J. Nutrition Edu.* **1980**, *12*, 14-18.
9. Burt, J.V.; Hertzler A.A.; *J. Nutrition Edu.* **1978**, *10*, 127-128.
10. Oliveria, S.A.; Ellison, R.C.; Moore, L.L.; Gillman. M.W.; Garrahie, E.J.; Singer, M.R. *Am. J. Clin. Nutr.* **1992**, *56*, 593-598.
11. Pliner, P. *J. Nutrition Edu.* **1983**, *15*, 137-140.
12. Pliner, P.; Pelchat, M.L. *Appetite* **1986**, *7*, 333-342.
13. Ritchey, N.; Olson C. *Ecology Food Nutrition* **1983**, *13*, 257-266.
14. Beauchamp G.K.; Cowart B.J. In *Sweetness*; Dobbing, J., Ed.; Springer-Verlag:Great Britain, 1987; pp 127-140.
15. Desor, J.; Greene, L.; Maller, O. *Science* **1975**, *190*, 686-687.
16. Desor, J.; Beauchamp, G. *Physiol. Behav.* **1987**, *39*, 639-641.
17. Fabsitz, R.R.; Garrison, R.J.; Feinleib, M.; Hjortland, M. *Behav. Genet.* **1978**, *8*, 15-24.
18. Falciglia, G.A.; Norton, P.A. *J. Am. Diet. Assoc.* **1994**, *94*, 154-158.
19. Krondl, M.; Coleman, P.; Wade, J.; Milner J. *Human Nutrition: Applied Nutrition* **1983**, *37A*, 189-198.
20. Rozin P.; Millman, L. *Appetite* **1987**, *8*, 125-134.
21. Greene, L.S.; Desor, J.A.; Maller. O. *J. Comp. Physiol. Psych.* **1975**, *89*, 279-284.
22. Cavalli-Sforza L.L. In *Genetic Variation and Nutrition;* Simopoulos A.P.; Childs, B., Eds.; Karger: Basel, Switzerland, 1990; pp. 35-48.
23. Pérusse L, Bouchard C. In *The Genetics of Obesity*, Bouchard, C., Ed.; CRC Press: Boca Raton, Florida, USA, 1994 ; pp. 125-134.
24. Bacon, A.W.; Miles .S.; Schiffman, S.S. *Physiol. Behav.* **1994**, *55*, 603-606.
25. Merton, B.B. *Acta Genet.* **1958**, *8*, 114-128.
26. Reed, D.R.; Bartoskuk, L.M.; Duffy, V.; Marino, S.; Price, R.A. *Chem. Senses* **1995**, *20*, 529-533.
27. Looy, H.; Weingarten, H.P. *Physiol. Behav.* **1992**, *52*, 75-82.
28. Duffy, V.B.; Weingarten, H.P.; Bartoshuk, L.M.; *Chem. Senses* **1995**, *20* (6), 688 abstract.

29. Kang, Y.S.; Cho, W.K.; Yurn, K.S. *Eugenics Quart.* **1967**, *14*, 1-6.
30. Drewnowski, A.; Henderson, S. A.; Shore, A. B. *Chem. Senses* **1997,** 22, 27-37.
31. Bartoshuk, L.M. *Chem. Senses* **2000**, *25* (4), 447-60.
32. Fuller, J.L. *J. of Heredity* **1974**, *65*, 33-36.
33. Stockton, M.D.; Whitney, G. *J. Comp. Physiol. Psychol.* **1974**, *86*, 62-68.
34. Lush, I.E. *Genet, Res. Camb.* **1989**, *53*, 95-99.
35. Phillips, T.J.; Crabbe, J.C.; Metten, P.; Belknap, J.K. *Clinical and Experimental Research* **1994**, *18*, 931-941.
36. Lush, I.E.; Hornigold, N.; King, P.; Stoye, J.P. *Genet. Res.* **1995**, *66*, 167-174.
37. Bachmanov, A.A.; Reed, D.R.; Ninomiya, Y.; Inoue, M.; Tordoff, M.G.; Price, R.A.; Beauchamp, G.K. *Mamm. Genome* **1997**, *8*, 545-548.
38. Blizard D.A.; Gudas, E.P.; Frank, M.E. *Chem. Senses* **1996**, *21*, 579.
39. Ninomiya, Y.; Funakoski, M. In *Mechanisms of Taste Transduction*; Simon, S.A.; Roper S.D., Eds.; CRC Press: Boca Raton, Florida, 1993.
40. Ninomiya, Y. Sako, N.; Katsukawa, H.; Funakoshi, M. In *Genetics of Perception and Communication*; Wysocki, C.J.; Kare, M.R., Eds.; Marcel Dekker: New York, 1991; pp 267-278.
41. Capeless, C.G.; Whitney, G. *Chem. Senses*, **1995**, 20: 291-298.
42. Abe, K.; Kusakabe, Y.; Tanemura, K.; mori, Y.; Arai, S. *FEBS Lett.* **1993a**, *316* (3), 253-256.
43. Abe, K.; Kusakabe, Y.; Tanemura, K.; Emori, Y.; Arai, S. *J. Biol. Chem.* **1993b**, *268* (16), 12033-12039.
44. Ansano-Miyoshi, M.; Kusakabe, Y.; Abe, K.; Emori, Y. *J. Biochem. (Tokyo)* **1998**, *124* (5), 927-933.
45. Hoon, M.A.; Ryba, N.J. *J. Dental Research* **1997**, *76*, 831-838.
46. Matsuoka, I., Mori, T., Aoki, J., Sato, T. and Kurihara, K. *Biochem. Biophys. Res. Commun.* **1993**, *194* (1), 504-511.
47. Naim, M.; Striem, B.J.; Tal, M. *Adv. Food. Nutr. Res.* **1998**, *42*, 211-243.
48. Hoon , M.A., Adler, E.; Lindemeier, J.; Ryba, N.J., Zucker, C.S. *Cell* **1999**, *96*, 541-551.
49. Li, X.; Inoue, M.; Reed, D.R.; Huque, T.; Puchalski, R.B.; Tordoff, M.G.; Ninomiya, Y.; Beauchamp, G.K.; Bachmanov, A.A. *Mamm. Genome* **2001**, *12*, 13-16.
50. Bachmanov, A.A.; Li, X.; Reed, D.R.; Ohmen, J.D.; Chen, Z.; Tordoff, M.G.; de Jong, P.J., Wu, C.; West, D.B.; Chatterjee, A.; Ross, D.A.; Beauchamp, G.K. *Chem. Senses* **2001**, 26, 925-933.
51. Li, X.; Bachmanov, A.A.; Li, S.; Chen, Z.; Tordoff, M.G.; Beauchamp, G.K.; de Jong, P.J.; Wu, C.; Chen, L.; West, D.B.; Ross, D.A.; Ohmen, J.D.; Reed, D.R. *Mamm. Genome*, in press.

52. Kitagawa M.; Kusakabe, Y.; Miura, H.; Ninomiya, Y.; Hino, A. *Biochemistry Biophysics Research Communication* **2001**, *283*: 236-242.
53. Max, M.; Shanker, Y.G.; Huan, L.; Rong, M.; Liu, Z.; Campagne, F.; Weinstein, H.; Damak, S.; Margolskee, R.F. *Nature Genetics* **2001**, *28*, 58-63.
54. Montmayeur, J.P.; Liberles, S.D.; Matsunami, H.; Buck, L.B. *Nature Neuroscience* **2001** *4*, 492-498.
55. Nelson, G.; Hoon, M.A.; Chandrashekar, J.; Zhang, Y.; Ryba, N.J.; Zuker, C.S. *Cell* **2001**, *106*: 381-390.
56. Sainz, E.; Korley, J.N.; Battey, J.F.; Sullivan, S.L *J. Neurochemi.*, **2001**, *77*, 896-903.

Chapter 5

Genetic Markers, Taste Responses, and Food Preferences

Adam Drewnowski

Nutritional Sciences Program, School of Public Health and Community Medicine, University of Washington, Seattle, WA 98195

Genetic sensitivity to the bitter taste of 6-*n*-propylthiouracil (PROP) has been linked with reduced acceptance of other bitter compounds and a dislike of some bitter foods. PROP tasters and "supertasters" were more likely to dislike black coffee, grapefruit juice, green tea, Brussels sprouts and some salad greens. Antioxidant phytochemicals such as flavonoids in citrus fruit, polyphenols in tea and red wine, glucosinolates in cruciferous vegetables, and isoflavones in soy products are almost always bitter. Many of these chemoprotective compounds have been linked to reduced risk of cancer and coronary heart disease. Consumer acceptance of these beneficial phytochemicals may be influenced by genetic taste factors.

Low concentrations of phenylthiocarbamide (PTC) and 6-*n*-propylthiouracil (PROP) taste bitter to some people but are tasteless to others *(1)*. The ability to taste dilute solutions of PTC/PROP is a heritable trait, thought to be determined by a dominant gene (2). Although phenotypic taste responses to

PTC/PROP are well-described in the literature, the gene responsible for this trait has not been described and its exact location is unknown.

The earliest studies used taste responses to PTC crystals or to PTC-impregnated filter paper to separate tasters from nontasters. PROP filter papers, used to test for genetic taste blindness, were distributed by the American Genetic Association as early as 1931. The proportion of PTC/PROP tasters in a Caucasian population was estimated at 70%. Later studies relied on the bimodal distribution of PROP detection thresholds, determined using the method of solutions *(2-4)*. Women were likely to be more PROP-sensitive than were men.

Following on Kalmus' observation *(2)* that PROP tasters showed a wide range of sensitivity to PROP, Bartoshuk (1994) suggested that the tasters group might include a smaller subsample of "supertasters", identified by low PROP thresholds and a high ratio of perceived PROP bitterness to the perceived saltiness of salt solutions. The speculation was that nontasters had two recessive alleles (tt); medium tasters were heterozygotes with one dominant allele (Tt); while supertasters had two dominant alleles (TT) *(5)*. Recent genetic studies in humans have provisionally linked the ability to taste PROP with a locus at 5p15 *(6)*.

PROP-tasting was associated with enhanced sensitivity to some, but not all, bitter tastes. Early studies showed that PTC/PROP tasters rated caffeine and quinine, though not urea, as more bitter *(7,8)*. PROP tasters also rated saccharin solutions, at concentrations found in diet soft drinks, as more bitter than did nontasters *(9)*. A more recent time-intensity study *(10)* found that PROP tasters gave higher aftertaste intensity ratings to a low concentration of caffeine (0.018mol/L) than did nontasters. One interpretation of such findings was that PTC/PROP tasters would avoid coffee and diet soft drinks sweetened with saccharin *(11)*. Later reports that PROP tasting was associated with enhanced oral burn of capsaicin, the active ingredient of hot peppers, led to the suggestion that PROP tasters might also avoid hot and spicy foods *(12)*. PROP tasters were also said to be more sensitive to the trigeminal irritation by ethanol, and reportedly avoided alcoholic beverages *(13)*.

Whether PROP-tasting influenced the perceived sweetness or reported liking for sugar solutions remains unclear. In some studies, low concentrations of sucrose and neohesperidin dihydrochalcone (NHDC), an intense sweetener, tasted sweeter to tasters than to nontasters *(9,14)*. PROP tasters were also reported to dislike sweet sucrose solutions *(15)*. However, other studies found no effect of PROP taster status on taste responses to sucrose solutions across a wider range of concentrations *(16)* or on self-reported preferences for sweet foods *(17)*.

Whether the ability to taste PROP predicts sensory response to dietary fats is another unresolved question. Two studies reported that PROP supertasters gave higher "fatness" ratings to a single sample of unsweetened heavy cream

(18) and to one high-fat salad dressing *(19)*. Other studies, based on more than one stimulus, found no effect of PROP taster status on the perceived sweetness, creaminess, or acceptability of sweetened dairy products *(20)*. There is no convincing evidence at this point that PROP tasters dislike high-fat foods.

Some studies have even made the claim that the ability to taste PROP protects against obesity, given that some PROP tasters appeared to have lower body mass indices than did nontasters *(18,19)*. However, those studies were limited to very small samples of college students. Epidemiological studies using regression models to account for covariates have failed to find a connection between PROP tasting and body mass.

Most researchers agree that PROP tasters are more responsive than nontasters to a number of other bitter compounds and are more likely to dislike some, but not all, bitter foods. Early studies on PROP tasters and nontasters focused on the perceived bitterness of urea, potassium benzoate, or quinine hydrochloride *(8,9)*, though sometimes with inconsistent results. However, these classic stimuli of taste psychophysics have no particular relevance to everyday food choices and eating habits. Instead, we chose to examine the perceived bitterness of dietary phytochemicals *(21)*, including those found in vegetables and fruit. Plant-based phenols and polyphenols, tannins, flavonoids, isoflavones, and glucosinolates are, almost without exception, bitter, acrid or astringent *(22)*. Clinical and epidemiological studies increasingly suggest that such phytochemicals have antioxidant and chemoprotective activity and successfully reduce the risk of cancer and other chronic disease *(23,24)*.

Increasing consumption of vegetables and fruit is a major dietary strategy for disease prevention *(25)*. According to the psychosocial literature on dietary choice, taste is the major barrier against the adoption of healthful eating habits *(17)*. Arguably, genetic taste factors that influence food preferences and food choices might reduce dietary exposure to substances known to affect cancer risk. However, since the ability to taste PROP does not confer increased responsiveness to all bitter compounds, it is important to know which compounds are perceived by PROP tasters as more bitter and which are not. A study of genetic taste markers would thus have implications for chronic disease prevention and public health *(22)*.

PROP Tasters and Nontasters

Individual sensitivity to PROP solutions is generally determined using the detection threshold procedure *(26)*. Such procedures have used a series of 15 PROP solutions, ranging in concentration from 1.0 x 10-6 mol/L to 3.2 x 10-3 mol/L PROP *(8)*. The highest concentration, solution no. 15, contained 0.5446 g/L PROP; the next concentration contained 0.3064 g /L, and so on *(2,3)*. In our studies of women (*16,17*), each participant was first presented with the least concentrated solution of PROP (solution 1), and then with increasingly higher solutions, until she reported detecting a taste distinct from that of water. She was then presented with two identical cups; one containing the detected concentration of PROP and the other containing deionized water. The water was at the same temperature and was stored in the same location as the PROP solution. She was asked to judge which of the two samples had the bitter taste *(4, 26-28)*. Wrong answers led to the presentation of the more concentrated PROP solution, while correct answers led to a second presentation of the same solution. Two consecutive correct answers at the same concentration led to presentation of less concentrated PROP solutions. Reversal points were defined as the concentration at which a series of correct responses turned to an incorrect response or vice versa *(26)*. Detection thresholds were based on a mean of five reversal points. Participants rinsed thoroughly with deionized water after tasting each PROP stimulus.

For bitterness intensity scaling, the same respondents tasted and rated 5 more solutions of PROP at concentrations of 0.032, 0.1, 0.32, 1.0, and 3.2 mmol/L (solutions 7, 9, 11, 13, and 15). Bitterness of each stimulus was rated using 9-point category scales, where 1 = "not at all bitter" and 9 = "extremely bitter". Respondents also ranked each stimulus along a 9-point hedonic preference scale *(29)*, that ranges from 1 = "dislike extremely" to 9 = "like extremely," with a neutral point at 5 ("neither like nor dislike").

Consistent with previous studies *(2)*, the distribution of PROP detection thresholds was bimodal. Figure 1 summarizes combined data from several studies for a total of 538 women, ranging in age from 18 to 80 years.

The antimode fell at solution 9. PROP tasters had thresholds below 0.1 mmol/L (solution 9) and nontasters had thresholds in excess of 0.2 mmol/L (solution 10). Cases with thresholds between 9 and 10 are usually rejected as unclassifiable (11).

Rating the intensity of 5 PROP solutions along a category scale is a faster and much less cumbersome procedure than the classic threshold detection method. As shown in Figure 2, summed bitterness intensity ratings for the 5 PROP solutions also showed a bimodal distribution. Summed bitterness ratings were inversely linked to the detection threshold ($r=-0.55$; $p<0.01$).

Summed bitterness intensity and summed hedonic ratings for the 5 PROP solutions, were strongly and inversely linked ($r=-0.83$; $p<0.01$). As shown in Figure 3, greater perceived bitterness was linked to a progressively increasing dislike of bitter taste. The same relationship between perception and hedonic response was obtained for tasters and nontasters. Consistent with past studies,

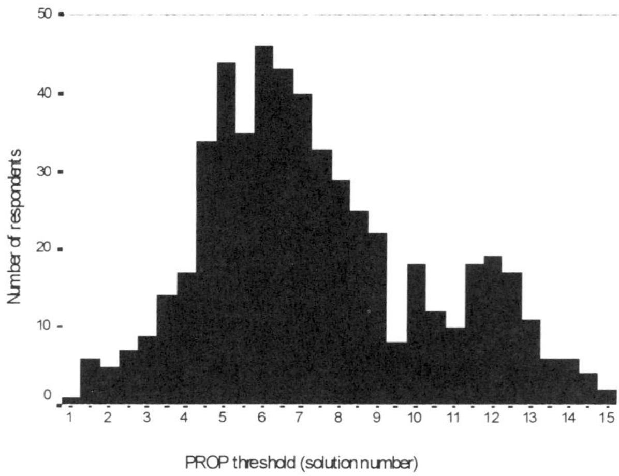

Figure 1. Distribution of PROP taste detection thresholds for 538 female respondents

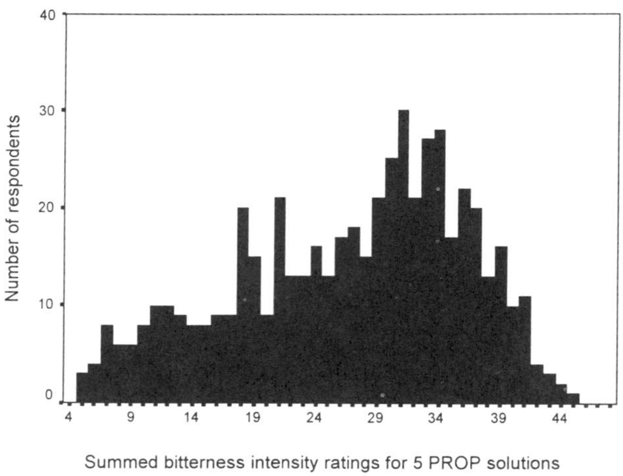

Figure 2. Distribution of summed bitterness ratings for 5 PROP solutions for 538 female respondents.

PROP tasting was unrelated to the perception of saltiness intensity of the 5 NaCl solutions.

In past studies, PROP "supertasters" were identified by a combination of PROP detection thresholds and intensity scaling of PROP relative to NaCl solutions (5, 11). Other studies, based on PROP-impregnated filter paper, identified those subjects who gave intensity scores of 8 or 9 on a 9-point scale as "high responders" or potential supertasters. We used summed intensity ratings for the 5 PROP solutions to assign respondents to 3 taste categories. Scores below 21 indicated nontasters, scores in the range 21 to 35 indicated regular tasters, while scores of 36 and above indicated supertasters. PROP responses for each group are summarized in Figure 4. As expected, bitterness intensity and hedonic curves were mirror images of each other, since increased perception of bitterness was associated with a greater dislike of PROP solutions. Furthermore, the shape of the bitterness intensity curve roughly indicated the location of the detection threshold for each taster group.

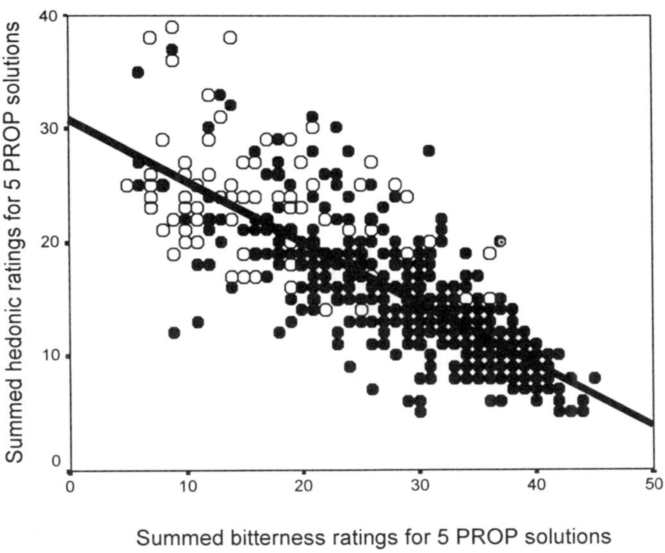

Figure 3. Summed hedonic ratings for PROP solutions as a function of summed bitterness. Data are shown separately for PROP tasters (closed circles) and PROP nontasters (open circles).

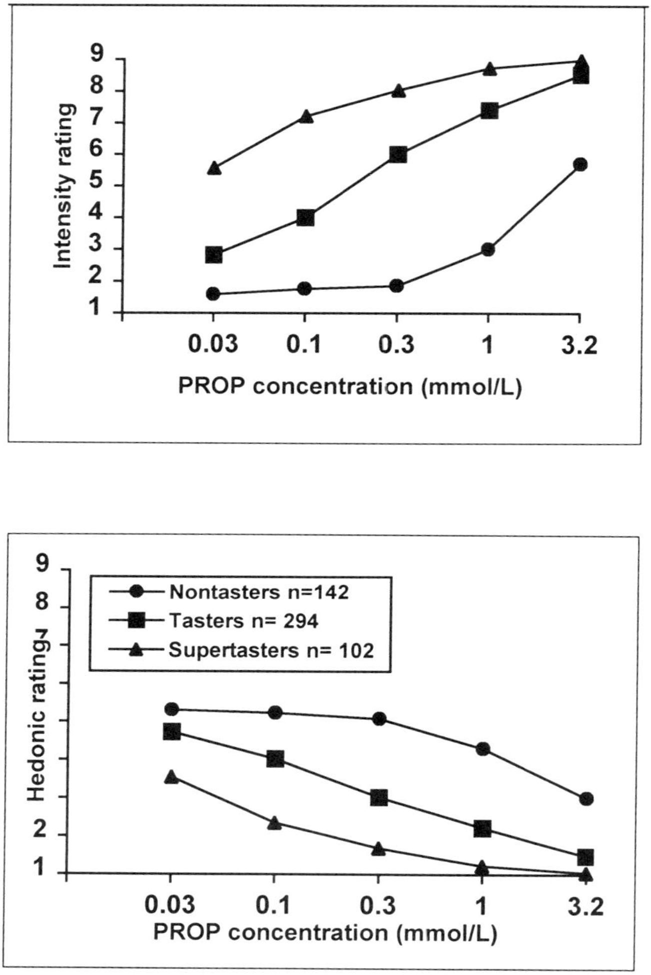

Figure 4. Bitterness intensity (top panel) and hedonic ratings (bottom panel) for 5 PROP solutions by PROP taster status for 538 women, aged 18-80 y.

PROP Tasting and Other Bitter Compounds

The ability to taste PROP was linked to higher bitterness ratings and lower preferences for naringin, a bioactive flavonoid that is the principal bitter component of grapefruit juice *(28)*. PROP supertasters gave significantly lower hedonic ratings to naringin solutions than did regular tasters or nontasters *(26)*. Naringin is not present in orange juice. A crosstab analysis of mean reported preferences for grapefruit juice showed that 19 out of 84 PROP tasters, as compared to only 3 out of 37 nontasters strongly disliked grapefruit juice (giving scores of 1 or 2 on a 9-point scale). No such differences by PROP taster status were observed for orange juice, oranges, or apples.

PROP tasting was also linked to an increased dislike of concentrated infusions of Japanese green tea *(30)*. In addition, PROP tasters gave lower self-reported acceptability ratings to soy foods, such as miso and tofu, on a food preference checklist *(30)*. Soy products, especially fermented ones, contain bitter isoflavones, genistein and dadzein *(31)*. Consistent with some past data PROP tasting was also associated with increased perceived bitterness of caffeine solutions. We had previously found *(17)* that PROP tasters gave lower acceptability ratings to coffee, instant coffee, and espresso. Respondents who expressed a decided dislike for coffee (ratings <3 on a 9-point scale) were, with only a few exceptions, tasters of PROP.

Table 1: Relationship between measures of PROP responsiveness and self-reported preferences for selected foods

Preference rating for	*PROP responsiveness*	
	Intensity	*Threshold*
Brussels sprouts	-0.18*	0.20*
Cabbage (raw)	-0.14	0.20*
Spinach (raw)	-0.11	0.18*
Coffee (regular)	-0.24*	0.27*
Coffee, Espresso	-0.26*	0.21*

SOURCE: From (ref. 7)

Another study, based on food preference checklists, examined food preferences by taste responsiveness to PROP. Respondents were 159 female nonsmokers, mean age 27.0 y. The food preference checklist (FPC) was based on 171 foods selected from all food categories. Subjects were asked to indicate how much they liked or disliked each food item using the 9-point hedonic preference scale. As expected, the more PROP-sensitive women reported lower preferences for coffee and espresso, but not for cocoa, tea or iced tea. In the same study, preferences for Brussels sprouts, cabbage and spinach were lower among tasters than among nontasters *(17)*. These data are shown in Table 1.

Phytochemicals and Bitter Taste

Many phytochemicals, including phenols and polyphenols, flavonoids, isoflavones, and glucosinolates, have antioxidant and anticarcinogenic effects, and a wide range of tumor-blocking activities *(23, 24)*. Most of these phytochemicals taste bitter. Phenolic compounds are directly responsible for the bitterness and astringency of many foods and beverages, including vegetables, fruits, legumes, tea, chocolate, coffee, and red wine *(31)*. Humans, conditioned through evolution to be wary of bitter plant-derived alkaloids and other bitter toxins, find excessive bitterness objectionable. Bitterness is the most commonly cited reason for the dislike for a particular food *(31)*, and lead to food avoidance *(26, 28, 32)*.

While many factors, including taste, may be responsible for food preferences, bitter taste is often the key reason for food rejection *(17)*. As documented above, PROP tasting was associated with greater perceived bitterness of selected flavonoids (naringin), catechins (green tea), isoflavones (soy products) and glucosinolates (Brussels sprouts).

How PROP tasting influences actual food choices is unclear. In early studies, based on food preference checklists, PTC/PROP tasters disliked cruciferous and green vegetables, rhubarb, sauerkraut, beer, coffee, and various sharp cheeses *(4, 33-36)*. However, other studies have failed to link PTC or PROP sensitivity with a consistent pattern of food dislikes *(37, 38)*. Studies on the consumption of cruciferous vegetables by elderly women showed only modest effects of PROP sensitivity on food choices *(39, 40)*. Increased sensitivity to the bitter taste of raw cruciferous vegetables, was not associated with a consistent pattern of food rejection. Vegetable consumption was low in tasters and nontasters alike *(40)*.

In a recent study *(21)*, we examined self-reported food preferences as a function of PROP taster status in a clinical sample of 326 female breast cancer patients and cancer-free controls. All patients were tested before (or shortly after) diagnosis and prior to any surgical, chemotherapy, or nutritional intervention. All completed the 171-item food preference checklist, that included grapefruit, grapefruit juice, lemons, oranges, and orange juice, as well as a variety of other vegetables and fruits *(29)*. Respondents were asked to indicate how much they liked or disliked each food using the 9-point hedonic preference scale. As shown in Figure 5, respondents who reported disliking

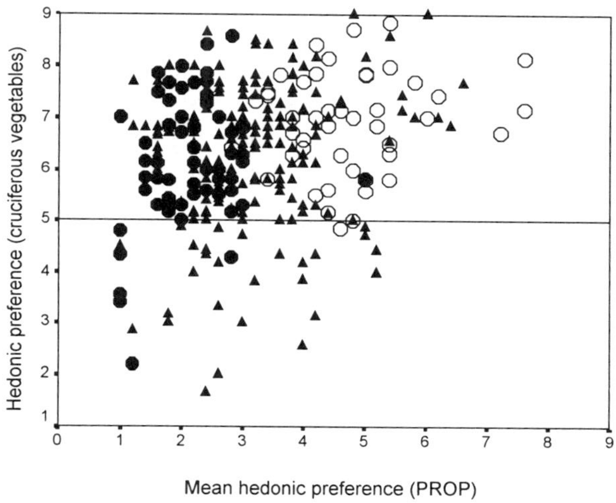

Figure 5. Scatterplot of mean hedonic ratings for 6-n-propylthiouracil (PROP) against mean preferences for cruciferous vegetables (broccoli, Brussels sprouts, cauliflower, cooked cabbage, raw cabbage, kale, and radishes.) ● = supertasters; ▲ = tasters; and O = nontasters. (Reproduced with permission from ref. 21. Copyright 2000 American Dietetic Association.)

cruciferous vegetables (rating <5) were tasters or supertasters of PROP, *(21)*. Preferences for sweet fruit were not affected by PROP taster status.

Debittering Processes

The sensation of bitter can be suppressed by cooking or by the addition of fat, sugar, or salt. The instinctive rejection of bitter may be immutable, since it has been crucial to survival. In general, humans reject those foods that are perceived as excessively bitter. Food industry routinely removes phenols and polyphenols, flavonoids, isoflavones and glucosinolates from plant foods through selective breeding and a variety of debittering processes. Many bioactive phytochemicals, currently studied in the laboratory, have long been treated as disposable bitter waste *(41)*. Studies now suggest that differences in PROP perception between tasters and nontasters can be reduced through the simple expedient of sweetening PROP solutions (42). Sweetening caffeine solutions eliminated taster and nontaster distinctions altogether, and no differences by PROP taster status were observed in the sensory evaluation of chocolate (42). Chocolate, composed of cocoa liquor, cocoa butter and sugar naturally balances the complex taste sensations of bitter, sweet, and fat.

Summary

Biologically active phytochemicals found in vegetables and fruit almost always have a bitter taste *(22, 31)*. Phenols and polyphenols, flavonoids and isoflavones, terpenes and glucosinolates are bitter phytochemicals that have recently been linked with cancer prevention in both animal and clinical studies *(24)*. Naringin, a bitter flavonoid found in grapefruit juice, acts as an antioxidant and inhibits tumor growth *(28)*. Bitter isoflavones in soybeans inhibit the growth of hormone-dependent and hormone-independent cancer cells, *in vitro (30)*. Isothiocyanates, or mustard oils, are reported to be effective phase II enzyme inducers in cancer prevention *(43)*. However, plant-based phytonutrients that are bitter, acrid, astringent or tear-provoking offer little in the way of eating pleasure and tend to be rejected by the consumer (41). Bitter taste is the most frequently cited reason for low acceptance of cruciferous and leafy green vegetables *(17)*.

Increasing fruits and vegetable consumption is the key dietary strategy for cancer prevention *(24, 44)*. Diets high in plant foods, notably cruciferous and green vegetables, allium vegetables, soy products, tomatoes, and citrus fruit, appear to confer a degree of protection against cancer *(44)*. Given that such diets are the cornerstone of current public health strategies for cancer prevention *(24, 45)*, the role of genetic taste markers in food acceptance or rejection needs to be better understood. If PROP taster status does predict the consumption of bitter vegetables, then inherited taste factors might pose a barrier to the adoption of plant-based diet. One possibility is that PROP tasters will avoid selected bitter foods altogether; an alternative possibility is that they will attempt to minimize bitter taste through the addition of fat, sugar, or salt. Genetic taste markers may thus affect dietary choices and the selection of healthful diets.

Acknowledgment

Supported by NCI grant CA-61680

References

1. Fox, A.F. *Proc. Natl. Acad. Sci. USA*. **1932,** 18, 33-44.
2. Kalmus, H. In *Handbook of Sensory Physiology;* Beidler, L. M., Ed.; Springer-Verlg: Berlin, 1971, pp 165-179.
3. Fisher, R. In *The Chemical Senes and Nutrition;* Kare, M.; Maller, O., Eds.; John Hopkins University Press: Baltimore, MD, 1967, pp. 61-81.
4. Fischer, R.; Griffin, F. *Drug Research*. **1964,** 14, 673-686.
5. Bartoshuk, L. M.; Duffy, V. B.; Miller, T. J. *Physiol. Behav.* **1994,** 56, 1165-1171.
6. Reed, D. R.; Nanthakumar, E.; North, M.; Bell, C.; Bartoshuk, L.M.; Price, R. A. *Am. J. Human Genet.* **1999,** 64, 1478-80.
7. Fisher, R. In *Gustation and Olfaction;* Ohloff, G.; Thomas, A. K., Eds.; Academic Press: London, 1971; pp. 187-237.
8. Hall, M. J.; Bartoshuk L. M; Cain, W. S.; Stevens, J.C. *Nature.* **1975,** 253, 442-3.
9. Bartoshuk, L. M. *Science.* **1979,** 205, 934-935.
10. Neely, G.; Borg, G. *Chem. Senses.* **1999,** 24,19-21
11. Bartoshuk, L. M. *Food Quality and Preference,* **1993,** 4, 21-32.
12. Karrer, T.; Bartoshuk, L. Physiology & Behavior, **1995,** 57, 421-9.
13. Pelchat, M. L.; Danowski, S. *Physiol. Behav.* **1992,** 51, 1261-6.
14. Gent, J. F.; Bartoshuk, L. M. *Chem. Senses.* **1993,** 7, 265-272.
15. Looy, H.; Weingarten, H. P. *Physiol. Behav.* **1992,** 52, 75-82.
16. Drewnowski, A.; Henderson, S. A.; Shore, A. B.; Barratt-Fornell, A. *Physiol. Behav.* **1997,** 62, 649-655.

17. Drewnowski A.; Henderson, S. A.; Levine., A.; Hann, C. *Public Health Nutrition.* **1999.**
18. Duffy, V. B.; Bartoshuk, L. M.; Lucchina, L. A.; Snyder D. J.; Tym, A. *AChemS XVIII Abstracts.* **1997.**
19. Tepper, B. J.; Nurse, A. C. *Physiol. Behav.* **1997,** 61, 949-954.
20. Drewnowski, A.; Henderson, S. A.; Barratt-Fornell, A. *Physiol. Behav.* **1998,** 63, 771-777.
21. Drewnowski, A.; Henderson, S.A.; Hann, C. S.; Berg, W.A.; Ruffin, M.T. *J. Am. Diet. Assoc.* **2000,** 100, 191-7.
22. Drewnowski, A.; Rock, C.L. *Am. J. Clin. Nutr.* **1995,** 62, 506-511.
23. Craig, W. J. *J. Am. Diet. Assoc.* **1997,** 97, S199-204.
24. Potter, J. P. *Food, Nutrition and the Prevention of Cancer: a Global perspective;* World Cancer Research Fund: Washington DC, 1997.
25. Steinmetz, K. A.; Potter, J. D. *J. Am. Diet. Assoc.* **1996,** 96, 1027-1039.
26. Drewnowski, A.; Henderson, S. A.; Shore, A. B. *Chem. Senses.* **1997,** 22, 27-37.
27. Bartoshuk, L. M.; *Psychology Today* **1988,** 14, 48-63.
28. Drewnowski, A.; Henderson, S. A.; Shore, A. B. *Am. J. Clin. Nutr.* **1997,** 66, 391-7.
29. Peryam, D. M.; Pilgrim, P. J. *Food Technology.* **1957,** 11, 9-14.
30. Akella, G. D.; Henderson, S. A.; Drewnowski, A. *Nutrition and Cancer.* **1998,** 29, 146-151.
31. Rousseff, R. L. In *Bitterness in Foods and Beverages;* Rousseff, R. L., Ed.; Elsevier, Amsterdam, **1990.**
32. Rozin, P.; Vollmecke, T. A. *Ann. Rev. Nutr.* **1986,** 6, 433,-456.
33. Fischer, R.; Griffin, F.; England, S.; Garn, S. M. *Nature.* **1961,** 191, 1328.
34. Glanville, E. V.; Kaplan, A. R. *Nature.* **1965,** 205,851-853.
35. Forrai, G.; Bankovi, G. *Acta Physiologica Hungarica,* **1984,** 64, 33-40.
36. Boyd, W. C. *Science.* **1950,** 112, 153.
37. Anliker, J. A.; Bartoshuk, L.; Ferris, A. M.; Hooks, L. D. *Am. J. Clin. Nutr.* **1991,** 54, 316-20.
38. Mattes, R.; Labov, J. *J. Am. Diet. Assoc.* **1989,** 89, 692-694.
39. Jerzsa-Latta, M.; Krondl, M.; Coleman, P. *Appetite.* **1990,** 15, 127-134.
40. Niewind, A.; Krondl, M.; Shrott, M. *Nutr. Res.* **1988,** 8, 13-20.
41. Drewnowski, A.; Gomez-Carneros, C. *Am. J. Clin. Nutr.* **2000,** 72, 1424-35.
42. Ly, A.; Drewnowski, A. *Chem. Senses.* **2001**, 26, 41-47.
43. Fahey, J. W.; Stephenson K. K.; Talalay, P. In *Functional Foods for Disease Prevention I. Fruits, vegetables, and teas;* Shibamoto, T.; Terao, J.; Osawa, T., Eds.; ACS Symposium Series 702; Washington, DC. **1998**; Vol. I, pp 16-22.
44. Steinmetz, K. A.; Potter, J. D. *J. Am. Diet. Assoc.* **1996,** 96, 1027-1039.
45. Havas, S.; Heimendinger, J.; Baranowski, T.; Nicklas, T. A.; Bishop, D.; Sorensen, G.; Beresford, S. A.; Cowan, A. *J. Am. Diet. Assoc.* **1994,** 94, 32-36.

Chapter 6

Clustering Bitter Compounds via Individual Sensitivity Differences: Evidence Supporting Multiple Receptor–Transduction Mechanisms

Jeannine F. Delwiche[1,2], Zivjena Buletic[2], and Paul A. S. Breslin[2]

[1]Department of Food Science and Technology, Ohio State University, 2015 Fyffe Road, Columbus, OH 43210

[2]Monell Chemical Senses Center, 3500 Market Street, Philadelphia, PA 19104

Although it had been well documented that people varied widely in their sensitivities to bitter compounds, the intercorrelation of these sensitivities remained unknown. By clustering bitter compounds representative of different chemical classes as a function of individual sensitivities, it was possible to infer the number and variety of potential bitterness transduction systems involved in bitter perception. Results indicated that bitter compounds could be grouped into two general groups, neither of which contains PROP (n-propylthiouracil). There are also subjects who possess diminished absolute sensitivity to bitter stimuli, but do not differ in their relative sensitivities to these compounds.

Compounds that elicit bitter-taste sensations are chemically diverse, and include both inorganic and organic compounds (1, 2). Compounds with bitter tastes have been categorized as acetylated sugars, alkaloids, amines, amino acids, carbamates, ionic salts, isohumulones, ureas/thioureas, phenols, etc. (1, 2). It is unlikely that a single bitter receptor can account for sensitivity to all classes of compounds that are perceived to be bitter tasting (1, 3). The possibility that classes of bitter transduction processes in the general population could be revealed by examining and correlating individual differences in sensitivities to bitter compounds was recently examined by Delwiche et al. (4).

Their approach was motivated by the observation that large individual differences in the bitterness of a variety of compounds have been reported. The most frequently studied variation among subjects' sensitivities to bitter chemicals has been for PROP (n-propylthiouracil), PTC (phenylthiocarbamide) and other antithyroid compounds that contain the N-C=S chemical group, (e.g., 5-12), although wide-ranging responses across individuals to other bitter compounds have been found as well (7). Despite numerous attempts to relate how strongly one perceives a particular concentration of PROP/PTC to how strongly one perceives particular concentrations of other bitter substances (including quinine, caffeine and urea), PROP/PTC has not proven to be a reliable indicator (7, 9, 13-17). Nevertheless, the possibility remains that individuals' sensitivities to other bitter compounds will correlate with each other.

The existence of more than one physiological transduction mechanism in the perception of bitterness is suggested by the substantial variation in sensitivities to different bitter-tasting compounds. Via primary neural recordings, Dahl et al. (18) attempted to infer the idiosyncratic distribution of transduction mechanisms for bitter taste stimuli in rat taste-receptor cells. They recorded single neuron responses to 10 bitter-tasting compounds in rats and summarized differences in evoked responses in multi-dimensional scaling space. The majority of physiological papers have only focused upon a small subset of bitter compounds, and many have used only a single compound - usually quinine (as noted by Dahl et al. in Ref 18). Similarly, many psychophysical studies have limited themselves to one or a few of the following bitter-tasting compounds: quinine, caffeine, urea, PTC, and PROP (e.g., 9, 10, 15, 17). The use of so few stimuli makes it difficult, if not impossible, to find covariance in perceived intensities across compounds. It may be possible to infer the number and variety of potential bitterness transduction systems for the compounds tested by adapting a strategy similar to that of Dahl et al. (18) and determining if compounds cluster together in humans as a function of sensitivities across individuals.

Applying a related approach, Yokomukai et al. (7) found that sensitivity to quinine sulfate and urea were unrelated, but individuals more sensitive to quinine than urea tended to find the bitterness of suprathreshold caffeine and sucrose octaacetate (SOA) to be greater than that of suprathreshold magnesium sulfate; individuals more sensitive to urea than quinine showed the reverse pattern. McBurney et al. (19) found that adaptation to quinine hydrochloride significantly decreased the perceived bitterness of SOA and caffeine, but not magnesium sulfate, while adaptation to urea significantly decreased the perceived bitterness of magnesium sulfate, but not SOA and caffeine. Similarly, Lawless (14) confirmed these findings for quinine, urea, and creatine.

Idiosyncratic Patterns of Perceived Bitter Intensity

In a recent extensive study by Delwiche et al. (4), the perceived intensities of bitter-tasting compounds reported by 26 individuals were examined for evidence of covariation in perceived bitter intensity. The subjects rated the bitter intensity of 11 chemically diverse compounds (typically described as bitter) using the LMS scale (20, 21). These compounds, and their concentrations, were as follows: 1.19×10^{-4}M quinine hydrochloride, 5.70×10^{-5}M sucrose octaacetate, 9.20×10^{-1}M urea, 1.37×10^{-4}M tetralone® (a mixture of iso-alpha-acids), 5.02×10^{-2}M L-phenylalanine, 2.69×10^{-2}M L-tryptophan, 1.09×10^{-2}M caffeine, 3.00×10^{-1}M magnesium sulfate, 4.99×10^{-7}M denatonium benzoate, 1.72×10^{-2}M (-)-epicatechin and 5.50×10^{-4}M n-propylthiouracil (PROP). In addition, these same subjects repeatedly ranked a subset of 9 of these compounds (all but epicatechin and PROP, due to expense and context effects, respectively) from weakest to strongest in bitterness, and a significant ranked order was determined for each individual (for more details on the psychophysical methodology, see Ref 4).

Pearson's product moment correlation coefficients were calculated for the rated intensities of all 11 compounds and Spearman's rho correlation coefficients were calculated for the bitter ranks of the 9 ranked compounds (see Table 1). Only three compound-pairs showed significant correlations in both data sets: caffeine/SOA, caffeine/tetralone®, and phenylalanine/tryptophan. No significant correlation was found between PROP and the ten remaining compounds for rated intensities ($p>0.05$, post-Bonferroni correction). This is especially striking since every other compound correlated significantly with at least one other compound, and the bitterness ratings of SOA correlated significantly with those of four other compounds (see Table 1, bottom). Similarly, with the ranked data, several significant correlations were found (see Table 1, top), but unlike with rated intensities, over half of the significant correlations were negative. However, it is important to note that every significant negative correlation found for the ranking data was a nonsignificant correlation for the ratings. The apparent discrepancies between the correlations based on ratings and those based on rankings can be explained by the fact that the ranking methodology is insensitive to absolute differences between subjects in perceived intensities, which allow negative correlations to be revealed. In contrast, with rating procedures some subjects consistently give higher ratings, and others lower ratings, to all bitter compounds, leading to positive correlations across subjects. As a consequence, negative correlations in the ranking data appeared as weakly positive correlations in the rating data.

Table 1. Correlation Coefficients of Bitter Ratings and Rankings

	DB	*Qui*	*SOA*	*Epi*	*U*	*Mg*	*Tetra*	*TRY*	*PROP*	*Caf*	*PHE*
DB	*	0.38	0.50	N/A	-0.50	-0.47	0.26	-0.44	N/A	0.51	-0.47
Qui	**0.63**	*	-0.09	N/A	-0.22	-0.51	-0.19	-0.05	N/A	0.04	-0.03
SOA	**0.64**	**0.67**	*	N/A	-0.47	-0.37	0.33	-0.53	N/A	**0.68**	-0.39
Epi	0.26	0.23	0.12	*	N/A	N/A	N/A	N/A	N/A	N/A	N/A
U	0.28	0.49	0.32	0.53	*	0.06	-0.54	0.46	N/A	**-0.61**	0.46
Mg	0.45	0.55	0.60	0.25	0.49	*	0.01	0.01	N/A	-0.34	-0.09
Tetra	0.55	0.54	**0.66**	-0.01	0.17	**0.70**	*	**-0.70**	N/A	**0.62**	**-0.60**
TRY	0.21	0.40	0.37	**0.71**	0.60	0.45	0.10	*	N/A	**-0.63**	**0.58**
PROP	0.24	0.26	0.39	0.24	0.43	0.34	0.52	0.33	*	N/A	N/A
Caf	0.53	**0.63**	**0.81**	0.09	0.46	0.52	**0.61**	0.31	0.37	*	-0.57
PHE	0.03	0.42	0.30	0.47	**0.65**	0.35	0.06	**0.68**	0.16	0.52	*

NOTE: Pearson's product moment correlation coefficients of bitter ratings (unshaded cells) and Spearman's Rho correlation coefficients of bitter ranking (shaded cells). Correlations in **bold** are significant at $p<0.05$. The following abbreviations were used: DB = denatonium benzoate, SOA = sucrose octaacetate, Qui = quinine hydrochloride, Epi = (-)-epicatechin, U = urea, Mg = magnesium sulfate, Tetra = tetralone®, TRY = L-tryptophan, PROP = n-propylthiouracil, Caf = caffeine, PHE = L-phenylalanine N/A = Not applicable.

(Source: Adapted from Reference 4. Copyright 2001 Psychonomic Society.)

These same data sets (4) were subjected to both unrotated principal components analyses, or PCA (shown in Figure 1), and cluster analyses. The placement of the compounds in both analyses (PCA and cluster analysis) of the bitter ratings reveals relationships that are largely in accordance with the correlation matrix. In both analyses, at least two main groupings of compounds were revealed, neither of which contained PROP. One of these groups consisted of phenylalanine, urea, tryptophan and more loosely epicatechin, while the second group consisted of the remaining compounds, excluding PROP. Analysis of the bitter rankings revealed that, paralleling the rating data, the compound

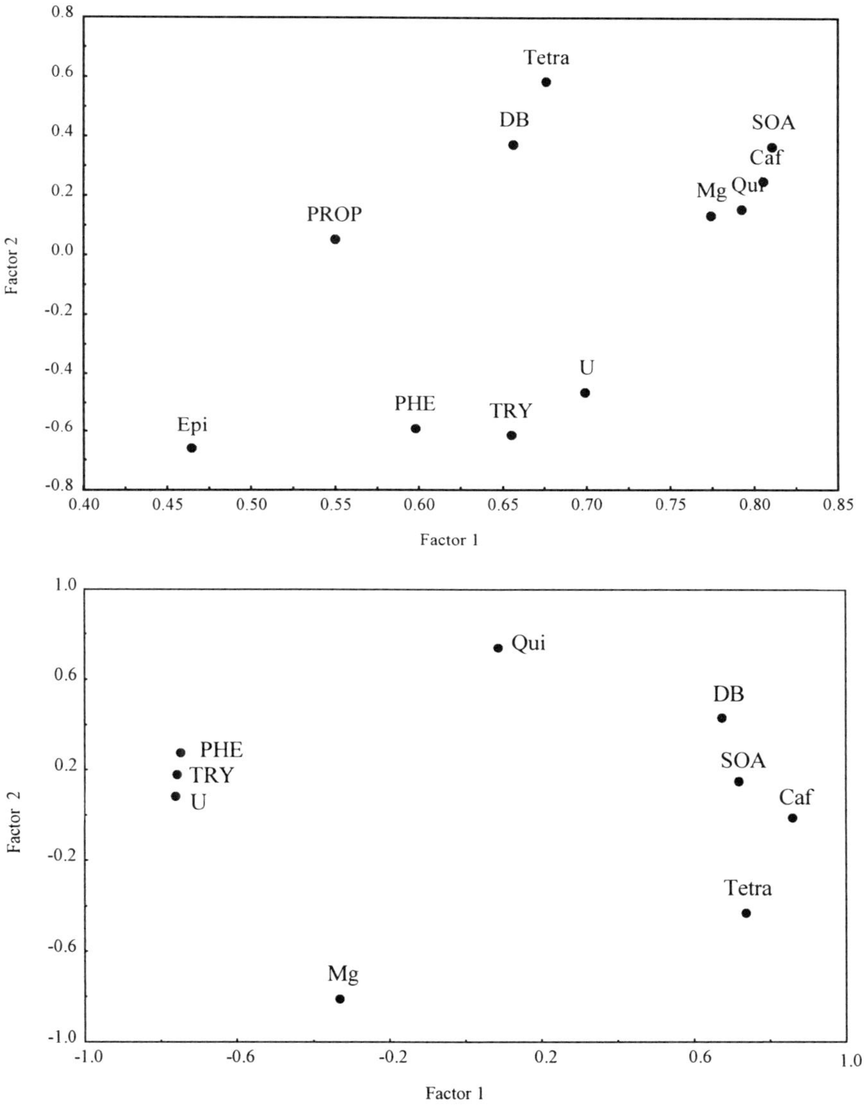

Figure 1. Unrotated PCA of Bitter Intensity Ratings (top) & Rankings (bottom)
NOTE: Abbreviations match those in Table 1.
(Source: Adapted from Reference 4. Copyright 2001 Psychonomic Society.)

groupings are largely in accordance with the correlation matrix of the rankings, and these compound groupings were similar to that found with the ratings. As with the ratings, two broad groupings of compounds were revealed in both analyses, the tightest of which consisted of phenylalanine, tryptophan, and urea while a looser group consisted of denatonium, SOA, caffeine and tetralone®. All significant negative Spearman's rho correlations were found between members of different clusters; no significant negative correlations were found between the compounds within a cluster. The totality of both the rating and ranking results suggested an especially close association of phenylalanine, tryptophan, and urea, as well as an association between caffeine, SOA, quinine, denatonium benzoate, and, less closely, tetralone®. The data also indicated that PROP is perceptually dissimilar to all the other compounds assessed. Other relationships (*i.e.*, that of epicatechin and magnesium sulfate) appeared to be more tentative (4).

The underlying factor(s) responsible for the groupings that were revealed in both the PCAs and cluster analyses are unknown. Several physical parameters examined (including molecular weight, percent carbon, percent nitrogen, percent oxygen, hydrophobicity, and pH) failed to correlate significantly to the factor loadings (4). A separation between the compounds which contain methyl groups and those which contain primary amines, despite tremendous differences in the chemical classes of the compounds contained within the two main compound groups, is evident in all analyses. While these observations of the chemical moiety differences do not fully account for the grouping of the compounds, they may lend insight into the types of transduction mechanisms involved in the perception of these bitter compounds and how these transduction mechanisms are likely to differ (4).

The Impact of PROP Sensitivity on Idiosyncratic Patterns

In the same study by Delwiche et al. (4), the impact of PROP sensitivity on idiosyncratic patterns of sensitivity to bitter compounds was also examined. The 26 subjects were divided into three groups: super-tasters ($n = 4$), medium tasters ($n = 18$) and non-tasters ($n = 4$). Repeated-measures ANOVA was performed with PROP status as an independent variable on the bitter intensity ratings and rankings (see Figure 2). The bitterness ratings (Figure 2, top) of the different PROP status groups were significantly different ($p<0.01$), although due to the low power associated with having only four members in two of the groups, Scheffe's test failed to reveal significant differences between the specific status groups for each particular compound. In contrast, no significant difference in bitterness rankings (see Figure 2, bottom) was found between the PROP status groups (4).

Visual inspection of Figure 2, top, indicates that individuals insensitive to PROP gave lower ratings to bitter compounds than did individuals more

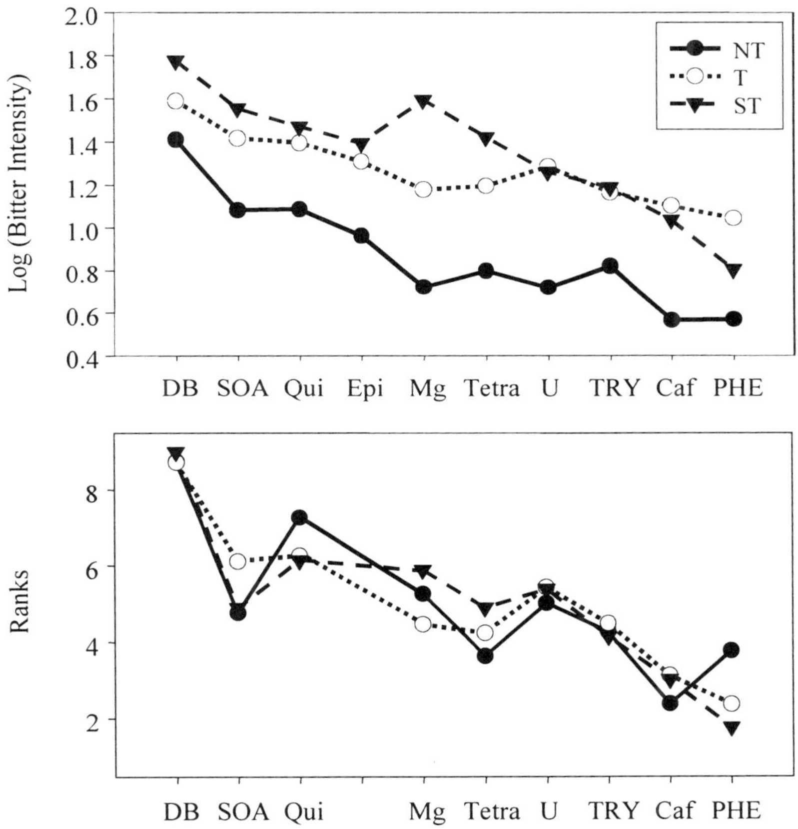

Figure 2: PROP Status Influence on bitter intensities (top) & rankings (bottom). NOTE: Non-tasters (NT) are indicated by closed circles, (medium) tasters (T) by open circles, and super-tasters (ST) by closed triangles. Abbreviations match those used in Table 1.
(Source: Adapted from Reference 4. Copyright 2001 Psychonomic Society.)

sensitive to PROP, mirroring the findings of other researchers (e.g., 14, 22). While these results initially may seem to contradict the lack of PROP correlations with other compounds reported above, the apparent discrepancy is due to the fact that only a very few individuals rated all compounds consistently lower than other subjects. In the ratings correlational analysis, the impact of these few subjects created weakly positive, but nonsignificant, correlations (as shown in Table 1). Only when the subjects are categorized such that these few insensitive individuals were separated from the others could the influence of PROP status be demonstrated.

There was, however, no significant difference between these same individuals in the rankings given to these same compounds. This seems to indicate that the individuals of each group differ only quantitatively, not qualitatively, suggesting that non-tasters have lower "system gain" for bitterness than do medium and super-tasters. The mechanism for this is unclear. Bartoshuk et al. (23) have suggested that this difference is perhaps related to the number of taste buds and/or taste pores, although increased or decreased numbers of taste buds might be expected to affect all taste qualities, not just bitterness. However, recent work (24) has demonstrated that while PROP-insensitive individuals tend to have fewer papillae than do PROP-sensitive individuals, papillae number is a poor predictor of perceived bitterness ratings across individuals. Regardless, the results of Delwiche et al. (4) indicate that general sensitivity to all bitter compounds is a factor that is independent from the issue of individual differences in responses to different compounds due to variation in transduction mechanisms. By making group comparisons, the general sensitivity effects can confound the measurement of sensitivities to individual compounds.

General Discussion

The work of Delwiche et al. (4) was guided by the hypothesis that bitter tasting compounds that cluster together as a function of individual differences in subjects' perceptual sensitivities are likely to share a common physiological mechanism somewhere between the mouth and the brain, most probably in the receptor cell transduction sequence. By using at least one representative from several of the less frequently used bitter chemical categories, this study maximized the chemical diversity of bitter stimuli, increasing the likelihood that such trends would be revealed. This work (4) resulted in evidence for reliable groupings of compounds into subclasses (illustrated in Figure 1): (1) denatonium benzoate, tetralone®, caffeine, SOA, and quinine; (2) urea, tryptophan, phenylalanine, and epicatechin; (3) magnesium sulfate; and (4) PROP. In general, the findings of this study are in good agreement with several different converging lines of multidisciplinary evidence, including recent findings on Tas2Rs and a variety of human psychophysical studies.

Molecular Studies Suggest Multiple Bitter Transduction Sequences

Recently, Adler et al. (25) and Matsunami et al. (26) identified a novel family of 40-80 putative G protein-coupled bitter receptors (Tas2Rs) that are expressed in selected taste receptor cells (~15%), suggesting bitter receptor heterogeneity. Organized into serial, head-to-tail clusters on human chromosomes 5, 7 and 12, the genes are associated with loci that influence bitter perception in humans. Physiological analysis of selected Tas2Rs, in an expression system (HEK-293 cells), by Chandrashekar et al. (27) provided compelling evidence that the proteins examined are likely bitter taste receptors (Tas2R-4 --denatonium & PROP sensitive in humans). And while Adler et al. (25) demonstrated that most or all these putative bitter taste receptors (which are co-localized with gustducin) tend to be co-expressed in the same cells, suggesting that a given taste cell will respond to many bitter taste stimuli, Caicedo and Roper (28) found evidence that different bitter stimuli may activate different subpopulations of bitter-sensitive taste cells.

Human Psychophysical Studies

Yokomukai et al. (7), Lawless (14), and McBurney et al. (19) interpreted their sensitivity and cross-adaptation data to suggest at least three separate, although perhaps overlapping, bitter transduction sequences. The three posited, largely in agreement with the findings of Delwiche et al. (4), were: one particularly sensitive to PTC and related compounds, one particularly sensitive to the bitterness of quinine, caffeine, and SOA, and one particularly sensitive to urea, magnesium sulfate and creatinine. Delwiche et al. (4), Yokomukai et al. (7) and Lawless (14) found PROP/PTC sensitivity to be independent of other bitter compounds. Delwiche et al. (4), Yokomukai et al. (7) and McBurney et al. (19) all suggest a tight relationship among quinine, caffeine, and SOA. In further agreement, Yokomukai et al (7) found indications of a negative correlation between the urea and caffeine, as was found in Delwiche et al. (4).

Breslin & Beauchamp (29) examined the effect of sodium on bitterness suppression and found differential suppression of bitterness across intensity matched concentrations of urea, quinine HCl, magnesium sulfate and caffeine. The effectiveness of suppression in bitterness by sodium across compounds separates the compounds into three classes (1:urea, 2: quinine and caffeine, and 3: magnesium sulfate, see Ref 29), matching the findings of Delwiche et al. (4).

Many studies on individual differences in human perception of PROP and PTC have been conducted (e.g., 5-10). Typically, subjects are divided into two (non-taster and taster) or three (non-taster, taster, and super-taster) groups (30). In agreement with Delwiche et al. (4), unless those who are highly insensitive are examined separately, how strongly one perceives a particular concentration of PROP has not been a reliable predictor of how strongly one perceives other bitter substances. In fact, individuals who are highly sensitive to one bitter compound can be quite insensitive to another (7).

While some studies have found a significant positive relationship between PROP or PTC sensitivity and quinine sensitivity (14, 15, 31, 32), several other studies have failed to find such a relationship (7, 9, 10, 13, 14, 16, 17, 22, 33, 34). Similarly, while Hall et al. (9) found that "sensitivity to the taste of PTC predicts sensitivity to caffeine," other studies failed to find a significant difference between tasters and non-tasters (15, 16, 22). No clear relationship between PROP/PTC and urea (7, 9, 16, 22), between PROP/PTC and denatonium benzoate (16, 22), or between PROP/PTC and SOA (16, 22, 35) has been found. Schiffman et al. (22) failed to find a significant difference between PROP taster groups for magnesium sulfate sensitivity, and Lawless (14) failed to find any relationship between sensitivity to PTC and creatine or creatinine. One study (36) found that PROP tasters rated isohumulone significantly more bitter than did non-tasters only when it was presented in water, but not when it was dissolved in beer.

Clearly, the relationship between the perception of PTC/PROP and other bitter compounds is not a simple one. Contradictions across studies could have arisen from the percentage of PROP/PTC insensitive individuals varying from one study to another; studies with a higher percentage of non-tasters would be more likely to find a significant relationship between PROP/PTC and other compounds than with those with a lower percentage. In addition, whether statistical analyses considered non-tasters as a separate category also impacts upon whether or not PROP/PTC status appears to influence sensitivity to other bitter compounds. Finally, the use of scales with ceilings and the use of designated intensity standards may minimize differences across subjects (37).

Correlations Among Papillae Density and Bitter Taste

The global insensitivity of the PROP non-tasters to bitter compounds, found in Delwiche et al. (4) and also noted by Lawless (14), could be explained by a reduced number of taste receptors, relative to that of PROP medium and super-tasters. Bartoshuk et al. (23) found a link between the perceived bitterness of PROP and the density of taste receptors on the anterior human tongue. Similarly, in mice, a relationship between SOA sensitivity and number of taste buds has been found (38). Since stimulation of an increasing number of taste buds in a population of fungiform papillae is known to give rise to progressively greater perceived taste intensities (39, 40), both mouse and human data suggest that the number of taste buds contributes to the absolute differences in taste sensitivities. However, as mentioned earlier, recent work (24) has demonstrated that papillae number is a poor predictor of perceived bitterness ratings across individuals.

Conclusions

Since sensitivity to any bitter compound may be affected by the type of transduction sequence involved for that compound, two compounds that share transduction components should be similarly affected by changes in these components. Significant correlations of sensitivities to certain compounds across individuals strongly suggest that those compounds share common transduction components. Several lines of evidence indicate that bitterness is transduced in humans via several heterogeneous interactions; among the 11 compounds employed by Delwiche et al. (4), compound groupings suggest four different transduction mechanisms, which is a high ratio of bitter compounds to transduction mechanisms. These individual differences in responses to different bitter compounds (suggesting variation in transduction mechanisms) can only be recognized when differences in global sensitivity to bitter compounds are not allowed to confound the former. Given the vast number of bitter tasting compounds and their structural diversity, it is perhaps not surprising that Adler et al. (25) and Matsunami et al. (26) have suggested the existence of 80 (and possibly more) bitter taste receptors. Their reports support the relatively large number of independent groupings found with the 11 stimuli used by Delwiche et al. (4).

It is important to point out that the clusterings discussed only relate to sensitivity to bitter compounds. Issues of bitterness quality or the possibility for multiple bitter sub-qualities are not addressed. For example, potentially all transduction interactions involve the same bitter-sensitive taste-receptor cells (as suggested in Ref 25), or converge onto the same primary neuron, in which case all compounds, regardless of their cluster identity or transductive mechanisms, would still give rise to the same quality of bitter taste.

Acknowledgments

The authors thank Dr. Alexander A. Bachmanov, Gary K. Beauchamp, Dr. Joseph G. Brand, Dr. Bruce Schneider, Dr. David V. Smith, and the anonymous reviewers for their thoughtful comments on earlier versions of this manuscript.

This work was funded by NIH grants R29 DC02995 (P.A.S.B) & F32 DC00384-02 (J.F.D).

References

1. Spielman, A.I.; Huque, T.; Whitney, G.; Brand, J.G. *Sensory Transduction*; Corey, D.P.; Roper, S.D., Eds.; Rockerfeller University Press: New York, NY, 1992; pp 308-324.
2. Belitz, H.D.; Wieser, H. *Food Reviews Intl* **1985**, 1, 271-354.
3. Brand, J.G. *Tasting and Smelling*; Beauchamp, G.K.; Bartoshuk, L., Eds.; Academic Press: San Diego, CA, 1997; pp 1-24.
4. Delwiche, J.F.; Buletic, Z.; Breslin, P.A.S. *Perception and Psychophysics* **2001**, *63*, 761-776.
5. Thorngate, J.H. III. *Modifying Bitterness: Mechanisms, Ingredients, and Applications*; Roy, G., Ed., Technomic Publishing: Valhalla, NY, **1997**; pp 139-160.
6. Bartoshuk, L.M.; Duffy, V.B.; Reed, D.; Williams, A. *Neuroscience and Biobehavioral Review* **1996**, *20*, 79-87.
7. Yokomukai, Y.; Cowart, B.J.; Beauchamp, G.K. *Chemical Senses* **1993**, *18*, 669-681.
8. Kemp, S.E.; Birch, G.G. *Chemical Senses* 1992, *17*, 151-168.
9. Hall, M.J.; Bartoshuk, L.M.; Cain W.S.; Stevens, J.C. *Nature* **1975**, *253*, 442-443.
10. Fischer, R.; Griffin, F. *Nature* **1963**, *200*, 343-347.
11. Barnicot, N.A.; Harris, H.; Kalmus, H. *Annals of Eugenics (London)* **1951**, *16*, 119-128.
12. Harris, H.; Kalmus, H. *Annals of Eugenics (London)* **1949,** *15*, 32-45.
13. Bartoshuk, L.M. *Science* **1979**, *205*, 934-935.
14. Lawless, H.T. *Chemical Senses and Flavour* **1979**, *4*, 249-258.
15. Leach, E.J.; Noble, A.C. *Chemical Senses* **1986**, *11*, 339-345.
16. Mela, D.J. *Chemical Senses* **1989**, *14*, 131-135.
17. Schifferstein, H.N.J.; Frijters, J.E.R. *Chemical Senses* **1991**, *16*, 303-317.
18. Dahl, M.; Erickson, R.P.; Simon, S.A. *Brain Research* **1997**, *756*, 22-34.
19. McBurney, D.H.; Smith, D.V.; Shick, T.R. *Perception and Psychophysics* **1972**, *11*, 228-232.
20. Green, B.G.; Dalton, P.; Cowart, B.; Shaffer, G.; Rankin, K.; Higgins, J. *Chemical Senses* **1996**, *21*, 323-334.
21. Green, B.G.; Shaffer, G.S.; Gilmore, M.M. *Chemical Senses* **1993**, *18*, 683-702.

22. Schiffman, S.S.; Gatlin, L.A.; Frey, A.E.; Heiman, S.A.; Stagner, W.C.; Cooper, D.C. *Neurobiology of Aging* **1994**, *15*, 743-750.
23. Bartoshuk, L.M.; Duffy, V.B.; Miller, I.J. *Physiology & Behavior* **1994**, *56*, 1165-1171.
24. Delwiche, J.F.; Buletic, Z.; Breslin, P.A.S. *Physiology & Behavior* **2001**, *74*, 1-9.
25. Adler, E.; Hoon, M.A.; Mueller, K.L.; Chandrashekar, J.; Ryba, N.J.P.; Zuker, C.S. *Cell* **2000**, *100*, 693-702.
26. Matsunami, H.; Montmayeur, J.-P.; Buck, L.B.. *Nature* **2000**, *404*, 601-604.
27. Chandrashekar, J.; Mueller, K.L.; Hoon, M.A.; Adler, E.; Feng, L.; Guo, W.; Zuker, C.S.; Ryba, N.J.P. *Cell* **2000**, *100*, 703-711.
28. Caicecdo, A.; Roper, S.D. *Science* **2001**, *291*, 1557-1560.
29. Breslin, P.A.; Beauchamp, G.K. *Chemical Senses* **1995**, *20*, 609-623.
30. Reed, D.R.; Bartoshuk, L.M.; Duffy, V.; Marino, S.; Price, R.A. *Chemical Senses* **1995**, *20*, 529-533.
31. Jefferson, S.C.; Erdman, A.M. *J. Home Economics* **1970**, *62*, 605-608.
32. Gent, J.F.; Bartoshuk, L.M. *Chemical Senses* **1983**, *7*, 265-272.
33. Bartoshuk, L.M.; Rifkin, B.; Marks, L.E.; Hooper, J.E. *Chemical Senses* **1988**, *13*, 517-528.
34. Frank, R.A.; Korchmar, D.L. *Physiology & Behavior* **1985**, *35*, 239-242.
35. Boughter, J.D., Jr.; Whitney, G. *Chemical Senses* **1993**, *18*, 445-448.
36. Mela, D.J. *Chemical Senses* **1990**, *15*, 485-490.
37. Bartoshuk, L.M.; Cunningham, K.E.; Dabrila, G.M.; Duffy, V.B.; Etter, L., Fast, K.R.; Lucchina, L.A.; Prutkin, J.M.; Snyder, D.J. *Tastes & Aromas*; Bell, G.A.; Watson, A.J., Eds.; University of New South Wales Press Ltd: Sydney, Australia, 1999; pp 12-22.
38. Miller, I.J., Jr.; Whitney, G. *Neuroscience Letters* 1989, *100*, 1-3.
39. Smith, D.V. *J. Experimental Psychology* **1971**, *87*, 163-171.
40. Arvidson, K. and Friberg, U. *Science* **1980**, *209*, 807-808.

Chapter 7

Neural Representation of Sweet Taste in the Cortex of the Monkey

Thomas R. Scott[1] and Carlos R. Plata-Salamán[2]

[1]College of Sciences, San Diego State University, San Diego, CA 92182
[2]Central Nervous System Research, R. W. Johnson Pharmaceutical Research Institute, Spring House, PA 19477

The development of a facility for recording the activity of single taste neurons in the unanesthetized macaque was driven by the need for an accurate neural model of human gustation. Rodents do not offer such a model. Rats and hamsters—nocturnal and occupying niches that do not offer long lines of sight—have not come to rely on vision for the majority of their sensory information as humans do. Rather, they are chemosensory animals. Their neuroanatomy weaves taste and viscerosensory afferents together even in the hindbrain (1). Rodents are satisfactory models as chemosensory specialists, just as cats long served as the neural model for visual animals. However, they are not adequate surrogates for human taste any more than the cat can reveal the cortical mechanisms of human color perception. Rodents continue to serve a central role in the study of gustatory neurophysiology as it pertains to the mechanisms of transduction or lower-order neural coding. But the revelation of higher-order processes requires study of an animal whose forebrain mechanisms more closely approximate those of the human. Indications from the peripheral level suggest that the chimpanzee would be the ideal lieutenant for humans (2-4). As in vision, however, the macaque is nearly as close, and much more plentiful and inexpensive.

The value of the primate model for human taste is revealed most clearly in the neural representation of sweet tastes. Chemicals that humans describe as predominantly sweet encompass a wide variety of molecular structures: mono-, di-, and polysaccharides, amino acids, dipeptides and individual and compound proteins, polyhydric alcohols, glycosides, sulfobenzamides, and salts of heavy metals. However, within each of these categories except the first, sweetness is only sparsely represented. Minor modifications of the structure of a sweet molecule often render it insipid or bitter. This implies what we now know to be the case: that specialized protein receptors linked to second messenger systems are responsible for the transduction of sweet tastes (5,6).

Given the idiosyncratic nature of this receptor-ligand relationship, the first question is whether various chemicals that taste sweet to humans generate neural activity in the macaque cortex, and whether the activity is similar to that evoked by sugars, implying sweetness. If so, then humans and macaques would appear to occupy the same range of sweet taste sensitivity. The more subtle issue is the nature of the relationship among these stimuli. Most sweet stimuli also evoke notes of other taste qualities in humans. In a neural model, the relationship among these notes is revealed in patterns of activity elicited by each stimulus. Therefore, an array of sweet stimuli serves as an effective probe to test the degree of similarity between human and macaque representations of this fundamental segment of the gustatory domain.

The afferent limb of the central taste system in the macaque is represented in Figure 1. Cranial nerves VII, IX and X (not shown) convey taste information from the mouth to the nucleus of the solitary tract (NST). From here, second-order fibers project rostrally toward the parabrachial nuclei, but do not terminate there, joining instead the central tegmental tract to distribute themselves within the parvicellular division of the ventroposterior medial nucleus in the thalamus (VPMpc). From the thalamus, third-order neurons project to the anterior insula (I) and frontal operculum (O) where gustation realizes its cortical representation (7). One projection from this primary taste cortex is to the central nucleus of the amygdala (Am) (8,9). From the amygdala, taste information could reach the basal forebrain and hypothalamus. Axons also project anteriorly from the insular-opercular cortex to end in the dysgranular caudolateral region of the orbitofrontal cortex (OFC) (10), termed "area G" (11). Fibers from this area join with those from a visual area immediately lateral, and from olfactory neurons just medial, to converge on cells yet more medial where they are hypothesized to interact to generate an appreciation of flavor.

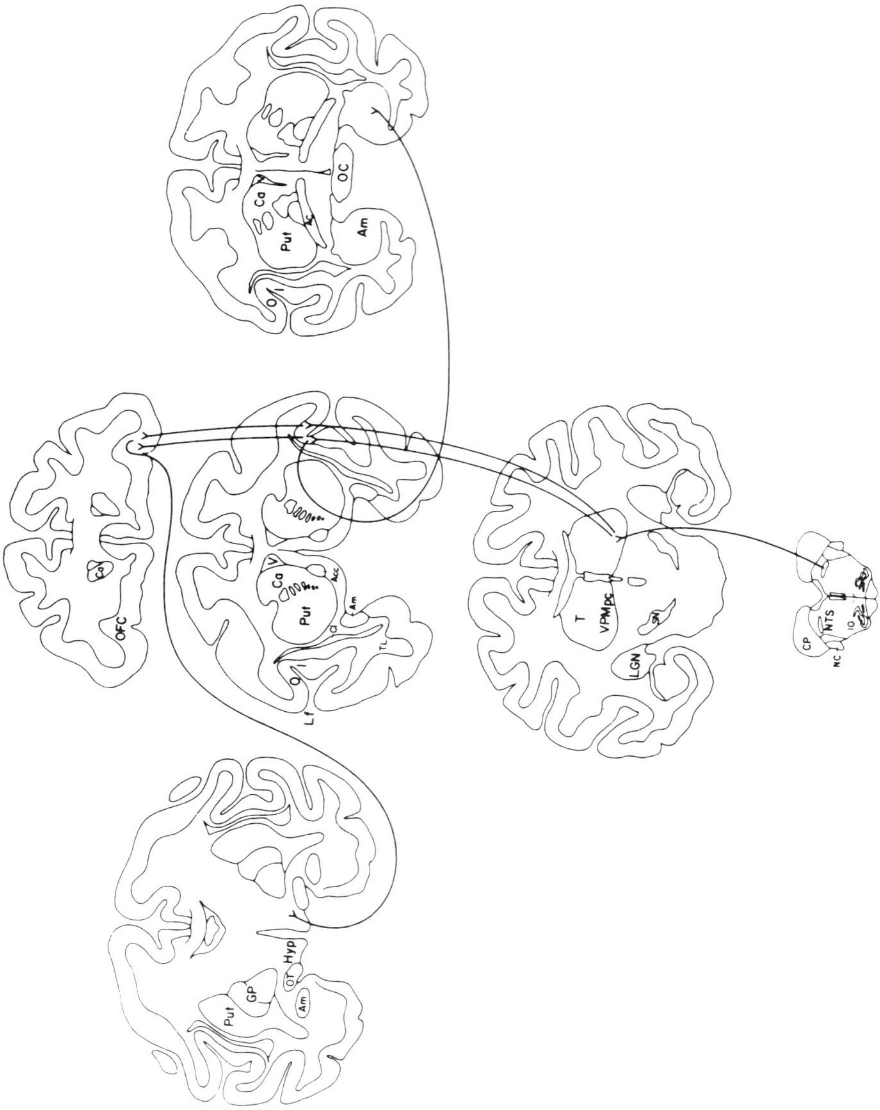

OC
Ca
Put
Am
Ac
O
I
OFC
Ca
Acc
Cl
TL
Lf
T
VPMpc
SN
LGN
NTS
IO
CP
NC
Hyp
GP
OT

Figure 1. Transverse sections summarizing the afferent limb of the central gustatory projections in the cynomolgus macaque. Abbreviations: **AC**, anterior commissure; **Acc**, nucleus accumbens; **Am**, amygdala; **Ca**, caudate; **Cl**, claustrum; **CP**, cerebellar peduncle; **GP**, globus pallidus; **Hyp**, hypothalamus; **I**, insula; **IO**, inferior olive; **Lf**, lateral fissure; **LGN**, lateral geniculate nucleus; **NC**, nucleus cuneatus; **NTS**, nucleus of the solitary tract; **O**, operculum; **OC**, optic chiasm; **OFC**, orbitofrontal cortex; **OT**, optic tract; **Put**, putamen; **SN**, substantia nigra; **T**, thalamus; **TL**, temporal lobe; **V**, ventricle; **VPMpc**, ventroposterior medial nucleus of the thalamus, pars parvicellularis.

Taste cells in the orbitofrontal cortex also send axons to the caudate nucleus (not shown) where they are distributed throughout the striatum, and to the lateral hypothalamus (Hyp) which communicates reciprocally with the central nucleus of the amygdala. From the amygdala, axons project back to the nucleus of the solitary tract (12), establishing the possibility of a lengthy and open circuit through which taste information, after cortical processing, could interact with areas involved with feeding, reward, and emotions.

Electrophysiological experiments have now been performed at each level of the macaque taste system: peripheral nerves (13-15), NST (16,17), thalamus (18), primary (19-21) and secondary (22,23) taste cortices, amygdala (24,25), hypothalamus (26) and caudate (27). Based on profiles of evoked activity, the most accurate neural representation of taste quality appears to exist in the primary taste cortex. That is, the relative similarities among the neural responses to a range of taste stimuli are most closely related to the relative similarities of those same stimuli as described by humans. Therefore, we will concentrate on the responses to sweet stimuli as evoked from primary cortical taste cells.

Primary taste cortex encompasses a volume of about 50 mm^3 in the anterior insula and frontal operculum (Figure 2). Gustatory neurons compose about 5% of its cells. Neurons intermingled with them respond during mouth movements, are activated by touch to areas in and around the mouth, discharge to the sight of an approaching taste stimulus, or when the macaque extends his tongue. Still, the combination of all these subpopulations only totals another 30% of the cortical cells. The majority of cells in primary taste cortex have not been characterized by the stimuli we have applied. It is likely that many are involved in an integrated series of activities that relates to the selection and the digestion of foods. Such activities would include somatosensory input from the hands, mouth and tongue (28) and visceral afferents (29-31). Motor activity would include the tongue and jaw movements mentioned above. It is also reported that electrical stimulation of the insular-opercular cortex in monkeys (32) and humans (33) causes visceromotor activity and sensations in the epigastrium. Thus, taste may not dominate primary gustatory cortex in the same manner as vision dominates V1, but may serve as one component of a system whose broader purpose is to choose and process nutrients (34).

The 22 stimuli we selected for an analysis of sweet taste are listed, along with other relevant data, in Table 1 (35). We used cran-raspberry juice (Ocean Spray) as an effective search stimulus since it has a complex and highly

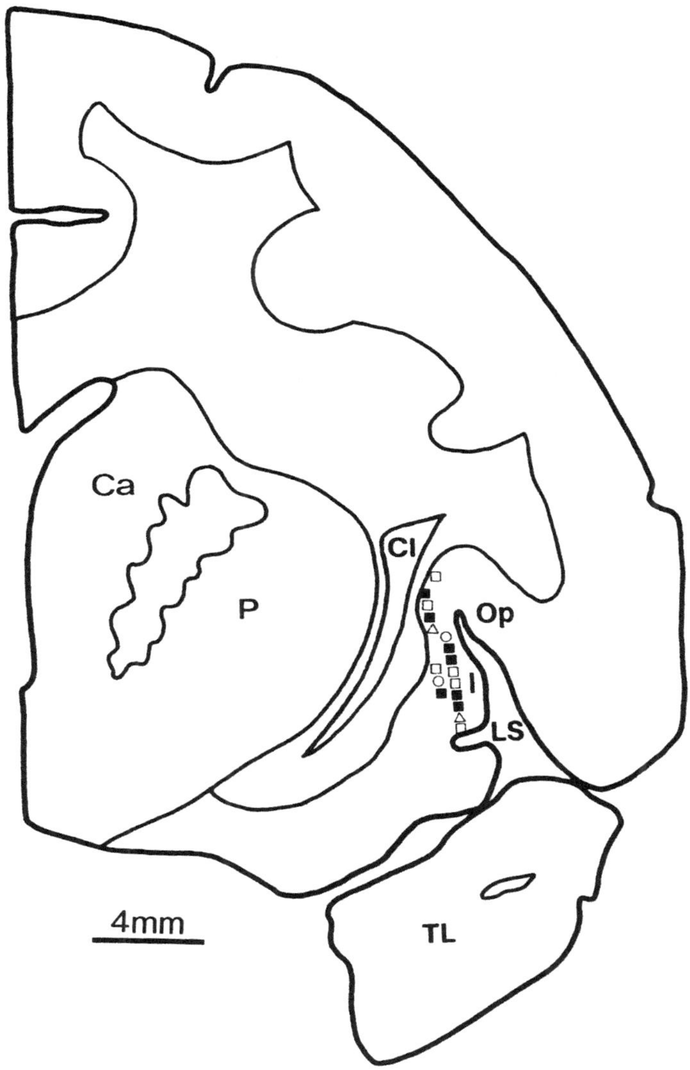

Figure 2. Tracing of a transverse section through the left hemisphere at the anteroposterior level of the anterior clinoid process of the sphenoid bone. Taste-responsive neurons encountered during two successful tracks through this section are labeled according to the basic stimulus that elicited the largest response. **Glucose**, filled squares; **NaCl**, open squares; **HCl**, triangles; **quinine HCl**, open circles. Abbreviations: **Ca**, caudate; **Cl**, claustrum; **I**, insula; **LS**, lateral sulcus; **Op,** frontal operculum; **P**, putamen; **TL**, temporal lobe.

Tab le I. Stimuli, Abbreviations, Concentrations, Molecular Weights, Chemical Groups

stimulus	*abbrev*	*conc*	*mw*	*chemical group*
distilled water	W	-------	18	-------------------
NaCl	N	0.3 M	58	-------------------
HCl	H	0.01 M	36	-------------------
quinine HCl	Q	0.001 M	361	-------------------
glucose	G	1.0 M	180	monosaccharide aldohexose
acesulfame K	ACE	0.7 %	163	oxathiazinone dioxide
aspartame	A	0.01 M	294	l-aspartyl-l-phenylalanine-1-methyl ester
Ca cyclamate	CC	1.2 %	198	calcium cyclohexyl-sulfamate
cranraspberry juice	C	20%	-----	-------------------
fructose	F	1.0 M	180	monosaccharide ketohexose
maltose	M	0.5 M	342	disaccharide
Monellin	MO	0.05%	10,700	protein
myoinositol	MYO	1.0 M	180	polyol
neohesperidin DHC	NEO	0.032%	613	dihydrochalcone glycoside
Polycose	PO	0.2 M	-----	glucose polymers
Na saccharin	SAC	0.09%	205	O-sulfobenzimide
sorbose	SE	1.0 M	180	monosaccharide ketohexose
sorbitol	SO	1.0 M	182	polyhydric alcohol
sucrose	S	0.5 M	342	disaccharide
stevioside	STE	0.18%	805	diterpene glycoside
L-tryptophan	TRP	0.03 M	204	l-amino acid
xylitol	XY	1.0 M	152	polyhydric alcohol
xylose	XE	1.0 M	150	monosaccharide aldopentose

palatable taste. The non-sweet stimuli—NaCl, HCl, and quinine HCl—were included to provide a broader context in which responsiveness to the 19 sweet stimuli could be evaluated.

We recorded the activity of 3,066 cells isolated during 66 recording tracks made in three male cynomolgus macaques over a period of nine months. Of these, 144 (5%) were responsive to taste. Many also had non-gustatory components, most notably touch or motor. When these caused major uncontrolled changes in a cell's activity, we rejected that neuron. Others were lost before we could apply the complete stimulus series, a process that required about 45 min. This left 47 taste cells, from whose responses we derive the following information.

Response characteristics

The mean spontaneous discharge rate of these cortical taste cells was 5.0 spikes/s. When 1.0 M glucose was placed on the tongue, the rate rose to a mean of 15.5 spikes/s, to 0.3 M NaCl, 11.4, to 0.001 M quinine HCl, 9.4, and to 0.01 M HCl, 6.7 spikes/s. Most cells responded best either to glucose (38%) or NaCl (36%), while few responded best to quinine HCl (15%) or HCl (11%).

Representation of taste quality

Our index of the relative similarity among these 22 taste stimuli was the correlation between the profiles of activity they elicited from the 47 taste cells. We calculated the correlation coefficient between the profiles of each possible pair of stimuli (22 x 21 / 2 = 231 coefficients), and organized them into a matrix. We subjected this to a multidimensional scaling routine (Alscal) to provide a spatial representation of relative similarity, and to a cluster analysis (36) to offer a quantitative indication of how the stimuli relate to one another.

Relationship between sweet and non-sweet stimuli

The two-dimensional space representing relative similarity among stimulus profiles is shown in Figure 3, while the dendrogram resulting from the cluster analysis appears in Figure 4. From the space, it is clear that sweet stimuli were quite distinct from the non-sweet chemicals. Quinine HCl (Q), HCl (H), and water (W)—toward the left side of the space—were all about equidistant from the sweet group, while NaCl (N) was isolated in the lower left. The

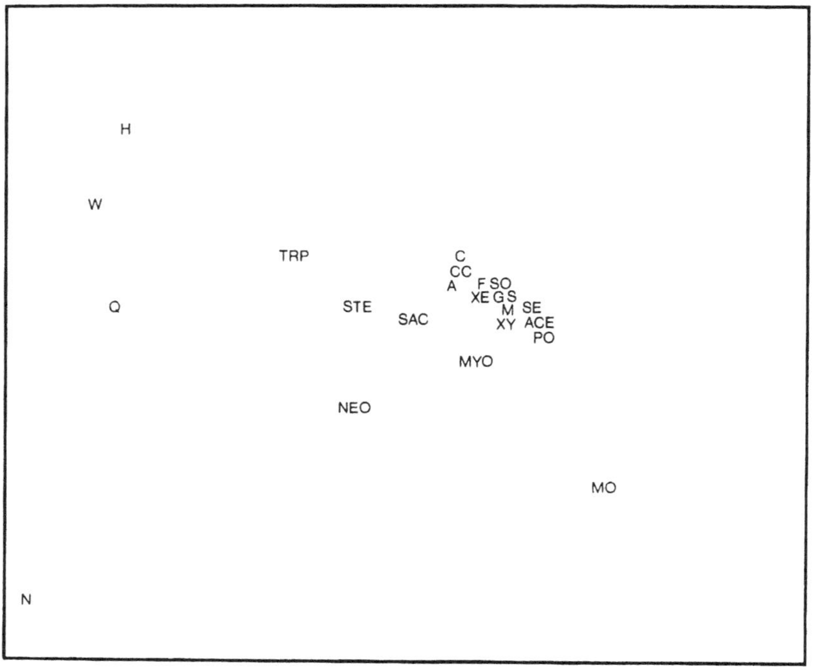

Figure 3. Two-dimensional space showing the distribution of taste qualities represented by this stimulus array. Sweet chemicals form a coherent cluster toward the right side of the space, from which stimuli that have sour or bitter notes extend out toward HCl, water, and quinine HCl.

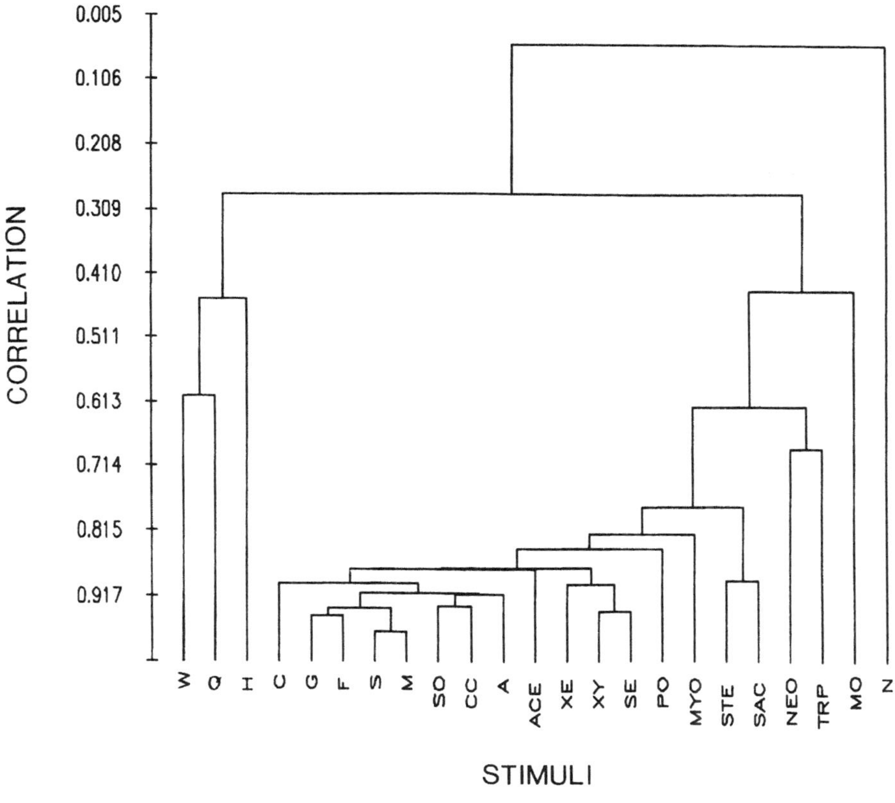

Figure 4. Dendrogram in which is represented the degree of similarity among the 22 tastants plus water. NaCl (N) is most isolated. Water (W), quinine HCl (Q) and HCl (H) are moderately well related among themselves and distinct from all other stimuli. The group of 19 sweet chemicals form a coherent cluster, described more fully in the text.

dendrogram shows the same relationship, with Q, H, and W all loosely affiliated and quite distinct from the sweet cluster in the center, while N was sharply separated from all other stimuli. Thus, the fundamental divisions that humans recognize, with acid (sour) and quinine (bitter) somewhat related and NaCl (salty) and glucose (sweet) separate and distinct, is reflected in the basic organization of the macaque's responses.

Relationship among sweet stimuli

Thirteen of the 19 sweet chemicals formed a coherent group in the taste space (Figure 3). At its center were the simple carbohydrates—glucose, fructose, sucrose and maltose—joined by sorbitol. The correlations (N = 10) among the profiles generated by these five chemicals were all greater than +0.92. Nearest this core were calcium cyclamate, aspartame, and cran-raspberry juice, with the profiles among all eight chemicals (N = 28) above +0.88. In the next concentric ring were acesulfame potassium, xylose, xylitol, sorbose, and Polycose. These 13 stimuli encompass taste qualities that are reliably discriminable to humans, though the correlations among their profiles remained moderately high, all exceeding +0.72 (N = 78).

Only the remaining six chemicals retreated from this central cluster: myoinositol, sodium saccharin, stevioside, neohesperidin DHC, L-tryptophan, and Monellin. The profiles evoked by myoinositol and sodium saccharin were still predominantly sweet in character, based on their high correlations (+0.70s and +0.80s) with those of the simple sugars and low (+0.20s and +0.30s) with non-sweet stimuli. Stevioside elicited a profile that was almost as similar to that of quinine HCl as to glucose, implying a distinctly bitter note to its sweetness, and causing its placement along a direct path toward quinine HCl in the space. The profile of neohesperidin DHC was more poorly correlated with those of the sugars, but this reduction was not replaced by increased correlations with non-sweet stimuli, suggesting a weaker sweet taste. Accordingly, neohesperidin DHC evoked less activity from the cortical neurons than any other stimulus except l-tryptophan (which is not sweet—see below) and Monellin.

Tryptophan lay away from the sweet group in the direction of water, HCl, and quinine HCl. We had inadvertently selected the l-isomer, which to humans tastes flat to bitter rather than sweet. Appropriately, then, the mean response was quite low and the profile correlated best with those of HCl and quinine HCl.

Most disparate among the sweet chemicals was Monellin. It evoked less activity than any other stimulus. Its profile correlated only modestly with those of the sugars, yet was totally unrelated to those of the non-sweet stimuli. Thus, although its relationship with sweet stimuli was weak, it had no hint of similarity to non-sweet chemicals, and so was relegated to a neutral corner of the space.

These same relationships are apparent in the dendrogram of Figure 4. There is a tight cluster of 12 sweet chemicals—from cran-raspberry juice (C) to sorbose (SE)—while the other seven sweet stimuli range progressively farther from this center, out to Monellin.

Neural responses and their relationship to human perception of sweetness

Schiffman et al. (37) conducted a psychophysical investigation of human sweet perception using 17 stimuli. Subjects gave descriptions of each chemical on scales of sweetness, bitterness, goodness, etc. They also tasted the chemicals in pairs and assigned a similarity rating to each pair (corresponding to a correlation between neural profiles in the monkey data). These ratings were used to generate a multidimensional space, analogous to the space of Figure 3. Fifteen of the sweet chemicals used by Schiffman et al. were also members of our stimulus set. Thus, even though the presence of different chemicals would cause some distortion of each space relative to the other, a comparison of the neurally and psychophysically-based spaces would still offer a useful means of determining the degree to which the subtle differences among sweet stimuli are common to macaques and humans.

We compared the organization of the spaces by defining the position of glucose as zero and measuring the distance from there to the locations of each of the common 14 stimuli. The correlation between these distances was +0.82.

As in the neural space, the simple carbohydrates and polyhydric alcohols were closely related in human perception. Among high intensity sweeteners, calcium cyclamate and aspartame generated the purest sweet taste in humans, and these (plus acesulfame potassium) were the only chemicals that elicited neural profiles in the macaque that correlated above 0.90 with those of the sugars.

L-tryptophan is described as flat to bitter, and stevioside and sodium saccharin are the most bitter among the sweet chemicals. Appropriately, in the neural space these three were situated away from the main sweet cluster, positioned in that same order progressively closer to the locations of HCl and quinine HCl.

Monellin was reported by humans to be rather sweet, moderately pleasant, and, among all stimuli, the only one to be totally devoid of bitterness. In the neurally-based space, Monellin is more isolated from other sweet stimuli than this description would imply. However it generated the lowest correlation of all stimuli with HCl and quinine HCl, and accordingly was the only stimulus on the opposite side of the main sweet cluster, away from the non-sweet stimuli.

Another point of comparison has to do with the temporal aspects of the responses. Of the 19 sweet stimuli we used, two—Monellin and neohesperidin—are noted for their slowly developing taste in humans. Accordingly, Monellin had the slowest-developing response among our stimuli, and neohesperidin DHC the third slowest (after sorbose).

Overall, there is a high level of agreement between the taste perceptions that humans report and the implied taste quality derived from response profiles in macaques. Among the 15 sweet stimuli used in both studies, 13 elicited profiles that would accurately predict the perception reported by human subjects. Only acesulfame potassium and Monellin presented discrepancies. The former evoked a neural profile that correlated well with those of the sugars and did not capture the degree of bitterness that humans report. This was the case even though our concentration (0.7 %) was twice that used by Schiffman et al. (38), and the bitterness of high intensity sweeteners is reported to increase with concentration. The latter generated a profile that did not correlate highly enough with those of the sugars to support human descriptions of its pure sweetness. Thus, with minor exceptions, the macaque offers an effective neural model for human gustation.

References

1. Hermann, G. C. and Rogers, R. C. (1985) *Convergence of vagal and gustatory afferent input within the parabrachial nucleus of the rat. Journal of the Autonomic Nervous System*, 13: 1-17.
2. Hellekant, G., Ninomiya, Y., DuBois, G.E., Danilova, V., and Roberts, T.W. (1996) *Taste in the chimpanzee. 1: The summated response to sweeteners and the effect of gymnemic acid. Physiology and Behavior*, 60: 469-479.
3. Hellekant, G., Ninomiya, Y., and Danilova, V. (1997) *Taste in the chimpanzee. 2: Single chorda tympani fibers. Physiology and Behavior*, 61: 829-841.

4. Hellekant, G., Ninomiya, Y., and Danilova, V. (1998*) Taste in the chimpanzee. III: Labeled-line coding in sweet taste. Physiology and Behavior*, 65: 191-200.
5. Herness, M. S. and Gilbertson, T.A. (1999*) Cellular mechanisms of taste transduction. Annual Review of Physiology*, 61: 873-900.
6. Lindemann, B. (1999) *Receptor seeks ligand: On the way to cloning the molecular receptors for sweet and bitter taste. Nature Medicine*, 5: 381-382.
7. Pritchard, T.C., Hamilton, R.B., Morse, J.R., and Norgren, R. (1986*) Projections of thalamic gustatory and lingual areas in the monkey, Macac fascicularis. Journal of Comparative Neurology*, 224: 213-228.
8. Aggleton, J.P., Burton, M.J., and Passingham, R.E. (1980) *Cortical and subcortical afferents to the amygdala of the rhesus monkey* (*Macaca mulatta*). *Brain Research*, 190: 347-368.
9. Mufson, E.J., Mesulam, M.-M. and Pandya, D.N. (1981) *Insular interconnections with the amygdala in the rhesus monkey. Neuroscience*, 6: 1231-1248.
10. Baylis, L.L., Rolls, E.T., and Baylis, G.C. (1995) *Afferent connection the caudolateral orbitofrontal cortex taste area of the primate. Neuroscience*, 64: 801-812.
11. Carmichael, S.T. and Price, J.L. (1993) *Sensory and premotor connections of the orbital and medial prefronal cortex of macaque monkeys. Journal of Comparative Neurology*, 363: 642-664.
12. Price, J.L. and Amaral, D.G. (1981) *An autoradiographic study of the projections of the central nucleus of the monkey amygdala. Journal of Neuroscience*, 1: 1242-1259.
13. Hellekant, G., Danilova, V., Roberts, T., and Ninomiya, Y. (1997a) *The taste of ethanol in a primate model. 1. Chorda tympani nerve response in Macaca mulatta. Alcohol*, 14: 473-484.
14. Hellekant, G., Danilova, V., and Ninomiya, Y. (1997b*) Primate sense of taste: Behavioral and single chorda tympani and glossopharyngeal nerve fiber recordings in the rhesus monkey, Macaca mulatta. Journal of Neurophysiology*, 77: 978-993.
15. Ogawa, H., Yamashita, S., Noma, A., and Sato, M. (1972) *Taste responses in the macaque monkey chorda tympani. Physiology and Behavior*, 9: 325-331.
16. Scott, T.R., Yaxley, S., Sienkiewicz, Z.J., and Rolls, E.T. (1986a) *Gustatory responses in the nucleus tractus solitarius of the alert cynomolgus monkey. Journal of Neurophysiology*, 55: 182-200.
17. Yaxley, S., Rolls, E.T., Sienkiewicz, Z.J., and Scott, T.R. (1985) *Satiety does not affect gustatory activity in the nucleus of the solitary tract of the alert monkey. Brain Research*, 347: 85-93.

18. Pritchard, T.C., Hamilton, R.B., and Norgren, R. (1989) *Neural coding of gustatory information in the thalamus of Macaca mulatta. Journal of Neurophysiology*, 61: 1-14.
19. Ogawa, H. (1994) *Gustatory cortex in primates: Anatomy and physiology. Neuroscience Research*, 20: 1-13.
20. Scott, T.R., Yaxley, S., Sienkiewicz, Z.J., and Rolls, E.T. (1986b) *Gustatory responses in the frontal opercular cortex of the alert cynomolgus monkey. Journal of Neurophysiology*, 56: 876-890.
21. Yaxley, S., Rolls, E.T., and Sienkiewicz, Z.J. (1990) *Gustatory responses of single neurons in the insula of the macaque monkey. Journal of Neurophysiology*, 63: 689-700.
22. Rolls, E.T., Sienkiewicz, Z.J., and Yaxley, S. (1989) *Hunger modulates the responses to gustatory stimuli of single neurons in the caudolateral orbitofrontal cortex of the macaque monkey. European Journal of Neuroscience*, 1: 53-60.
23. Rolls, E.T., Yaxley, S., and Sienkiewicz, Z.J. (1990) *Gustatory responses of single neurons in the orbitofrontal cortex of the Macaque monkey.Journal of Neurophysiology*, 64: 1055-1066.
24. Scott, T.R., Karádi, Z., Oomura, Y., Nishino, H., Plata-Salamán, C.R., Lénárd, L., Giza, B.K., and Aou, S. (1993) *Gustatory neural coding in the amygdala of the alert macaque monkey. Journal of Neurophysiology*, 69: 1810-1820.
25. Yan, J. and Scott, T.R. (1996) *The effect of satiety on taste cells in the amygdala of the alert macaque. Brain Research*, 740: 193-200.
26. Karádi, Z., Oomura, Y., Nishino, H., Scott, T.R., Lénárd, L. and Aou, S. (1992) *Responses of lateral hypothalamic glucose-sensitive and glucose-insensitive neurons to chemical stimuli in behaving rhesus monkeys. Journal of Neurophysiology*, 67: 389-400.
27. Karádi, Z., Faludi, B., Lénárd, L., Czurko, A., Niedetsky, C., Vida, I.,and Nishino, H. (1995) 28.28. *Complex functional attributes, Brain Research Bulletin*, 37: 157.
28. Preuss, T.M. and Goldman-Rakic, P.S. (1989) *Connections of the ventral granular frontal cortex of macaques with perisylvian premotor and somatosensory areas: Anatomical evidence for somatic representation in primate frontal association cortex. Journal of Comparative neurology*, 282:293-316.
29. Mesulam, M.-M. and Mufson, E.J. (1982) *Insular of the old world monkey. III. Efferent cortical output and comments on function. Journal of Comparative Neurology*, 212: 38-52.
30. Mufson, E.J., Mesulam, M.-M., and Pandya, D.N. (1981) *Insular interconnections with the amygdala in the rhesus monkey. Neuroscience*, 6: 1231-1248.

31. Mufson, E.J. and Mesulam, M.-M. (1982) *Insular of the old world monkey. II. Afferent cortical input and comments on the claustrum Journal of Comparative Neurology*, 212: 23-37.
32. Wall, P.D. and Davis, G.D. (1951) *Three cerebral cortical systems affecting autonomic function. Journal of Neurophysiology*, 14: 508-517.
33. Penfield, W. and Raulk, M.E. (1955) *The insula. Further observations on its function. Brain*, 78: 445-470.
34. Scott, T.R. and Plata-Salamán, C.R. (1999) *Taste in the monkey cortex. Physiology and Behavior*, 67: 489-511.
35. Plata-Salamán, C.R., Scott, T.R., and Smith-Swintosky, V.L. (1993) *Gustatory neural coding in the monkey cortex: The quality of sweetness. Journal of Neurophysiology*, 60: 482-493.
36. Wishart, D. (1978) *CLUSTAN*. Edinburgh: Edinburgh University Program Library Unit.
37. Schiffman, S.S., Reilly, D.A., and Clark, T.B. (1979) *Qualitative differences among sweeteners. Physiology and Behavior*, 23: 1-9.
38. Schiffman, S.S., Booth, B.J., Losee, M.L., Pecore, S.D., and Warwick, Z.S. (1995) *Bitterness of sweeteners as a function of concentration. Brain Research Bulletin*, 36: 505-513.

Chapter 8

Age-Related Chemosensory Losses: Effect of Medications

Susan S. Schiffman, Jennifer Zervakis, Brevick G. Graham, and Holly L. Westall

Department of Psychiatry, Duke University Medical Center, Durham, NC 27710

Abstract. Significant losses in taste perception can occur with advancing age, and these losses can contribute to inadequate food intake leading to malnutrition and weight loss. Both experimental data and clinical reports suggest that medications play a major role in age-related chemosensory changes. Over 250 drugs have been reported clinically to affect the sense of taste. Taste impairments from medications include: ageusia (absence of taste), hypogeusia (diminished sensitivity of taste), and dysgeusia (distortion of normal taste). The sites of action for most pharmaceutical compounds that induce taste losses are not known, but medications can act at several levels including peripheral receptors, chemosensory neural pathways, and/or the brain. Extensive research has shown that drugs are secreted into the saliva, and salivary levels of many drugs are high enough to exert adverse effects on taste sensations either by modifying taste transduction mechanisms or by producing a taste of their own.

Extensive shifts in the demographics of the world's population are currently taking place with dramatic increases in both the number and percentage of elderly persons (1). Currently over 400 million people in the world are 65 years of age or older (2), and by the year 2025, the elderly population is expected to reach 1.121 billion people (3). Older individuals have special sensory problems such as taste and smell impairments that must be taken into account in order to provide optimum care to this expanding segment of the population. Altered taste and smell perception in the elderly is a consequence of normal aging, certain disease states (especially Alzheimer's disease), medications, surgical interventions, and/or environmental exposure (4,5).

Taste and smell are chemical senses that respond to molecules in foods, liquids, and air; they also play a fundamental role in food selection and digestion. Taste and smell sensations trigger digestive secretions (e.g. salivary, gastric, pancreatic, and intestinal secretions) that prepare the body for the absorption of nutrients (6). Chemosensory signals serve as indicators of a food's nutritional value from learned association of a food's taste/smell with its postingestive effects. This learned association of taste/smell with the metabolic consequences of eating enables food intake and meal size to be modulated in anticipation of nutritional needs. Thus, impairment of taste and smell can have a significant impact on nutritional status in the elderly (4,5,6).

The current literature suggests that a significant proportion of the elderly experience taste and smell losses during their lifetimes (4,5,7) with chemosensory decrements becoming noticeable after the age of 60 years. Taste changes are less severe than smell until medication use increases (4,5). The medical terms used to classify altered taste and smell sensitivity include: ageusia (no taste sensation), hypogeusia (decreased taste sensation), dysgeusia (distorted taste sensation), anosmia (no sensation of smell), hyposmia (decreased sensation of smell), and dysosmia (distorted smell sensation).

Age-related losses for so-called basic tastes such as NaCl (salty), sucrose (sweet), citric acid (sour), and quinine HCl (bitter) using whole mouth stimulation are small in elderly who suffer from no diseases and take no prescription medications. However, far greater losses occur in those elderly taking medications but who otherwise live active, normal lives (5,8). The mean detection thresholds for elderly individuals taking an average of 3.4 medications were significantly elevated compared to a young cohort, i.e. 11.6 times higher for sodium salts; 2.7 times higher for sweeteners; 4.3 times higher for acids; 7.0 times higher for bitter compounds; 2.5 times higher for amino acids; 5.0 times higher for glutamate salts. The average loss across these taste qualities is 5.41.

The burden of chemosensory disorders from medications is exaggerated in a geriatric population compared to the young due to the disproportionate use of prescription and nonprescription drugs by older individuals. Community-dwelling elderly over the age of 65 take an average of 3.3 ± 0.4 medications (9), and the number increases significantly to 7 drugs or more for elderly living in retirement and nursing homes (5). An overview of data on prescription drug usage in the United States and United Kingdom indicates that the elderly account for 25 to 39% of the prescription drug costs and up to 40% of nonprescription drugs dispensed; yet the elderly account for only 10-18% of the population (10).

Neither the sites of action nor the complicated cascades of cellular events by which medications induce taste changes are well understood. However, medications can impact taste perception at several levels of the nervous system including the peripheral receptors, chemosensory neural pathways, and/or the brainstem and brain. Drugs can produce tastes of their own when they are secreted into the saliva (or build up over time in taste tissues) at concentrations that are greater than taste detection thresholds. In this paper, three experiments are described that deal with the impact of drugs (or other bitter tasting compounds) at the periphery as if they were in saliva. First, the taste thresholds, taste quality, and the effect of a drug on other tastes were assessed for 62 different drugs. Second, the effect of the taste of a bitter drug on neuroendocrine secretions was evaluated. Third, methods to reduce bitter taste were explored.

Experiment 1—Effect of topical application of drugs to the lingual surface

The purpose of these experiments was to determine taste thresholds, taste quality, and the effect of the drug on other tastes for the sixty-two drugs given in Table 1. Detection and recognition thresholds were found using methods described by Schiffman and colleagues (11-14). At a concentration 4 times above the detection threshold, subjects were asked to rate the taste quality of the drug using 14 adjectives: overall intensity, sweet, sour, salty, bitter, metallic, cooling, hot, spicy, burning, anesthetic, astringent, medicinal, and minty/menthol. Thresholds were determined for both young and elderly subjects. Thresholds for HIV drugs were also tested in HIV-infected patients. The effect of topical application of each drug on other tastes was also assessed to mimic the situation in which the drug is secreted into the saliva. The perceived intensities of suprathreshold concentrations of NaCl, KCl, $CaCl_2$, sucrose, quinine HCl, citric acid, capsaicin (pungent), WS-3 (n-ethyl-p-menthane-3-carboxamide) which has a menthol-like taste, and $FeSO_4$ (metallic) were compared when the tongue was treated topically with a drug (near threshold concentrations) or a water control for 3 to 4 minutes. A summary of the results of these studies is given in Table 1. In the first column the drug and its predominant taste at a concentration four times the detection threshold are given. In second column, the detection threshold for young subjects is listed. Thresholds for elderly subjects were higher on average than those for younger subjects. The symbol "n/a" in the threshold column indicates that the drug either had no taste or was so insoluble in water that the threshold could not be determined. The next columns indicate that changes in intensity of certain taste qualities for other taste compounds occurred when the tongue was treated with a drug. A single arrow indicates that the change in intensity was less than 20% while a double arrow indicates that the change was greater than 20%.

The results of this series of experiments revealed that the taste detection thresholds for these drugs ranged from as low as 2.9 μM for saquinavir (Invirase) to 24 mM for didanosine. All of the drugs, which had a taste, were bitter, sour, or metallic at a concentration 4 times the detection threshold. Finally, most of the drugs modified the taste of other compounds to different degrees. This nonuniform alteration of other tastes by drugs may be responsible, in part, for the dysgeusia that is often experienced by elderly who are taking medications.

Table 1: Drugs that interfere with the taste system

Classification/Drug	*Threshold (in mM)*	*Effect of topical application on intensity or major taste quality (↑↓ refers to changes of less than 20%, ⇑⇓ to changes greater than 20%)*								
Psychotropic		NaCl	KCl	$CaCl_2$	Sucrose	QHCl	Citric Acid	Capsaicin	WS-3	$FeSO_4$
Amitriptyline HCl (Bitter)	0.155 ± 0.042	⇓	⇓	⇓	⇓	⇓	⇓	⇓	⇓	
Clomipramine HCl (Bitter)	0.122 ± 0.028	⇓	⇓	⇓	⇓	⇓		⇓		
Desipramine HCl (Bitter)	0.161 ± 0.049	↓					⇓	⇓	⇓	⇑
Doxepin HCl (Bitter)	0.143 ± 0.021	⇓	⇓	⇓	⇓	⇓	⇓	⇓	↓	⇑
Imipramine HCl (Bitter)	0.125 ± 0.024	⇓	⇓	⇓	↓	↓	⇓	↓	⇓	
Trifluoperazine HCl (Bitter)	0.065 ± 0.012	⇓	⇓	⇓		↓	⇓			⇑
Cardiovascular	*Threshold (in mM)*	*NaCl*	*KCl*	*$CaCl_2$*	*Sucrose*	*QHCl*	*Citric Acid*	*Capsaicin*	*WS-3*	*$FeSO_4$*
Captopril (Sour, bitter)	0.132 ± 0.150		↓		↓					
Diltiazem HCl (Bitter)	0.142 ± 0.03			⇑ burn	⇑		⇑		⇓	
Enalapril Maleate (Sour, bitter)	0.107 + 0.03						⇓			

Continued on next page

Table 1. Continued

Classification/Drug	Threshold (in mM)	Effect of topical application on intensity or major taste quality (↑↓ refers to changes of less than 20%, ⇑⇓ to changes greater than 20%) NaCl	KCl	$CaCl_2$	Sucrose	QHCl	Citric Acid	Capsaicin	WS-3	$FeSO_4$
Ethacrynic acid (Bitter)	0.1859 ± 0.056						⇑			⇑
Hydrochloro-thiazide	n/a			⇑ bitter	⇓					⇓
Labetalol HCl (Bitter)	0.182 ± 0.04			↓		↓		⇑		
Mexiletine HCl (Bitter)	0.463 ± 0.34	⇓		↓	↓	↓		⇓	⇑	⇓
Nifedipine	n/a	n.s.								
Procainamide HCl (Bitter)	0.438 ± 0.10		↓	↓				⇑		
Propafenone HCl (Bitter)	0.048 ± 0.03		⇓			⇓	⇓			
Propranolol HCl (Bitter)	0.209 ± 0.07							⇑		
Analgesic/anti-inflammatory	*Threshold (in mM)*	*NaCl*	*KCl*	*$CaCl_2$*	*Sucrose*	*QHCl*	*Citric Acid*	*Capsaicin*	*WS-3*	*$FeSO_4$*
Dexamethasone (Bitter)	0.0902± 0.024			⇓						
Diclofenac sodium (Bitter)	1.008 ± 0.220				↓		⇓	⇓	⇑	
Fenoprofen calcium (Metallic, sour)	1.111 ± 0.388									⇓
Hydrocortisone (Bitter)	0.0928 ± 0.02	ns								

Ibuprofen	n/a								↑ bitter	
Ketoprofen	n/a						⇑ bitter			
Nabumetone	n/a	⇓	⇓	⇓ sour	↓					
Sulindac	n/a		↓							
Antihistamines	*Threshold (in mM)*	*NaCl*	*KCl*	*$CaCl_2$*	*Sucrose*	*QHCl*	*Citric Acid*	*Capsaicin*	*WS-3*	*$FeSO_4$*
Chlorpheniramine maleate (Bitter)	0.085 ± 0.02								⇓	
Terfenadine	n/a			⇑						
Sedatives and antiemetics	*Threshold (in mM)*	*NaCl*	*KCl*	*$CaCl_2$*	*Sucrose*	*QHCl*	*Citric Acid*	*Capsaicin*	*WS-3*	*$FeSO_4$*
Prochlorperazine (Bitter)	0.103 ± 0.045						⇑			
Promethazine (Bitter)	0.079 ± 0.025	⇓	↓	⇓	⇓		⇑ bitter			
Antimicrobial	*Threshold (in mM)*	*NaCl*	*KCl*	*$CaCl_2$*	*Sucrose*	*QHCl*	*Citric Acid*	*Capsaicin*	*WS-3*	*$FeSO_4$*
Ampicillin (Bitter)	1.458 ± 0.392	⇓		⇓			⇓ bitter		⇑	
Atovaquone	n/a							⇓		
Dapsone	n/a		⇓				⇓			
Enoxacin (Metallic, bitter)	0.040 ± 0.014				↓		⇓			

Continued on next page

Table 1. Continued

Classification/Drug	*Threshold (in mM)*	*Effect of topical application on intensity or major taste quality (↑↓ refers to changes of less than 20%, ⇑ ⇓ to changes greater than 20%)*								
Ethambutol 2HCl (Bitter)	0.247 ± 0.055		⇓ bitter	↓					⇓	
Griseofulvin	n/a	n.s.								
Lomefloxacin HCl (Bitter)	0.379 ± 0.102						⇑	⇑		
Ofloxacin (Bitter)	0.387 ± 0.155		⇑				⇑			
Pentamidine isethionate (Bitter)	0.062 ± 0.014	↓	⇓	⇓						
Pyrimethamine	n/a			⇑ burn						
Sulfamethoxazole (Sour, bitter)	0.639 ± 0.119			↓ bitter	↑					
Tetracycline HCl (Sour-bitter)	0.061 ± 0.010		↓	↑ bitter						
Trimethoprim (Bitter)	0.264 ± 0.118	n.s.								
Others (indication	*Threshold (in mM)*	*NaCl*	*KCl*	*$CaCl_2$*	*Sucrose*	*QHCl*	*Citric Acid*	*Capsaicin*	*WS-3*	*$FeSO_4$*
Acyclovir (antiviral)	n/a		↓ bitter					⇓		

Allopurinol (reduces serum & urinary uric acid)	n/a	n.s.								
Baclofen (muscle relaxant) (Bitter-metallic)	3.50 $\pm$ 0.723								↓	
Buspirone HCl (antianxiety) (Bitter)	0.269 $\pm$ 0.147 mM			↑ bitter	↓	⇓		⇑		
Cyclobenzaprine HCl (antispasmotic, muscle) (Bitter)	0.349 $\pm$ 0.156	⇓	⇓	⇓	⇓			⇑		
Dicyclomine HCl (antispasmotic, GI) (Bitter)	0.0354 $\pm$ 0.010		⇓	⇓						
Gemfibrozil (antilipidemic)	n/a		⇓							⇑
Levamisole (immunomodulator) (Bitter)	0.449 $\pm$ 0.148			⇑			↑			
Methimazole (antithyroid agent) (Bitter)	2.111 $\pm$ 1.07	n.s.								
Pentoxifylline (blood viscosity modulator) (Bitter)	0.930 $\pm$ 0.261	⇑		⇓						
Phenytoin (antiepileptic)	n/a								⇑	
HIV drugs	*Threshold (in mM)*	*NaCl*	*KCl*	*$CaCl_2$*	*Sucrose*	*QHCl*	*Citric Acid*	*Capsaicin*	*WS-3*	*$FeSO_4$*

Continued on next page

Table 1. Continued

Classification/Drug	*Threshold (in mM)*	*Effect of topical application on intensity or major taste quality (↑↓ refers to changes of less than 20%, ⇑⇓ to changes greater than 20%)*								
Didanosine (Bitter)	24.0 ± 4.22		↓	⇓					⇓ burn	
Lamivudine (Bitter)	4.36 ± 1.37		⇓	⇓				↓	↓	
Nevirapine	n/a	↓				⇓				
Stavudine (Bitter)	5.99 ± 1.436		⇑	⇓						
Zidovudine (Bitter)	2.15 ± 0.60		⇓					⇓		
Indinavir (Crixivan) (Bitter)	0.237 ± 0.013					⇓	↓	⇑		⇓
Nelfinavir	n/a				⇓	⇓			⇓	
Ritonavir (Bitter)	0.0702 ± 0.039	↓	↓							
Saquinavir mesylate (Bitter)	0.0029 ± 0.0004	↓				⇓ burn	⇑	⇑		

Experiment 2—Elevation of norepinephrine from bitter taste of a drug

Little is currently known about acute physiological side-effects caused by brief unpleasant taste sensations. In this study, subjects were given a single brief exposure to bitter taste and oral irritation to determine their effects on neuroendocrine levels (specifically norepinephrine, epinephrine, and cortisol) in humans. Norepinephrine, epinephrine, and cortisol were investigated because they are associated with stress and arousal. Secretion of norepinephrine and epinephrine involves the adrenal-medullary catecholamine system while the release of cortisol involves the pituitary-adrenal cortical system.

Nine subjects participated individually in one 4 1/2 hour session in the Clinical Research Unit at Duke University Medical School. Four oral stimuli were tested: water (control), carbonated water, Invirase (a bitter drug used to treat HIV infection), and capsaicin (component in chili pepper). Invirase (0.09 mM) and capsaicin (100 ppm) had strong perceived intensities but were not harmful to the subjects. Three stimuli were delivered as a 10 ml liquid bolus (water, carbonated water, and Invirase dissolved in deionized water); capsaicin was delivered on filter paper discs. The taste stimuli were presented in a fixed order: water, carbonated water, bitter water (Invirase), and capsaicin.

A saline lock was placed in a peripheral arm vein sixty minutes prior to the beginning of the study. Then a baseline blood draw was performed followed by delivery of 10 ml of the first oral stimulus (water) squirted into the mouth. Subjects swished the sample throughout the oral cavity and then expectorated. Additional blood samples were taken at 5 minutes, 10 minutes, and 25 minutes after stimulus administration. There was a twenty-five minute break between each test stimulus. Three more test stimuli were then tested in a manner similar to the control stimulus (water), i.e. blood draws were taken immediately before, and 5, 10, and 25 minutes after administration of the stimulus. The subjects' pulse and blood pressure were taken immediately before stimulus administration, and 25 minutes after stimulus administration. Subjects also rated the perceived overall intensity and pleasantness of each stimulus. The blood concentrations of epinephrine, norepinephrine, and cortisol were analyzed for each blood sample. Catecholamines were separated by high performance liquid chromatography (HPLC) and quantitated electrochemically (15). Plasma cortisol was measured by specific radioimmunoassay using HPLC-purified tritiated tracer, 3H cortisol antiserum, and standards purchased from ICN Biochemicals. Bound and free cortisol were separated by dextran-coated charcoal; assay sensitivity was 5 pg/tube.

An analysis of variance was performed to determine is there were differences between stimuli in their effects on neuroendocrine levels. The effect of the stimuli was significant for norepinephrine ($F = 3.59$, $p<0.05$). The results for norepinephrine are shown in Figure 1. Plasma levels increased after a single

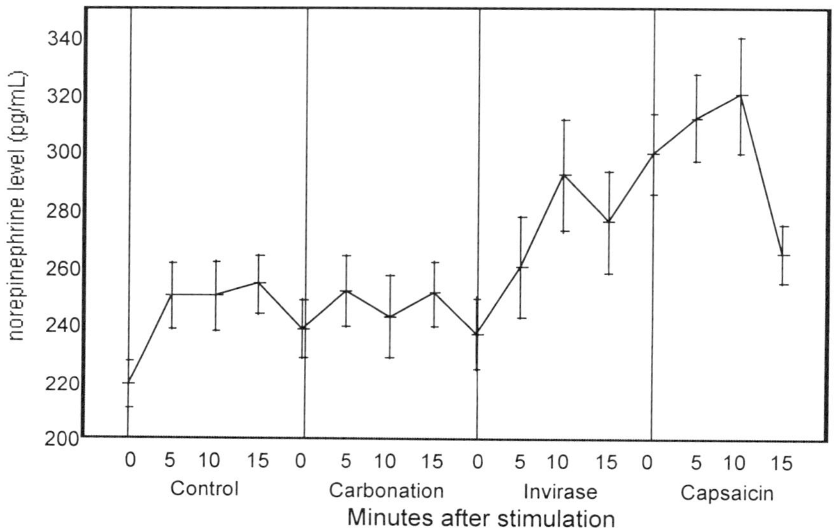

Figure 1. Levels of norepinephrine in the blood after a single taste of water, carbonated water, the bitter-tasting drug Invirase, and the oral irritant capsaicin respectively. The tastes of Invirase and capsaicin both elevated norepinephrine.

presentation of the bitter-tasting drug Invirase and did not return to baseline after the 25 minute rest period. A comparison of norepinephrine levels for the two stimuli Invirase and capsaicin indicated norepinephrine levels were higher than for the control and carbonation groups ($F = 6.02$, $p < 0.05$). There were no significant changes for epinephrine (mean = 20.46 pg/ml $\pm$ 7.90) or cortisol (mean = 75.13 ng/ml $\pm$ 15.18) over stimuli. However, cortisol did significantly decrease with repeated blood draws within a stimulus ($F = 8.31$, $p < 0.05$). Blood pressure, heart rate, and respiration rates were statistically equivalent when measured before and 25 minutes after presentation of a stimuli. Mean overall intensity ratings for the stimuli were: water (1.0 $\pm$ 0.82, very weak), carbonated water (4.10 $\pm$ 1.45, moderate), Invirase (5.20 $\pm$ 1.03, moderate strong), and capsaicin (6.40 $\pm$ 1.10, strong). Mean pleasantness ratings were: water (4.30 $\pm$ 0.67, neither pleasant nor unpleasant), carbonated water (3.50 $\pm$ 1.35, slightly unpleasant), Invirase (1.40 $\pm$ 0.52, very unpleasant), and capsaicin (2.00 $\pm$ 1.94, moderately unpleasant).

The main finding was that a single presentation of the bitter drug (Invirase) and an oral irritant (capsaicin) elevated plasma norepinephrine levels. However, neither epinephrine nor cortisol were altered by a single presentation of these stimuli. These findings are consistent with previous research on stress and arousal which suggests that at the early stages, there is an increase in norepinephrine (fight hormone which provides the drive to meet the challenge with action). As anxiety and uncertainty regarding staying in control grows, epinephrine (fight, anxiety) and ultimately cortisol (helplessness, depression) tend to rise (16). Further studies must be performed to determine if sustained exposure to unpleasant tastes of drugs elevates epinephrine or cortisol as well as norepinephrine. Many medications taken by patients are extremely bitter, and could induce trigeminal/taste stimulation that not only increases stress hormones but also metabolic rate or even wasting. It is important to know what, if any, physiological consequences and effects there are from tasting bitter foods/drugs or carbonated beverages, especially for clinical populations in which wasting is a concern.

Experiment 3- The Effect of Bitter Inhibitors on bitter taste perception of urea, quinine HCl, magnesium chloride, and caffeine.

Little is known about methods for bitterness reduction and inhibition of drugs in part because the range of compounds and mechanisms involved in bitter taste transduction are so diverse. Roy (17) recently reviewed some of the current methods available for bitterness reduction. The purpose of this study was to assess eleven compounds given in Table 2 for their efficacy as bitter blockers of the bitter tastes of caffeine, $MgCl_2$, quinine HCl, and urea. The intensity of each bitter compound mixed with a potential bitter blocker was

Table 2: Average percent suppression of bitterness for each blocker in rank order (Table gives details on which blocker was most effective with each bitter compound; numbers in each % change column with the same superscript are statistically equivalent)

Caffeine	%Change	$MgCl_2$	%Change	QHCl	%Change	Urea	% Change
Cysuc	-37.82^{a}	NaCl	-28.02^{a}	Suc	-19.96^{a}	Suc	-32.12^{a}
Suc	-37.47^{a}	Suc	-27.25^{a}	Cysuc	-8.668^{b}	Cysuc	$-29.43^{a,b}$
NC201	-27.35^{b}	Cysuc	$-23.12^{a,b}$	Hpβcd	$-6.271^{b,c}$	NaCl	$-22.77^{b,c}$
Hpβcd	-25.20^{b}	Hpβcd	$-15.43^{b,c,}$	NaCl	$-3.072^{b,c,d}$	Poly	$-18.96^{c,d}$
Poly	-24.59^{b}	βcyclo	$-15.15^{b,c}$	Malt	$-0.028^{b,c,d}$	Hpβcd	$-12.12^{d,e}$
MAG	-22.32^{b}	Poly	-13.88^{c}	βcyclo	$1.836^{c,d}$	Cypha	$-4.063^{e,f}$
NaCl	$-20.67^{b,c}$	MAG	-13.86^{c}	Cypha	$1.877^{c,d}$	βcyclo	$-3.271^{e,f}$
Malt	$-17.73^{b,c,d}$	NC201	$-10.44^{c,d}$	Emalt	4.525^{d}	MAG	-1.008^{f}
βcyclo	$-13.28^{c,d}$	Emalt	$-9.553^{c,d}$	Poly	4.923^{d}	Emalt	0.129^{f}
Cypha	$-12.16^{c,d}$	Cypha	$-7.523^{c,d}$			NC201	0.864^{f}
Emalt	11.08^{d}	Malt	-4.085^{d}			Malt	3.398^{f}

Abbreviations for bitter blockers: *N*-(4-cyanophenyl)-*N'*-[(sodiosulfo)methyl]urea (NC201), ±2-(4-methoxyphenoxy) propionic acid (Cypha), monoammonium glycyrrhizinate (MAG), β-cyclodextrin (βcyclo), hydroxypropyl β–cyclodextrin (Hpβcd), polydextrose (Poly), maltol (Malt), ethyl maltol (Emalt), sodium chloride (NaCl), sucrose (Suc), and a mixture of sucrose and ±2-(4-methoxyphenoxy) propionic acid (Cysuc). Concentrations of bitter blockers used: NC201 (4mg/ml), Cypha (500 ppm), MAG (456 ppm), βcyclo (1.5%w/v), Hpβcd (10%w/v), Poly (50%w/v), Mal (80 ppm), Emalt (50 ppm), NaCl (0.3M), sucrose (60%w/v), and Cysuc (60%w/v Suc and 500 ppm Cypha). Concentrations of four bitter compounds that were mixed with the potential bitter blockers: urea (0.178 M to 1.35M), quinine HCl (8.68μM to 2.11mM), $MgCl_2$ (0.0104 M to 0.333 M), and caffeine (2.94 mM to 48.3 mM).

evaluated by 15 adult subjects, consisting of 8 tasters and 7 non-tasters of phenylthiocarbamide (PTC). The subjects also evaluated the intensity of the same bitter compound alone as a control. All combinations of bitter compounds and potential blockers were assessed with the exception of quinine HCl, which was not tested with *N*-(4-cyanophenyl)-*N'*-[(sodiosulfo)methyl]urea or monoammonium glycyrrhizinate due to solubility problems at suprathreshold levels. Quinine HCl (0.156 mM) and sucrose (7%) were used as mid-point references on 100 mm visual analog scales for bitterness and sweetness respectively.

N-(4-cyanophenyl)-*N'*-[(sodiosulfo)methyl]urea is an analog of suosan which was discovered by Muller and colleagues (18) at the Nutrasweet company (code NC00201). Cypha (±2-(4-methoxyphenoxy)propionic acid) is a GRAS (generally regarded as safe) compound that was patented as a sweetness reducing agent (US patent 4,567,053). Monoammonium glycyrrhizinate is the salt of glycyrrhizinic acid which is extracted from licorice. Cyclodextrins have been discussed as bitter blockers by Roy (17). Polydextrose is a polymer of dextrose that has little taste; anecdotal evidence suggests it may block bitter taste. Maltol ($C_6H_6O_3$) and ethyl maltol ($C_7H_8O_3$) are marketed commercially as compounds that minimize bitterness and enhance sweetness. Sucrose is used as a pharmaceutical excipient to reduce bitter taste. Sodium salts have also been shown to reduce some bitter tastes (19).

The average percent suppression of bitterness for each potential blocker is given in Table 2. The potential blockers that were the most effective overall in reducing the bitterness ratings were sucrose alone and NaCl. The bitter blocking potential of the other inhibitors varied with the structure of the bitter compound.

Final comment

Medications applied topically to the tongue can produce a taste of their own and alter the quality and intensity of other tastants in both young and elderly individuals as well as HIV-infected patients. Furthermore, brief exposure to a bitter-tasting drug can elevate norepinephrine levels. Blockers can be added to bitter solutions to reduce the bitter taste. Further research is necessary to determine the mechanisms by which drugs alter taste perception at the peripheral and central levels.

Supported by NIA AG00443 & AG00029, and NCRR GCRC MO1RR30

References

1. American Association of Retired Persons (AARP). *A profile of older Americans;* AARP: Washington, DC, 1995, pp 1-13.

2. Davies, A. M. *World Health Forum* 1989, 10, 299-306; discussion 306-321.
3. US Senate Special Committee on Aging (in conjunction with the American Association of Retired Persons, the Federal Council on the Aging, and the Administration on Aging). *Aging America, Trends and Projections 1985-1986 Edition*; United States Senate: Washington, DC, 1986, pp. 8-28.
4. Schiffman, S. S. *N. Engl. J. Med.* 1983, 308, 1275-1279, 1337-1343.
5 Schiffman, S. S. *JAMA* 1997, 278, 1357-1362.
6. Schiffman, S. S.; Warwick, Z. S. In *The Science of Food Regulation: Food Intake, Taste, Nutrient Partitioning, and Energy Expenditure;* Bray, G. A.; Ryan, D. H., Eds.; Louisiana State University Press: Baton Rouge, LA, 1992, pp 293-312.
7. Doty, R. L., Shaman, P., Applebaum, S. L, Giberson, R., Siksorski, L., and Rosenberg, L. *Science* 1984, 226:1441-1443.
8. Schiffman, S. S. *Crit. Rev. Food Sci. Nutr.* 1993, 33, 17-26.
9. Lewis, I. K.; Hanlon, J. T.; Hobbins, M. J.; Beck, J. D. *Spec. Care Dentist.* 1993, 13, 171-176.
10. Atkin, P. A.; Veitch, P. C.; Veitch, E. M.; Ogle, S. J. *Drugs Aging* 1999, 14, 141-152.
11. Schiffman, S. S.; Zervakis, J.; Suggs, M. S.; Shaio, E.; Sattely-Miller, E. A. *Physiol. Behav.* 1999, 66, 183-192.
12. Schiffman, S. S.; Zervakis, J.; Shaio, E.; Heald, A. E. *Nutrition* 1999, 15, 854-859.
13. Schiffman, S. S.; Zervakis, J.; Heffron, S.; Heald, A. E. *Nutrition* 1999, 15, 767-772.
14. Zervakis, J.; Graham, B. G.; Schiffman, S. S. *Physiol. Behav.* 2000, 68, 405-413.
15. Kilts, C. D.; Gooch, M. D.; Knopes, K. D. *J. Neurosci. Methods* 1984, 11, 257-273.
16. Henry, J. P. *Acta Physiol. Scand. Suppl.* 1997, 640, 10-25.
17. Roy, G. *Trends Food Sci. Technol.* 1992, 3, 85-91.
18. Muller, G. W.; Culberson, J. C.; Roy, G., Ziegler, J.; Walters, D. E.; Kellogg, M. S.; Schiffman, S. S.; Warwick, Z. S. *J Med. Chem.* 1992, 35, 1747-1751.
19 Breslin, P. A.; Beauchamp, G. K. *Chem. Senses* 1995, 20, 609-223.

Olfactory Mechanisms and Flavor Release

Chapter 9

Olfactory Receptor Neuron Signaling

Stuart Firestein

Department of Biological Science, 923 Fairchild, MC 2438, Columbia University, New York, NY 10027

Olfactory receptor neurons are the primary sensory cell in a system designed to detect and discriminate between a large and diverse array of chemical ligands called odors. The mechanisms employed for this task are a model of biological engineering.

Introduction

The vertebrate olfactory system is perhaps the most capable chemical detector on the face of the planet. It is able to detect and discriminate between an immense and diverse array of small chemical ligands that we typically call odorants. The odors, which may number more than 10,000, are found in many chemical classes and vary in structure from the simplest of aliphatic short chain organic molecules to highly complex aromatics with multiple functional groups. In many instances a change of no more than a single carbon atom can give rise to different odor qualities, while in other cases apparently unrelated chemicals produce similar odor perceptions.

Remarkable as these capabilities may seem from an engineering point of view the olfactory system accomplishes its sensory tasks with biological mechanisms that are common to many signaling systems. In the past 15 years, the application of modern techniques in physiology, imaging, molecular biology and genetics has given rise to a consensus model of olfactory signaling that bears striking similarities to other signal transduction systems in the brain and to other sensory systems, most notably visual transduction. This article will describe the current consensus model of olfactory signaling.

Location and Morphology of Olfactory Receptor Neurons

The olfactory sensory neuron (OSN) is the primary sensory neuron of the olfactory pathway. While located peripherally within the olfactory epithelium, a thin tissue lining the bones of the upper or caudal recesses of the nasal cavity (Figure 1), these cells are true central nervous system neurons, possessing axons and deriving embryonically from neural tissue. The olfactory epithelium consists of only three major cell types: the olfactory sensory neurons (OSNs), sustentacular cells (a type of glial or supporting cell), and basal cells (see later). There are between 6 and 10 million OSNs in the epithelia of different mammals. The simple structure of this tissue, its peripheral location, and the presence of only a single type of neuron, makes this an experimentally convenient system for physiological, molecular and biochemical studies.

The OSN itself is also a morphologically simple cell. The soma is oval shaped, and from it's distal pole a single rather thick dendrite projects to the surface of the epithelium. This dendrite ends in a knob-like swelling from which emanate up to 20 or so very fine and long cilia. These cilia project into the lumen of the nasal cavity being protected by thin layer of watery mucus. From the proximal pole of the soma a thin, unmyelinated axon projects to the olfactory bulb, the site of the first synapse in the system. The axons from several hundred neurons fasiculate together entering the olfactory bulb through the cribiform plate. Once in the bulb the axons find their way, by as yet unknown mechanisms, to specific targets within structures known as *glomeruli*. The glomeruli are small spherical knots of neuropil consisting mostly of incoming OSN axons and dendrites from the main input-output cells of the olfactory bulb, the *mitral cell*. There are two remarkable features of the glomerular connections: 1) all of the ORNs expressing a particular odor receptor send their axons to the same glomerulus; and 2) there is a tremendous convergence at the level of the glomeruli, with several thousand ORNs projecting to a single glomerulus and several hundred synapses being made on any of the 10-25 mitral cells innervating a single glomerulus.

Like all sensory receptors the OSN is highly compartmentalized, both functionally and structurally. The soma is important for normal cellular

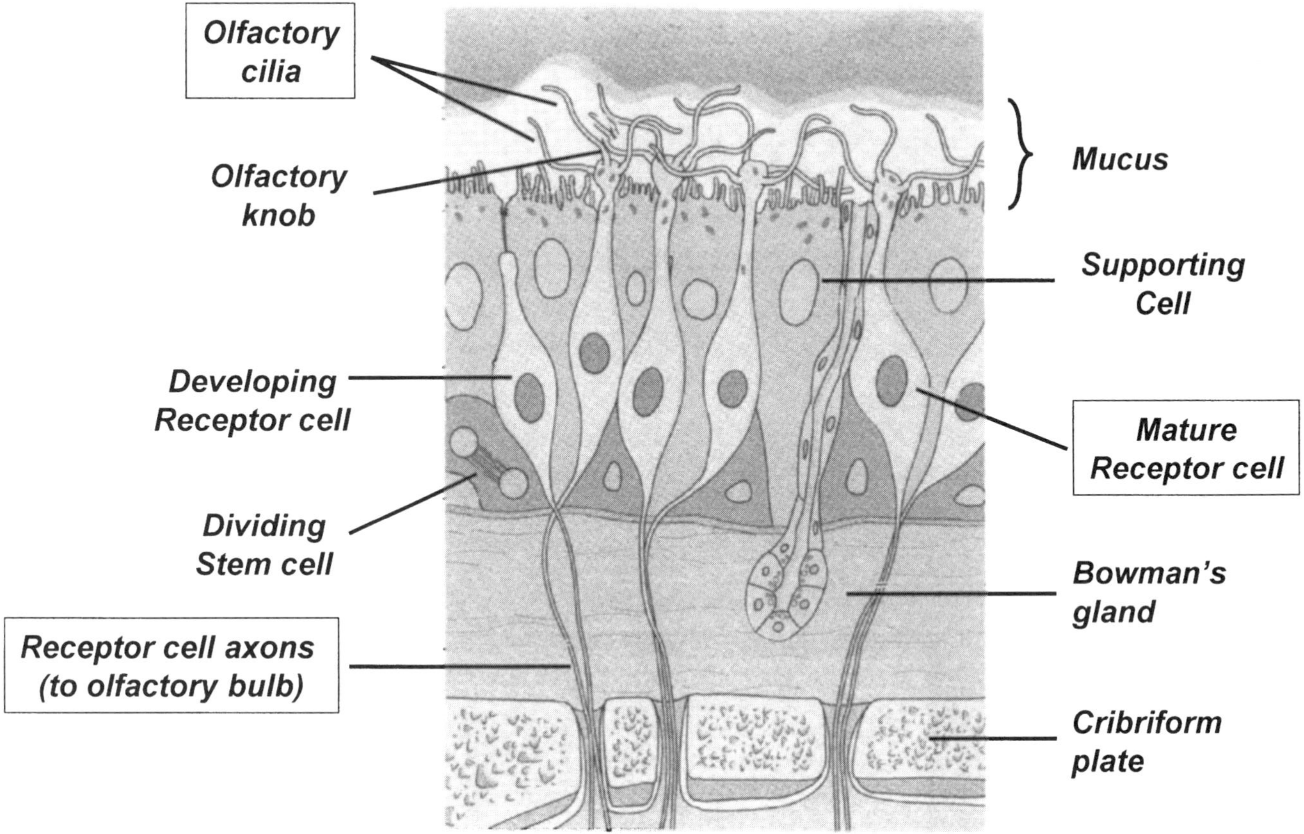

Figure 1. Schematic of the mammalian olfactory epithelium.

functions, the axon supports the propagation of action potentials centrally, and the cilia are the site of sensory transduction.

Expression of a large family of odorant receptors by olfactory receptor neurons

Cloning of odor receptors.

In 199, Linda Buck and Richard Axel successfully cloned the first odor receptor gene (*1*) , thus providing a molecular basis for understanding odor coding. Three remarkable results emerged from these studies.

1. The receptors belonged to the super family of receptors that are characterized structurally by possessing seven transmembrane domains and functionally by coupling to a G-protein for signaling (these receptors are often called GPCRs, for G-Protein Coupled Receptors). Although this was expected from earlier work that characterized the second messenger signaling system (see below) it provided important confirmation and also meant that the odor receptors were closely related to receptors mediating phototransduction (rhodopsin), neurotransmission (biogenic amine receptors) and many other cellular processes.

2. The odor receptors comprised a large gene family of nearly 1000 separate genes, each coding for a distinct receptor. This makes the odor receptors the largest gene family in the nervous system. If, as is believed, the entire mammalian genome consists of 50-100,000 genes then odor receptors comprise nearly 1-2% of the entire genome. Because the genes for G-protein receptors are typically intronless, the odor receptors are encoded by individual genes; i.e. there is no alternative splicing to make new receptors.

3. The structure of the receptors revealed that while they shared many common motifs there was a "hyper-variable" region in which the sequences encoding the 2nd to 6th transmembrane domains diverged. Since this is the region that has been implicated in ligand binding in other members of the GPCRs, this provided a rationale for understanding the large number of odors that could be discriminated by the olfactory system. Each of the thousand receptors differed in the precise structure of that region of the molecule between the 2nd and 6th

alpha helices, and this presumably gives rise to their differential binding properties. Although some combinatorial algorithm for recognizing thousands of odors must still be invoked, a significant amount of discriminatory power rests with the initial steps of olfactory transduction (see below).

Patterns of receptor expression

Following the cloning of the odor receptor gene family a series of elegant *in situ* hybridization studies were undertaken by several laboratories to determine the organization of gene expression in such a large family (*2*) . Receptors can be grouped into sub-families based on sequence similarity, and members of sub-families were found to be expressed in one of four wide bands, or zones, that stretch in the rostral-caudal direction across the epithelium (Figure 1). Any one receptor gene was found to be expressed in only one of the four zones, but within that zone its expression appeared to be random. The functional significance of the zonal organization remains obscure as it does not neatly overlay any known pattern of odor sensitivity across the epithelium; it may be a developmental epi-phenomenon. Each labeled receptor was expressed in about 0.1% (i.e., 1/1000) of the cells, leading to the belief that each cell expresses only a single odor receptor type. This remains to be definitively shown, but all existing data is consistent with this pattern of receptor gene expression. How an OSN decides which of the thousand receptors to express is unknown.

Signaling in olfactory receptor neurons

Second messenger pathway

Following the initial binding of odorant with the molecular receptor a cascade of biochemical events is initiated that results in the depolarization (e.g., change in the membrane potential to a more positive voltage) of the receptor neuron membrane and the generation of an action potential (Figure 2). The steps in this cascade are now reasonably well understood. Odor receptors are coupled to a G-protein specific to olfactory receptors, called G_{olf}, but which is of the G_s type of G-protein. The activated alpha subunit of Golf binds to Adenylyl Cyclase, catalyzing the production of cAMP from ATP. Cyclic AMP

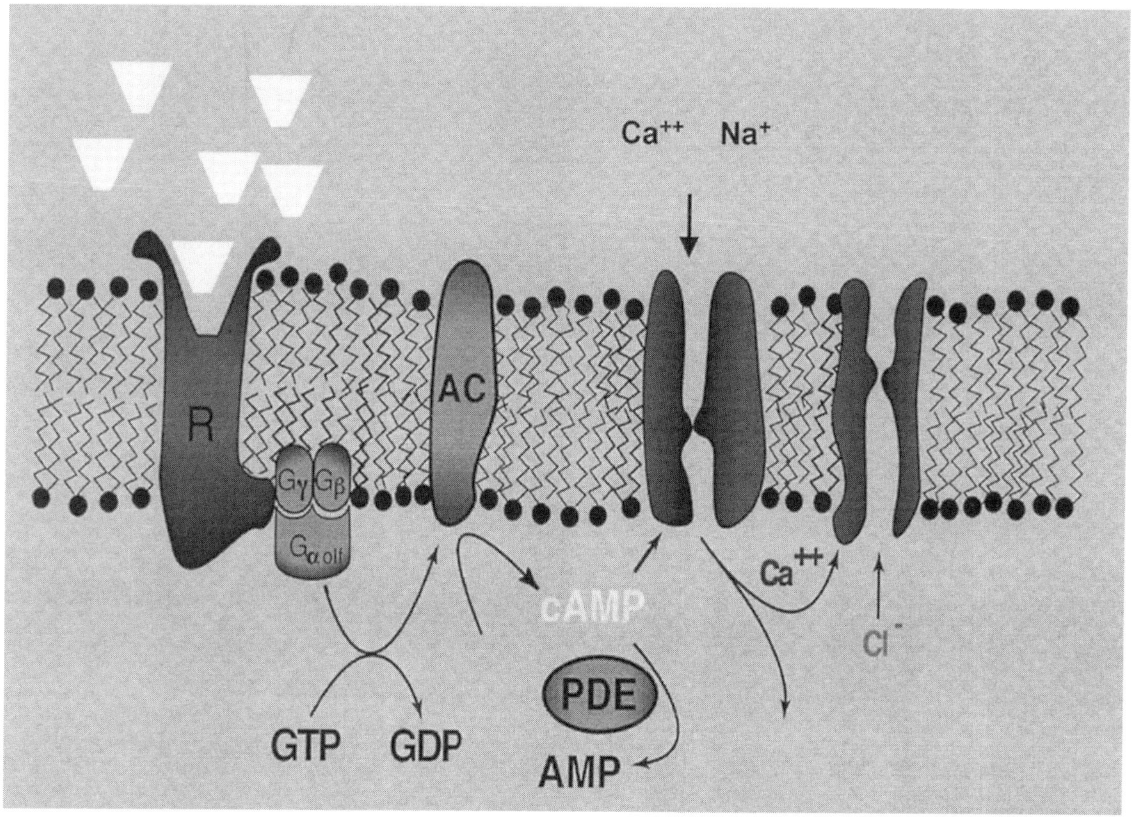

Figure 2. Elements of the second messenger transduction pathway.

binds directly to an ion channel, increasing its open probability and allowing positive ions to flow across the membrane into the cell. This serves to depolarize the membrane by a few tens of millivolts, sufficient to activate Na and K currents in the soma membrane which act to generate an action potential. That action potential is propagated down the axon to the synaptic terminal where it induces transmitter release at the OSN-mitral cell synapse. Each of the several steps in the transduction cascade represents an amplification of the initial signal. Each activated receptor generates several dozen or more G_{olf} alpha subunits, each of which can activate an Adenylyl Cyclase. The adenylyl cyclase in turn produces as many as 1000 molecules of cAMP per second, and those cAMP's open ion channels that allow several hundred thousand ions to flow per second. There is some evidence that the OSN can, in the limit and under ideal circumstances, respond to a single odor molecule using this signal amplification scheme. Indeed one model, based on physiological recordings, suggests that OSNs count molecules of odor bound over time as a measure of concentration. That is the OSN really detects "flux" (number of molecules over time) rather than the more static parameter, concentration (molecules per volume) (*3*).

A unique and interesting feature of the olfactory transduction pathway is the addition of a chloride channel at the last step. The cAMP gated channel is permeable to calcium ions as well as Na and K, and this calcium is also able to bind to and activate a calcium-dependent chloride channel. Normally chloride channels act to inhibit a cell by driving the membrane potential more negative as Cl- ions flow into the cell. However, olfactory neurons maintain an unusually high intracellular Cl- concentration so that when these channels open the Cl- ions flow out of the cell, leaving behind a net positive charge on the membrane. Thus they add to the odor induced depolarization and act to amplify the signal through the cAMP gated channels.

Adaptation

Calcium ions flowing in through the cAMP gated channels play an additional and critical role in olfactory adaptation. As with all sensory input, if the stimulus is maintained for long periods of time the perception of it decreases. Thus a strong odor perceived upon entering a room is, after some time, no longer noticed, even though the odor is still present (as evidenced by the reaction of new people entering the room). In olfaction this process appears to occur in the olfactory sensory neuron, at the earliest stages of detection. Calcium ions entering through the cAMP sensitive channels bind to the modulatory molecule *calmodulin* and together the Ca-calmodulin complex interacts with the ion channel causing it to close, even in the presence of cAMP

(*4*). Thus this negative feedback loop acts to turn off the electrical response in a cell that is signaling the presence of a particular odor.

Other, longer term mechanisms may also be operating in adaptation to odors. Special enzymes are known to specifically phosphorylate activated (i.e., bound) odor receptors, causing them to be less effective in activating the G-protein. These processes are thought to operate over longer time periods than the calcium feedback loop described above, and may even result in the internalization of receptor protein. These desensitization mechanisms are common to many types of G-protein coupled receptors and were first described in B2-Adrenergic receptors and Rhodopsin.

How are different odors discriminated?

This remains one of the more perplexing questions in olfactory science. Information from physiological, biochemical, behavioral and molecular experiments does not always converge, and the differing techniques do not allow easy comparisons. Physiological recordings from individual neurons suggest that most OSNs respond to more than one odor . Biochemical measurements of second messenger accumulation and physiological recordings of whole tissue responses also indicate that most odors are able to induce responses in many cells and cell populations. On the other hand, molecular biological data consistently indicates that each neuron expresses only one of the thousand different odor receptors. Taken together these data suggest a family of receptors that are relatively promiscuous in their ligand selectivity, and a large repertoire of ligands, many of which are likely to activate more than one receptor. Odor discrimination then requires not only receptor-ligand interactions but also some set of combinatorial algorithms at the system level for deciphering patterns of activation at the periphery. This will be discussed below.

Genetic basis of olfactory discrimination

The gene structure of the odor receptor family provides a molecular rationale for odor discrimination at the first step. Although all members of the olfactory gene family share certain amino acid motifs there is significant variability in certain regions. In particular, there appears to be a hyper-variable region stretching from transmembrane domains 2 thorough 6. From previous studies in Beta-Adrenergic and other monoamine receptors it had been postulated that the major binding pocket in GPCRs lay within the transmembrane barrels, in an aqueous filled pocket about a third to a half of the way through the membrane. Thus the region of greatest variability among the family of odor receptors correlates with the presumptive binding pocket of the GPC receptors.

The odor receptors can be grouped into sub-families based on nucleotide sequence similarity. Receptors are generally considered members of the same subfamily if they share 85% or greater sequence similarity. In fact, among the receptors with known sequences there are instances in which two receptors differ by only 3 nucleotide changes resulting in a single amino acid difference. Although it seems likely that receptors with closely related sequences recognize chemically or structurally related molecular ligands, this has not yet been shown.

The receptive field of olfactory receptors

Until recently no particular receptor gene had been paired with its cognate ligands. For reasons that remain obscure odor receptor genes do not express well in heterologous cell systems making it difficult to achieve functional expression of odor receptors under conditions favorable for ligand binding studies. Recently one receptor, the rat I7 receptor, was expressed using an adenovirus vector to drive overexpression in rat olfactory epithelium (*5*). The functional receptor was paired with a homologous series of aliphatic aldehydes with carbon chains of 7, 8, 9 and 10. The best recognized odor was octyl aldehyde (C8) followed by heptyl (C7), nonyl (C9) and finally decyl (C10) aldehydes. There was no activation by aldehydes with 6 or fewer carbons or with 11 or more carbons. Thus in this case, a single odor receptor is responsive to a series of closely related molecules, but is capable of clearly distinguishing between molecules that are different by only a single carbon atom (C7 vs. C6, or C10 vs. C11).

This sort of "molecular receptive field" is similar to that recorded from mitral cells of the rabbit olfactory bulb in response to the presentation of a similarly homologous series of odors to the rabbit nose. Kensaku Mori and his colleagues undertook an extensive investigation (*6*) of the responses in single mitral cells that innervate a single glomerulus and are presumably receiving inputs from a single population of receptor neurons, all expressing the same receptor (see below). They demonstrated that mitral cells appear to be tuned to structurally related molecules, in much the same way as was reported for the I7 receptor.

Most recently, utilizing the adenovirus expression system and the identified I7 odor receptor we have taken an approach based on the principles of medicinal chemistry and explored the chemical space around octyl aldehyde for other agonists or antagonists at this receptor. Reasoning that there were a large number of available compounds that differed in specific characteristics from octyl aldehyde it seemed possible to gain some appreciation for the structure of the binding pocket for this receptor.

In particular we utilized molecules that differed in saturation, carbon chain length, functional group, or structure to explore the conditions which activated the I7 receptor. From these data several interesting conclusions arise. First the aldehyde functional group is critical for activity. Exchanging it for any other functional group, even in an otherwise identical molecule, completely removes function. Thus 8 carbon saturated aliphatic ester, alcohols, ketones, carboxylic acids, halides and formiates were all ineffective. Second the length of the molecule is an important determinate for activity. Using unsaturated aldehydes, which occupy primarily the extended conformer, as a molecular ruler we were able to determine that only molecules with lengths between 8 and 12 angstroms could induce receptor activity (Figure 3). These dimensions correlate with compounds possessing a carbon chain of between 7 and 11 carbons. Thirdly, the carbonyl end of the molecule was more critical in determining activity than the tail of the molecule. Indeed all sorts of additions and substitutions from C4 back were reasonably well tolerated, whereas double bonds and methyl groups at the C2 or C3 carbons generally reduced or abolished activity (Figure 4). In particular the presence of both a methyl group at C2 or C3 and a double bond at C2-C3 caused the methyl group to be held in the plane of the aldehyde resulting in a steric hindrance to binding. Details of these interactions can be found in Araneda et al (*7*).

Significant work remains to be done in this area - both in the areas of odor chemistry, where molecules with small changes (i.e., the addition of a double bond) could be synthesized and investigated for changed discriminability, and the odor receptors themselves where the particular residues critical to odor binding and activation could be identified by site directed mutagenesis.

Summary

Olfaction is a primary sense in most animals, mediating all of the major behaviors- feeding, rearing, aggression, sex - and is of more importance in humans than generally credited due to its largely unconscious role in our everyday lives. The olfactory system of vertebrates is capable of recognizing a diverse and numerous array of odorous chemicals, making it the most effective

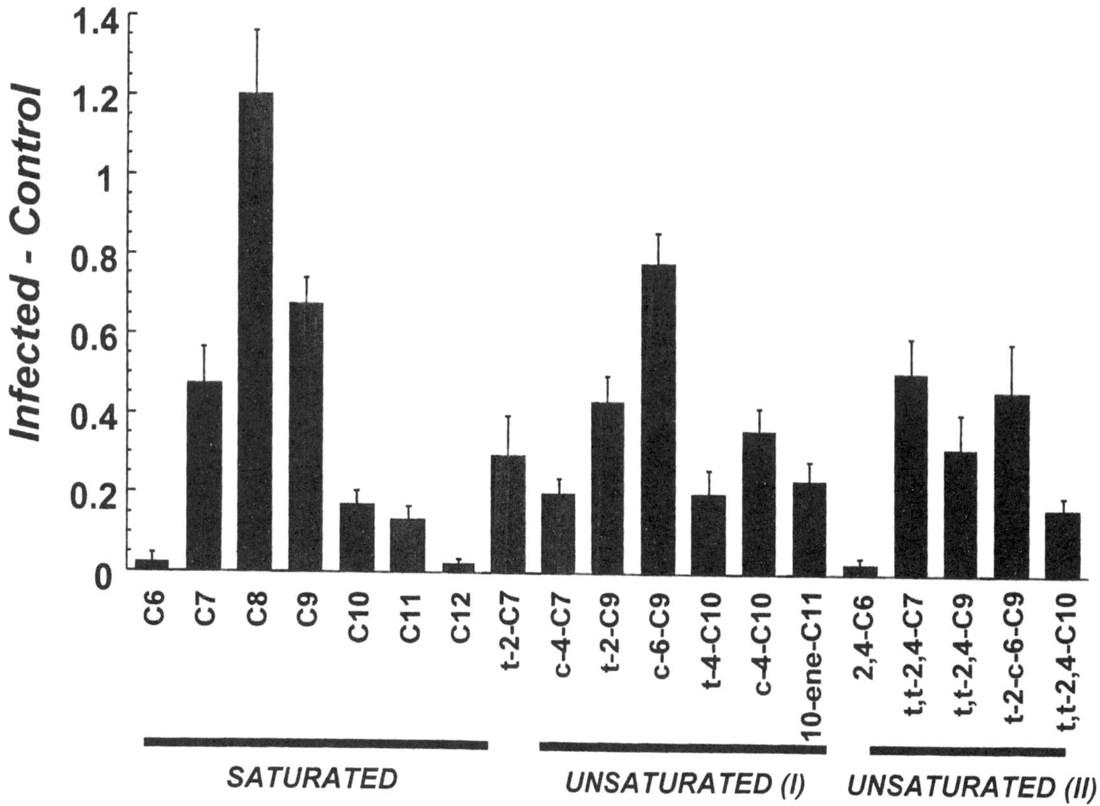

Figure 3. Responses of cells with the I7 receptor to a series of saturated and unsaturated aliphatic aldehydes with carbon chains from 6-12.

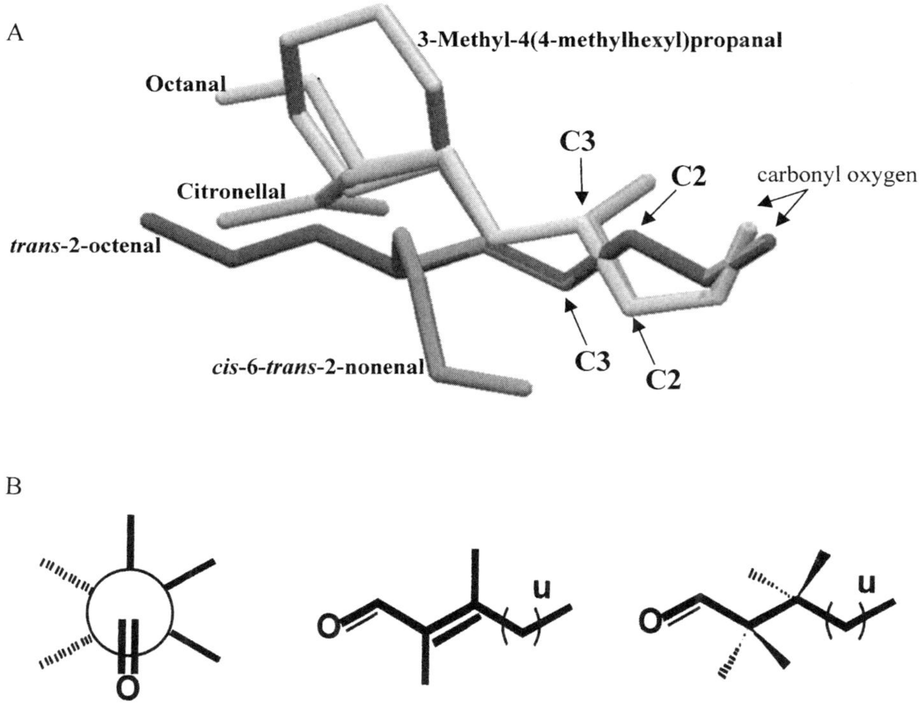

Figure 4. A. An overlay of some agonists of the rat I7 receptor are shown. As the models illustrate, the lowest energy conformer of each of these molecules occupies both extended and bent conformations. This demonstrates that in the absence of a double bond at C2-C3, molecules that adopt a bent conformation are capable of activating the receptor contrary to molecules like citral which adopts a bent conformation but is unable to activate the receptor. B. Newman projections showing that the double bond in C2-C3 forces the methyl group into the plane of the aldehyde group causing a possible steric hindrance.

chemical sensor on the planet. The process of olfactory transduction makes use of common molecular and cellular mechanisms, including membrane bound receptors and second messenger amplification systems. Signaling from the periphery to the brain appears to impart a spatial code onto the olfactory stimulus set, creating a "molecular receptive field" for each receptor. Processing in higher centers is not yet well studied, but, as with studies of the peripheral system, should be of considerable interest for the insights that may be provided regarding general issues of signal generation and processing in the nervous system.

References

1. Buck, L.; Axel, R. *Cell* **1991**, *65*, 175-187.
2. Ressler, K. J.; Sullivan, S. L.; Buck, L. B. *Cell* **1993**, *73*, 597-609.
3. Firestein, S.; Picco, C.; Menini, A. *J. Physiology* **1993**, *468*, 1-10.
4. Kurahashi, T.; Menini, A. *Nature* **1997**, *385*, 725-729.
5. Zhao, H.; Ivic, L.; Otaki, J. M.; Hashimoto, M.; Mikoshiba, K.; Firestein, S. *Science* **1998**, *279,* 237-242.
6. Imamura, K.; Mataga, N.; Mori, K. J. Neurophysiol. **1992***,* *68*, 1986-2002.
7. Araneda R. C.; Kini A. D.; Firestein, S. *Nature Neuroscience* **2000** 3(*12*), 1248-1255

Chapter 10

Olfactory Psychophysics

Richard L. Doty

Smell and Taste Center, University of Pennsylvania Medical Center, 3400 Spruce Street, 5 Ravdin Pavilion, Philadelphia, PA 19104-4283

Within the last 20 years, standardized means for assessing olfactory function have become generally available. In this paper, I describe numerous psychophysical test procedures that have been employed to measure olfactory function. Emphasis is placed on practical psychophysical tests used to assess olfactory function in academic, industrial, and clinical settings. Issues related to test reliability are discussed, along with variables known to influence test scores.

Introduction

Psychophysics, in a strict sense, is the quantitative study of relationships between stimuli and psychological sensations. Psychophysical methods form the backbone of the development of 20th Century experimental psychology, and today are commonly used to evaluate olfactory function in humans in academic, clinical, and industrial enterprises. Such measures are used, for example, to elucidate structure activity relationships and to produce consumer products of higher acceptibility to the public.

Psychophysical procedures can be divided into two categories -- threshold procedures and suprathreshold procedures. Threshold procedures can be further subdivided into tests of detection sensitivity, recognition sensitivity, and differential sensitivity. Suprathreshold tests include measures of odor identification, discrimination, memory, and perceived odor intensity. The present review provides a brief synopsis of modern psychophysical tests used for olfactory assessment; the reader is referred elsewhere for more extensive and detailed treatments of this topic [1-4].

Threshold Procedures

The formal development of threshold methodology is attributed to Fechner [5], whose 1860 treatise, *Elemente der Psychophysik*, described and operationalized the "classical" methods for measuring sensory thresholds.[1] While this was the first compendium and systematization of threshold methods, the measurement of thresholds -- albeit in a less defined form -- had been practiced by others before Fechner [6]. For example, Valentin described, in 1848, a procedure in which he mixed a given volume of odorous gas with a volume of air 100 times as large [7]. The resulting mixture was then again similarly diluted 100 fold. Each resulting mixture was successively diluted, allowing for a series of geometrically decreasing concentrations of odorant that could be sampled. This general procedure was employed in the 19th Century to determine the minimum level of warning agents (e.g., mercaptans) needed to be

---

[1] The three methods involved were the method of limits, the method of constant stimuli (also termed the method of right and wrong cases), and the method of average error. It should be noted that the general concept of a threshold or limen was present in ancient times, although the primordium of the modern sensory threshold concept came from the theorizing of such mathematicians and philosphers as Gottfried Wilhelm von Leibniz (1646-1716), the co-discover of integral and differential calculus, and Johann Friedrich Herbart (1776-1841), perhaps best known for his so-called mental calculus [65]. Herbart, who was not an empiricist, taught that the mind consists of two parts -- the conscious and the unconscious -- and that ideas cross back and forth between the boundary (threshold) of these two parts. Some ideas are dominant over others and, thus, keep other ideas from reaching the threshold of consciousness. Herbart's concepts were employed nearly a Century later by Sigmund Freud, who made them a cornerstone of psychoanalytic theory.

added to natural gas [8] and forms the general basis for modern air-dilution olfactometry. Modern olfactometers – instruments specifically designed to provide accurate odorant concentrations for testing the sense of smell – capitalize on advances in airflow control technology (e.g., mass flow controllers, rotameters, permeation tubes), and provide the most accurate means for quantifying stimuli to be presented to subjects [3]. Unfortunately, most of these devices are impractical and expensive, not only to purchase, but to operate and maintain.

A more practical approach to presenting stimuli of varying concentrations was developed by Passy [9]. Passy successively diluted his stimuli in alcohol, rather than air, and provided the resultant mixtures in closed containers. Although his method suffered from the use of a diluent that had an odor, Zwaardemaker, the great Dutch olfactory scientist, viewed Passy's method as "a great step in the history of olfactometry" [10]. Subsequent workers have employed diluents without as severe of problems as those posed by alcohol, including water, mineral oil, and propylene glycol (e.g., [11-13].

Definitions of Threshold

The lowest concentration of an odorant that an individual can reliably detect (usually defined as that concentration where detection is midway between chance and perfect detection) is termed his or her detection or absolute threshold. At very low concentrations no odor quality can be discerned -- only that something is present that differs from air or the comparison diluent blank or blanks. In modern olfactory detection threshold testing, the subject is asked to report which of two or more stimuli (i.e., an odorant and one or more blanks) smells strongest, rather than to simply report whether an odor is perceived or not. Such "forced-choice" procedures are less susceptible to contamination by response biases (e.g., the conservatism or liberalism in reporting the presence of an odor under uncertain conditions) than non-forced-choice procedures. In addition, they are more reliable and produce lower threshold values [14]. The instructions provided to a subject are critical in measuring a detection threshold, since if the subject is instructed to report which stimulus produces an odor, rather than which stimulus is stronger, a spuriously high threshold value may result (odor quality is present only at higher perithreshold concentrations).

The recognition threshold is defined as the lowest concentration where odor quality is reliably discerned. Unfortunately, it is nearly impossible to control criterion biases in recognition threshold measurement. Thus, in a forced-choice situation, guesses are not randomly distributed among alternatives, potentially leading to a spuriously low recognition threshold for the preferred alternative. A classic example of this problem comes from taste psychophysics, where some subjects report "sour" much more frequently than the other primary qualities in the absence of a clearly discernible stimulus, resulting in a spuriously low sour-taste recognition threshold measure [15].

A third type of threshold is the differential threshold, the smallest amount by which a stimulus must be changed to make it perceptibly stronger or weaker. This is also termed a difference threshold or a "just noticeable difference" (JND). The size of the increment in odorant concentration (ΔI) required to produce a JND increases as the comparison concentration (I) increases, with the ratio approximating a constant; i.e., $\Delta I/I = C$. This phenomenon was described by Weber in 1834 and was termed "Weber's Law" by Fechner in 1860 [5]. Although differential thresholds have been measured in olfaction, they have received little practical application.

Modern Odor Detection Threshold Measurement Procedures

Two types of detection threshold procedures have received the most use in the last two decades: the ascending method of limits procedure (AML) and single staircase procedure (SS). In the AML procedure, odorants are presented sequentially from low to high concentrations and the point of transition between detection and no detection is estimated. In the SS method, the concentration of the stimulus is increased following trials on which a subject fails to detect the stimulus and decreased following trials where correct detection occurs. An average of the up-down transitions ("reversals") is used to estimate the threshold value. In both the AML and SS procedures, the direction of initial stimulus presentation is made from weak to strong in an effort to reduce potential adaptation effects of prior stimulation.

It has been generally believed that threshold measures exhibit considerable variability. For example, one study of 60 subjects reported that intersubject variation was as great as 5 log units [16]. Another study, employing a non-forced-choice ascending threshold procedure, found intersubject variation on the order of 16 log units [17]. Stevens et al. [18] obtained 60 threshold values over the course of 30 days from three subjects (20 for butanol, 20 for pyridine, and 20 for β-phenyl-ethylmethylethylcarbinol). The intrasubject variability across test days was found to be as great as intersubject variability on a given test day. This suggested to these authors that the large individual differences observed in threshold values are not a reflection of big differences among stable threshold values of subjects, but reflect large day-to-day fluctuations in the test measures. Unfortunately, Stevens et al. used a comparatively crude and relatively unreliable single AML procedure to establish thresholds for all three chemicals, likely accounting for the large within subject variability. The stability or reliability of a threshold measure is predictably related to the number of trials presented. Procedures with more trials, such as the SS procedure, produce less variable and more reliable measures than simple AML procedures [14] and negate the notion that, within an individual, thresholds vary markedly from day to day.

Several threshold tests are currently available commercially. One relies upon felt tip marker-like dispensing agents [19], one upon bottles into which strips of blotter paper are dipped [17], and two which employ squeeze bottles [20; 21]. The most recent test developed for commercial use, the The Smell Threshold Test™ (STT), is pictured in Figure 1. This test employs 17 dilution steps of the odorant phenyl ethyl alcohol, spanning a wide concentration range (-10.00 to -2.00 $\log_{10}$ units in half-log steps). Versions of this have been employed at the University of Pennsylvania Smell and Taste Center over the course of nearly two decades [14; 22-31].

Suprathreshold Procedures

Psychophysical tests employing clearly discernible stimuli are generally classified as suprathreshold (i.e., above threshold) tests. Some suprathreshold tasks capitalize on the fact that psychological attributes, such as strength, pleasantness, and quality, can be assigned to such stimuli. In general, suprathreshold procedures overcome a number of the problems of threshold tests [e.g., the need for painstaking mixing of the dilution series, care in minimizing stimulus contamination (particularly at lower odorant concentrations), and long test sessions] and measure different elements of the perceptual process.

The large number of suprathreshold tasks precludes their mention here, and the reader is referred elsewhere for a more comprehensive review of such tasks [3]. For the present purposes, I discuss tasks that fall into three general categories: (i) scaling & magnitude estimation tasks; (ii) identification tasks; and (iii) memory tasks.

Scaling and Magnitude Estimation Tasks

Numerous procedures have been developed to establish the nature of the relationship between stimulus concentration and perceived intensity and other attributes (e.g., pleasantness). In general, scientists have found that logarithmic or power functions usually fit odorant concentration/perceived intensity data quite well, although this is not the case for some other attributes. In general, the

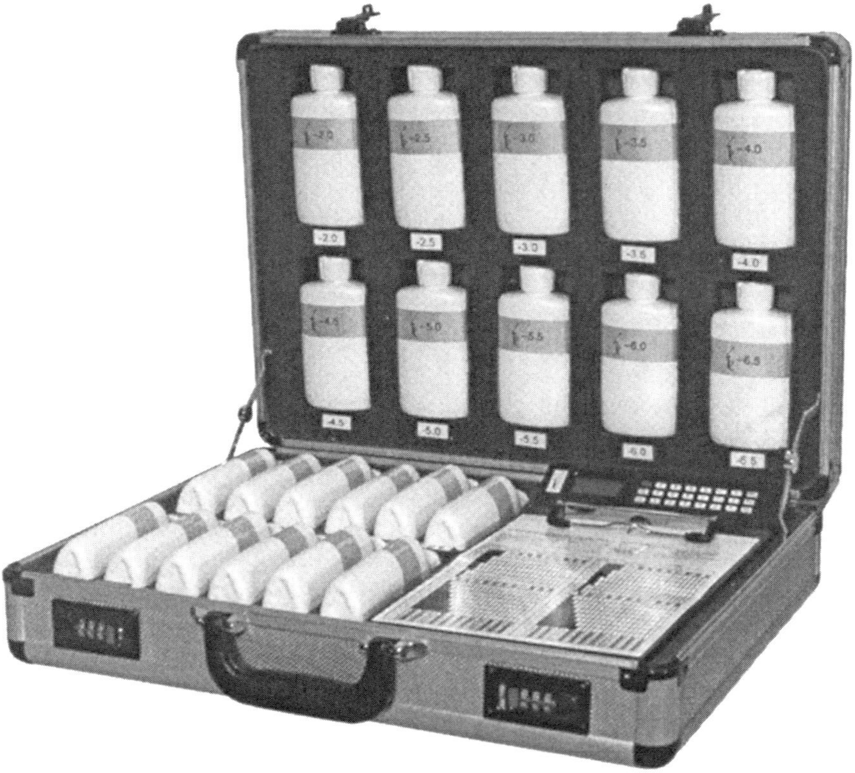

Figure 1. A modern commercially-available detection threshold test, "The Smell Threshold Test™." In this test, standardized forms are used with randomized presentation orders to present stimuli in a single staircase procedure. Environmental temperature is recorded, and the geometric mean of the last of four staircase procedures serves as the threshold estimate. Photograph courtesy of Sensonics, Inc., Haddon Hts., NJ USA.

nature of the scaling task largely dictates the function obtained between the stimulus concentration and the intensity measure under consideration.

As in the other senses, psychophysicists and psychometricians have sought to develop psychological scales with ruler-like properties whenever possible (i.e., so-called ratio scales, where distances along the scale have ratio properties and a true zero point). Among the leaders in this field was the late S.S. Stevens of Harvard University [32]. The degree to which such scales can be developed is debatable, however, as the judments of odor intensity are relative, being influenced by both subject idiosyncracies and contextual factors (e.g., a moderately intense odor is reported to be more intense when presented with weak comparision stimuli than with strong comparision stimuli). Fortunately, for most practical purposes, neither the exact form of the underlying psychophysical function nor the influences of stimulus context are of great concern, so long as the test procedures are standardized and it can be demonstrated that the responses on the scaling tasks are reliable and differentiate between the groups being evaluated.

Rating & Magnitude Estimation Scales

Rating scales can be used to estimate the relative amount of a psychological attribute perceived by a subject. In chemosensory tasks, two types are popular: category scales, where the relative amount of sensation is signified by indicating which of a series of discrete categories best describes the sensation, and line scales (also termed visual analog or graphic scales) where the stength of the sensation is indicated by the patient placing a mark along a line that has descriptors (termed anchors) located at its extremes (e.g., very weak – very strong). The reader is referred elsewhere to discussions of the properties of rating scales, including the influence of category number on their psychometric properties (e.g., [1,33]).

A popular procedure for examining the association between odorant concentration and perceived intensity is termed magnitude estimation. In the most common version of this procedure, the subject assigns numbers to stimuli in relation to their relative intensity. For example, if a number of 50 is used to indicate the intensity of one concentration of an odorant, a concentration that smells four times as intense would be assigned a number of 200. If another concentration is perceived to be half as strong as the initial stimulus, it would be assigned the value of 25. The examinee can assign any range of numbers to the stimuli, so long as they reflect the relative magnitudes of the perceived intensities. In some cases, a standard for which a number has been preassigned (often the middle stimulus of the series) is presented to the subject in an effort to make the responses more reliable. In other cases, the individual is free to choose any number system he or she wishes, so long as the numbers are made proportional to the magnitude of the attribute (the so-called "free modulus method"). The important point for assessing the function relating stimulus concentration to perceived intensity is that the absolute values of the numbers are not important, but only the ratios between them.

To obtain an index of suprathreshold function, magnitude estimation data are most commonly plotted on log-log coordinates (log magnitude estimates on the ordinate and log odorant concentrations on the abscissa) and the best fit line is determined using linear regression. The resulting function, $\log \psi = n \log \phi + \log k$, where ψ = perceived intensity, k = the Y intercept, ϕ = stimulus concentration, and n = the slope, can be represented in its exponential form as a power function, $\psi = k \log \phi^n$, where the exponent n is the slope of the function on the log-log plot. In olfaction, n varies in magnitude from odorant to odorant, but is generally less than 1, reflecting a negatively accelerated function on linear-linear coordinates. As noted elsewhere [1], numerous modifications have been made in this equation in an attempt to take into account such factors as threshold sensitivity and adaptation (see, e.g., [34]).

Although the standard magnitude estimation procedure allows for a determination of the slope of the function relating odorant concentration to perceived intensity, the y intercept in such studies does not accurately signify information about the absolute perceived intensity of the stimuli, since it is dependent upon the subject's choice of numbers or the standard assigned by the experimenter. In an attempt to gain additional information from the function's ordinate position, investigators have employed the method of magnitude matching, which provides, at least theoretically, information about the perceived overall intensity of stimuli from the absolute position of the magnitude estimation fucntion and corrects, to some degree, for differences among subjects in number usage (for a detailed discussion of this procedure, see Marks et al [35]). In the most common application of this method, judgments of the intensity of sensations from two modalities (e.g., loudness, odor intensity) are made on a common magnitude estimation scale. Under the assumption that subjects experience stimuli on one of the continua (i.e., loudness) in a similar manner, differences among their loudness ratings would be expected to reflect differences in number usage. The odor intensity continuum can then be adjusted accordingly. Such normalizations allow, theoretically, for a direct comparision of scale values across subjects. Thus, if the adjusted odor intensity magnitude value for one subject is 10 and for another subject is 20 at the same odorant concentration level, the second subject is presumed to experience twice the odor intensity as the first.

Odor Identification Tests

A straight-forward means of assessing olfactory function is to have a subject report the identity of a number of smells. The proliferation of easy-to-use commercially available tests of odor identification have, in fact, significantly increased our understanding of the sense of smell in humans. The most popular of such tests, the 40-odor University of Pennsylvania Smell Identification Test (UPSIT; known commercially as the Smell Identification Test™ or SIT) [36; 37], has been estimated to have been administered to nearly 200,000 persons over the last decade (Figure 2). This highly reliable self-administered test (test-retest r = 0.94 [38]) employs norms based upon nearly 4,000 persons, and is now available in English, Spanish, French, and German language versions [37; 39]. This test was the impetus for a massive smell function survey sent to nearly 11 million subscribers of the National Geographic Magazine in 1986 [40]. Shorter versions of this test include the 12-odor Brief-Smell Identification Test™ (B-SIT; also known as the Cross-Cultural Smell Identification Test™ [4; 41], and the 3-odor Pocket Smell Test™ (PST)[14; 42].

The essence of the UPSIT and other such tests is that the subject smells an odor and then identifies the smell from a list of forced-choice alternatives. The choice list is essential, as many odors are not readily identifiable without such a list. Malingering can be determined, as by chance alone approximately 25% of the stimuli in a four alternative test should be correctly identified and probabilities of obtaining fewer correct responses by chance can be quite low (e.g., obtaining a score of 0 on the UPSIT and not avoiding the correct responses is around 1 in 100,000). Major non-clinical findings of the last decade and a half derived from the UPSIT include the following: first – women, on average, have a better sense of smell than men, and this superiority is noticable as early as four years of age and is culture independent [41; 43-46]; second, there is a substantial genetic influence on the ability to identify odors [46; 47]; third, major loss of olfactory function occurs after the age of 65 years, with over half of those between 65 and 80 years of age, and over three-quarters of those 80 years of age and older, having such loss [41; 44; 48]; fourth, women – on average -- retain the ability to smell longer than men [44]; fifth, the decrement in olfactory function associated with smoking is present in past smokers and recovery to presmoking levels, while possible, can take years, depending upon the duration and amount of past smoking [49]; and sixth, olfactory function is compromised in urban residents and in workers in some industries, including the paper and chemical manufacturing industries [50-57].

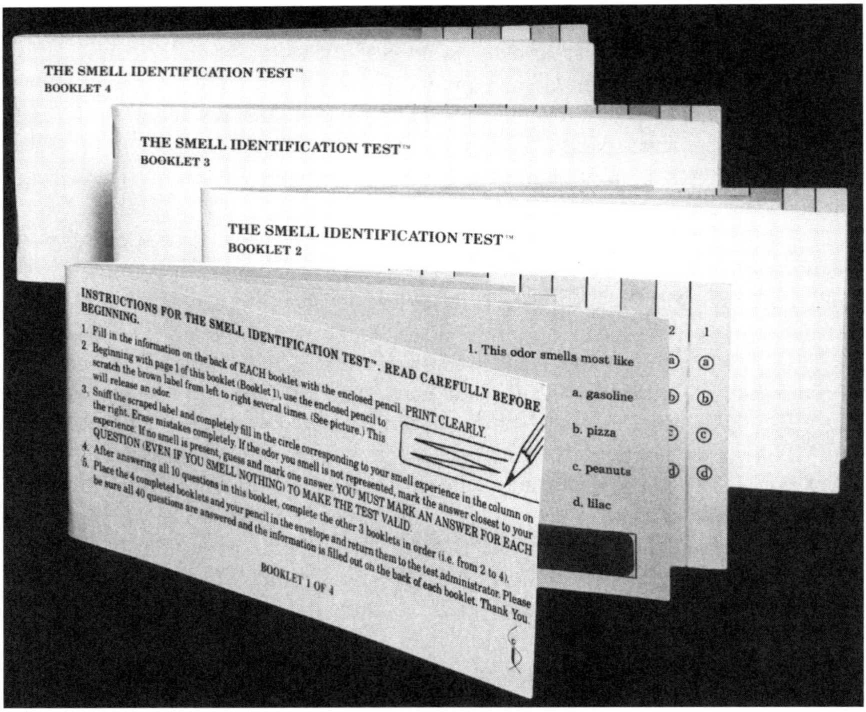

Figure 2. The widely-used University of Pennsylvania Smell Identification Test, a 40-odorant forced-choice self-administered test of olfactory function. This test is the most widely used test of olfactory function in the world, being available commercially as "The Smell Identification Test™) (Sensonics, Inc., Haddon Hts., New Jersey, USA).

The reader is referred elsewhere to the many clinical studies employing the UPSIT and related tests [58].

Odor Memory Tests

A large number of paradigms have been developed for assessing odor memory [59], although only one memory test is available commerically [60] (Figure 3). In a basic odor recognition memory test, a subject is required to smell an odorant or a small set of odorants (termed the target or inspection stimulus or stimuli) and to select, after various set intervals of time, that odorant or set of odorants from foils. Repeated trials may be performed at one or more retention intervals for each of several stimuli or sets of stimuli. In an effort to minimize the rehearsal of verbal labels reflecting the odor qualities during the delay intervals, the examinee is sometimes asked to perform an unrelated task during the retention period, such as counting backward. The proportion of correct responses at each retention interval is typically the dependent measure of such tests.

A conceptual problem with an odor memory test is whether, in fact, it is the odor that is being remembered or whether it is the concept that is being remembered. Thus, if a man is asked to smell an odor which reminds him of a lemon, and is asked a day later to choose that same odor from a series of foils, he may have recalled that he had smelled a lemon (i.e., he remembers the concept of having smelled a lemon) even though he can't conger up the actual lemon odor smell. Thus, when he runs across the lemon odor amongst some foils, he correctly identifies it as the odor smelled the day before. It is very difficult to find sets of odors that do not have verbal or conceptual referents to employ in odor memory tests.

Reliability of Psychophysical Olfactory Tests

It is an axiom of test measurement theory that a test cannot be valid if it is not reliable (although it is possible to have a reliable but invalid test). Although the reliability of many olfactory tests has not been determined, reliability coefficients are available for the most widely used psychophysical olfactory test measures.

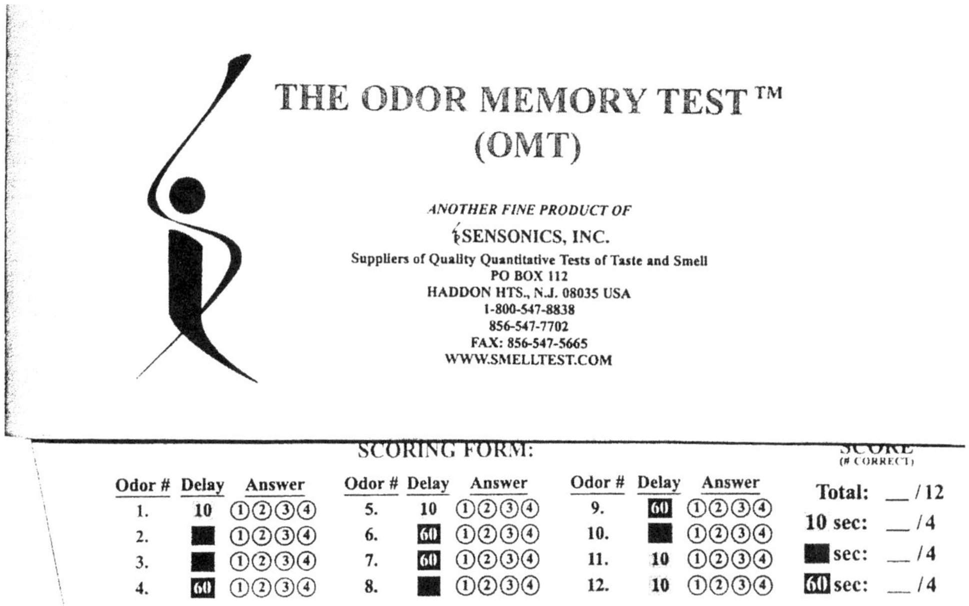

Figure 3. The Odor Memory Test[1]. In this 12-item test, an inspection stimulus is presented and a delay period enforced, following which four stimuli are presented, one of which is the inspection stimulus. The subject is required to identify the previously smelled inspection stimulus. 10-, 30-, and 60-second retention intervals are employed in this standardized test, although other retention intervals can also be used. Photograph courtesy of Sensonics, Inc., Haddon Hts., New Jersey, USA.

In the case of olfactory threshold measures, reliability is higher for forced choice than for non-forced choice procedures. Reliability is also higher for threshold procedures that collect more perithreshold information (e.g., staircase procedures). Cain and Gent [61], in a study of 32 subjects ranging in age from 22 to 59 years, found the correlation between single ascending series butanol thresholds determined for the left and right sides of the nose (which they used as a reliability measure) was, at best 0.68 and as low as 0.30 when the butanol threshold was the first in a series of four threshold tests. Costanzo (1986) evaluated the test-retest reliability of the ascending butanol threshold procedure in 16 subjects aged 17-52 years. The reliability coefficient for the left side of the nose was 0.45 and that right 0.08 [14]. Hummel et al. [62] reported the test-retest reliability of a single ascending butanol threshold test in a 104 subjects to be 0.36. In contrast, the test-retest reliability of a threshold test that employs a single initially ascending staircase procedure and seven staircase reversals (the last four of which were used as the threshold estimate) established in a study of 57 normal subjects spanning a wide age range was 0.88 [14]. However, when only the first reversal of this test was evaluated for its reliability, the reliability coefficient ($r = 0.45$) was within the range of reliability coefficients reported for single ascending series thresholds, emphasizing the importance of collecting more than a single reversal in a detection threshold measure. In an empirical assessment of reliability of threshold tests, it was determined that the number of reversals included in the threshold measure is positively related to the test's reliability, and reliability increased as the number of threshold reversals increased. The Spearman-Brown Prophesy Formula [63] provided an excellent fit ($r^2 = .984$) to the relationship between the number of reversals included and the reliability coefficient.

The reliability of tests of odor identification are similarly related to their length by the Spearman-Brown Prophesy Formula. In one study, the reliability of the 40-item UPSIT was found to be 0.922, whereas the relability of the one-, two-, and three-booklet fractions of the UPSIT was found to be 0.752, 0.855, and 0.890, respectively. In another study, the reliability of the 12-item version of the UPSIT (i.e., the Brief Smell Identification Test™) was found to be 0.71, a value not significantly different from the 0.75 value for observed in the earlier study of the reliability of a single UPSIT booklet.

There are surprisingly few studies of the reliability of suprathreshold rating or scaling measures, although generally they are believed to be very reliable. In one study, the test-retest reliability of the mean of suprathreshold ratings given to a range of above threshold concentrations of amyl acetates was 0.76; the reliability of the slopes of functions fitted to these data was 0.68 [14]. The test-retest reliability of a multiple target odor memory task was reported to be 0.62 for the left side of the nose, 0.77 for the right side of the nose, and 0.69 for both sides combined. Analogous reliability coefficients for a single target odor memory task were 0.70, 0.64 and 0.72 (all ps < 0.001)[64].

Summary

In this paper, the basic olfactory tests that are most commonly used today for assessing the ability to smell are briefly reviewed. Presently, tests of odor detection and identification are most commonly used to assess olfactory function, although other types of olfactory tests, including those of odor memory, are also employed. The reliability of olfactory tests using few trials is generally suspect, as there is a strong relationship between the number of items or trials in an olfactory test and its general reliability.

References

1. Doty, R.L. In *Smell and Taste in Health and Disease*; Getchell, T.V.; Doty, R.L.; Bartoshuk, L.M.; Snow, J.B., Jr., Eds.; Raven Press: New York; 1991; pp. 175-203.
2. Doty, R.L.; Smith, R.; McKeown, D.A.; Raj, J. *Percept Psychophys* **1994**, *56*, 701-707.
3. Doty, R.L.; Kobal, G. In *Handbook of Olfaction and Gustation*; Doty, R.L., Ed.; Marcel Dekker: New York, 1995; pp. 191-225.
4. Doty, R.L., Marcus, A. and Lee, W.W. *Laryngoscope* **1996**, *106*, 353-356.
5. Fechner, G.T. *Elemente der Psychophysik*, Breitkopf & Harterl: Leipzig, 1860.
6. Boring, E.G.; *Sensation and perception in the history of experimental psychology*; Appleton-Century-Crofts, Inc.: New York, 1942
7. Valentin, G.; *Lehrbuch der Physiologie des Menschen*; Braunschweig, 1848.
8. Fischer, E.; Penzoldt, F.; *Sitzungsber Phys Med Soz Erlangen* **1886**, *18*, 7-10.
9. Passy, J. *Comp.Rend.Soc.Biol.* (Paris) **1892**, 44, 84-88, 239-243.
10. Zwaardemaker, H. *L'Odorat*; Paris, Doin, 1925.

11. Toulouse, E.; Vaschide, N. (1893) *C.R.Acad.Sci.Biol.* **1893**, *51*, 381-383.
12. Proetz, A.W. *Ann. Otol. Rhinol. Laryngol.* **1924**, *33*, 275-278.
13. Ventrom, D. and Amoore, J.E. *J. Food Sci.*, **1968**, *33*, 264-265.
14. Doty, R.L.; McKeown, D.A.; Lee, W.W. ; Shaman, P. (1995*) Chem Senses* **1995**, *20*, 645-656.
15. Weiffenbach, J.M. (1983) *Percept. Psychophys.* **1983**, *33*, 251-254.
16. Brown, K.S.; MacLean, C.M.; Robinette, R.R. *Human Biology*, **1968**, *40*, 456-472.
17. Yoshida, M. *Bull. Fac. Sci. Eng. Cho. Univ.*, **1984**, *27*, 343-353.
18. Stevens, J.C.; Cain, W.S.; Burke, R.J. *Chem. Senses* **1988**, *13*, 643-653.
19. Kobal, G.; Hummel, T.; Sekinger, B.; Barz, S.; Roscher, S.; Wolf; *Rhinology* **1996***, 34*, 222-226.
20. Amoore, J.E. and Ollman, B.G. *Rhinology* **1983**, *21*, 49-54.
21. Doty, R. L. *Odor Threshold Test* TM *Administration Manual.* Sensonics, Inc.: Haddon Hts., New Jersey, 2000.
22. Betchen, S.A. ; Doty, R.L. *Chem. Senses* **1998** *23*, 453-457.
23. Deems, D.A; Doty, R.L. *Trans. Penn. Acad. Ophthalmol.Otolaryngol.* **1987**, *39*, 646-650.
24. Doty, R.L.; Shaman, P.; Dann, M. *Physiol. Behav.* **1984***, 32*, 489-502.
25. Doty, R.L.; Gregor, T.P.;Settle, R.G. *Chem. Senses* **1986**, *11*, 259-264.
26. Doty, R.L., Reyes, P.F. and Gregor, T. *Brain Res. Bull.* **1987***, 18*, 597-600.
27. Doty, R.L.; Deems, D.A.; Stellar, S. *Neurology* ***1988****, 38*, 1237-1244.
28. Doty, R.L.; Deems, D.A.; Frye, R.E.; Pelberg, R.; Shapiro, A. *Arch. Otolaryngol. Head Neck Surg.* **1988***, 114*, 1422-1427.
29. Doty, R.L.; Riklan, M.; Deems, D.A.; Reynolds, C.; Stellar, S. *Ann. Neuro.* **1989**, *25*, 166-171.
30. Pierce, J.D.J.; Doty, R.L.; Amoore, J.E. *Percept. Motor Skills* **1996**, *82*, 451-458.
31. Smith, R.S.; Doty, R.L.; Burlingame, G.K.; McKeown, D.A. *Percept. Psychophys.* **1993**, *54*, 649-655.
32. Stevens, S.S. In *Sensory Communication*; Rosenblith, W.A., Ed.; MIT Press: Cambridge, 1961.
33. Anderson, N.H. *Psychological Review* **1970**, *77*, 153-170.
34. Overbosch, P. *Chem. Senses* **1986**, *11*, 315-329.
35. Marks, L.E. *Percept. Psychophys.* **1988**, *43*, 511-525.
36. Doty, R.L., Shaman, P. and Dann, M. *Physiol. Beha.v* **1984**, *32*, 489-502.
37. Doty, R.L. *The Smell Identification Test*TM *Administration Manual*, , Sensonics, Inc., Haddon Hts., New Jersey, 1995, pp. 1-57

38. Doty, R.L.; Frye, R.E.; Agrawal, U. *Percept. Psychophys.* **1989**, *45*, 381-384.
39. Doty, R.L.; Shaman, P.; Kimmelman, C.P.; Dann, M.S. *Laryngoscope* **1984**, *94*, 176-178.
40. Gibbons, B. *National Geographic Magazine* **1986**, *170*, 324-361.
41. Liu, H.C.; Wang, S.J.; Lin, K.P.; Lin, K.N.; Fuh, J.L.; Teng, E.L. *Physiol. Behav.* **1995**, *58*, 1251-1255.
42. Solomon, G.S.; Petrie, W.M.; Hart, J.R.; Brackin, H.B.,Jr. Journal of *Neuropsychiatry & Clinical Neurosciences* **1998**, *10*, 64-67.
43. Doty, R.L. In *Clinical Measurement of Taste and Smell*; Meiselman, H.L.; Rivlin, R.S., Eds.; MacMillan: New York, 1986, pp. 377-413,
44. Doty, R.L.; Shaman, P.; Applebaum, S.L.; Giberson, R.; Siksorski, L.; Rosenberg, L. *Science* **1984**, *226*, 1441-1443.
45. Doty, R.L.; Applebaum, S.; Zusho, H.; Settle, R.G. *Neuropsychologia* **1985**, *23*, 667-672.
46. Segal, N.L.; Topolski, T.D.; Wilson, S.M.; Brown, K.W.; Araki, L. *Physiol. Behav.* **1995** *57*, 605-609.
47. Segal, N.L.; Brown, K.W. and Topolski, T.D. *Acta Genet. Med. Gemellol.* **1992**, *41*, 113-121.
48. Ship, J.A. ; Weiffenbach, J.M. J. *Gerontol.* **1993**, *48*, M26-M32.
49. Frye, R.E.; Schwartz, B.S.; Doty, R.L. *JAMA* **1995**, *263*, 1233-1236.
50. Rose, C.S; Heywood, P.G. ; Costanzo, R.M. *J. Occup. Med.* **1992**, *34*, 600-605.
51. Schwartz, B.S.; Ford, D.P.; Bolla, K.I.; Agnew, J.; Bleecker, M.L. *Amer. J. Psychiat.* **1991**, *148*, 751-756.
52. Bolla, K.I.; Schwartz, B.S.; Stewart, W.; Rignani, J.; Agnew, J.; Ford, D.P. *Amer. J. Indust. Med.* **1995**, *27*, 231-246.
53. Schwartz, B.S.; Ford, D.P.; Bolla, K.I.; Agnew, J.; Rothman, N.; Bleecker, M.L. *Amer. J. Indust Med.* **1990**, *18*, 697-706.
54. Schwartz, B.S.; Doty, R.L.; Monroe, C.; Frye, R.; Barker, S. *Amer. J. Pub. Health* **1990**, *79*, 613-618.
55. Ahman, M.; Holmstrom, M.; Cynkier, I.; Soderman, E. *Occupational & Environmental Medicine* **1996**, *53*, 112-117.
56. Hirsch, A.R.; Zavala, G. *Occupational & Environmental Medicine* **1996**, *56*, 284-287.
57. Corwin, J.; Loury, M.; Gilbert, A.N. *Journals of Gerontology Series B, Psychological Sciences & Social Sciences* **1995**, *50*, 179-186.
58. Doty, R.L. *Ann. Rev. Psychol.* **2001**, *52*, 423-452.
59. Schab, F.R. *Psychol. Bull.* **1991**, *109*, 242-251.
60. Doty, R. L. *The Odor Memory Test™ Administration Manual.* Sensonics, Inc.: Haddon Heights, New Jersey, 2000.

61. Cain, W.S.; Gent, J.F. *J. Exp.: Hum. Percept. Perform.* **1991**, *17*, 382-391.
62. Hummel, T.; Sekinger, B.; Wolfe, S.R.; Pauli, E.; Kobal, G. *Chem. Senses* **1997**, *22*, 39-52.
63. Guilford, J. P. *Psychometric Methods,* McGraw-Hill:New York, 1954.
64. Bromley, S.M; Doty, R.L. *Cortex* **1995**, *31*, 25-40.
65. Herbart, J.F. *Psychologie als Wissenschaft, neu gegrundet Auferfahrung*, Metaphysik, und Mathematik., Unzer, Konigsberg, 1824.

Chapter 11

Dose–Response Curves of Odor and Taste Stimuli: Influence of Sweetening Agents

Christine Vuilleumier, Isabelle Cayeux, and Maria Inés Velazco

Corporate R&D Division, Firmenich S.A., P.O. Box 239, CH–1211 Geneva 8, Switzerland

Abstract

In perfumery, the dose-response curves of odorant compounds may be determined by olfactometers and evaluation panels. The olfactometers that we have developed enable us to provide a range of perfectly controlled concentrations of compounds in the gas phase. We are thus able to represent the perceived intensity (orthonasal) as a function of the logarithm of the gas phase concentration. This treatment allows a straightforward comparison of different compounds or mixtures of compounds. Similarly, for flavor compounds, it is possible to characterize a perception dose-response curve (both orthonasal and retronasal) by both sniffing and tasting samples prepared at varying concentrations. Dissolution of the sample in a solvent and tasting of these solutions introduce further complexity owing to partitioning between the various different phases. These data are plotted as a function of the logarithm of the different concentration levels in the liquid phase. We have conducted a comparative study of the two olfactory airways (orthonasal and retronasal) in order to study the difference between the observed dose-response curves. In addition, to determine the influence of sweetening agents on odor and taste, we have included a study of the effect of sucrose and aspartame. The test sample is a lemon flavor, constituted of (±)-linalool, (+)-limonene and citral, compounds which we also studied individually. Whether or not this influence is a result of the various phase partitions or a consequence of the "cortical" integration of olfactory and taste contributions is also discussed.

Introduction

The influence of sweetening agents on flavor release and perception is a widespread phenomenon. For example, when sucrose is replaced by a sweetener to lower the calorie content of a soft drink, although the sweetness is generally maintained at the same level, the flavor quality and intensity of the product is often modified ([1-5]). This change could be due to both the mechanism of sensory perception and/or interactive effects of the flavor components. In 1996, Nahon *et al.* [6] reviewed several types of interactions (psychological, chemical, medium related, and sensory receptor related) and techniques for their measurement. Most of the published work involves measurement of the interactions between sweeteners and flavors by descriptive analysis. The same authors, in 1998 [7-8], showed that a solution flavored with an orange aroma containing aspartame is perceived as more sour, chemical and possessing an aftertaste, in comparison with solutions containing sucrose. The discovery that aspartame enhances longlastingness has been reported by several authors [2, 9-10]. Several studies also describe the chemical reactions of sweeteners with volatile compounds; for example, aspartame is reported to react with aldehydes and other carbonyl compounds [11-13]. Besides chemical interactions, flavor release could vary depending on the sweetener formulations. In this context, whereas Von Sydow *et al.* [14] noted that the addition of sucrose to a blueberry juice gave only minor changes in the headspace composition, Nahon *et al.* [7-8] found significant changes in the release of 15 selected volatile compounds between a flavored solution with sucrose (10% w/w) and the aqueous control; however, there was no observed change between the same flavored solution with added sodium cyclamate and the aqueous control. Deibler *et al.* [15] reported that many volatile components of soft drink (headspace) are affected by a variety of factors such as sweeteners, acidity, pH and temperature, and that none of the components behaved identically to change in these factors. Psychological effects rather than receptor or chemical interactions are also often described. Taste and smell behave independently, though perception of them can often be confused in cases when olfactory stimulation evokes sensations of taste. Some authors [14, 16] have examined the dependence of the flavor on sweetness, and others have proposed an additive model which integrates taste with smell [17]. Our investigation treats these interactions in a different manner. The sensory methodology is not the usual descriptive analysis but instead the establishment of dose-response curves, whereby we measure the dependency of the perceived intensity versus concentration. This method has already been applied to human olfaction and gustation, and has resulted in various mathematical treatments [18-19]. From such data, we generally observe that, between the non-perception and the saturation levels, the perception intensity of a chemical is proportional to the logarithm of its concentration [20]. We have chosen to work with simple systems, consisting of 3 pure chemicals and a model mixture containing only these chemicals, in 3 different media: water, an aqueous sucrose solution, and an aqueous aspartame solution. In order to describe and compare orthonasal and retronasal perception, perceived intensity is measured by sniffing, then by tasting from a cup. This process is quantified by analytical measurements for the liquid-gas partition in the cup, as well as in the mouth when tasting. An additional olfactometric study was effected to determine the perceived intensity versus the gas phase concentration. The aim of this approach was an understanding of the changes in perceived intensity by comparison and interpretation of the sniffing and tasting results.

General approach

This study was performed on a set of 3 flavor ingredients which give, when mixed in adequate proportions, a lemon-like odor (lemon flavor). These ingredients are:

(±)-linalool

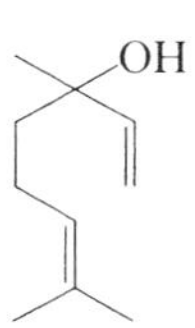

(+)-limonene

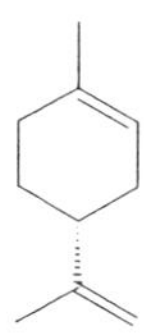

citral

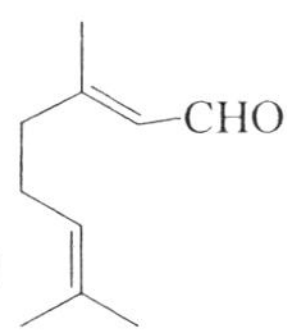

The composition (w/w) of the lemon flavor was: (+)-limonene (94.7%), citral (5%, *cis*-isomers 30-45%; *trans*-isomers 45-60%), and (±)-linalool (0.3%).

The 3 media used were:

- Water (*Henniez* Swiss mineral water),
- Water containing sucrose 5% (w/w),
- Water containing aspartame 0.042% (w/w).

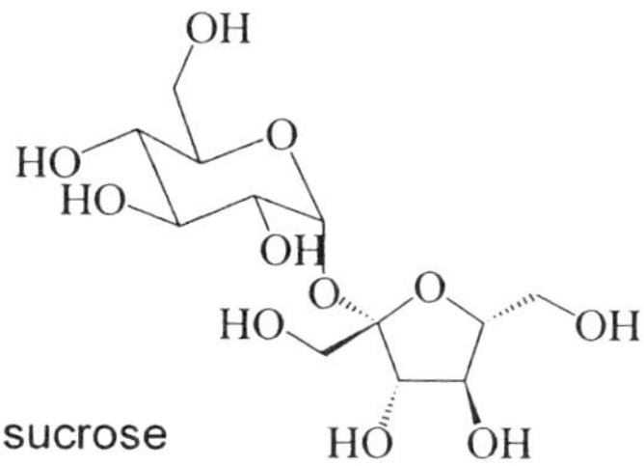

sucrose

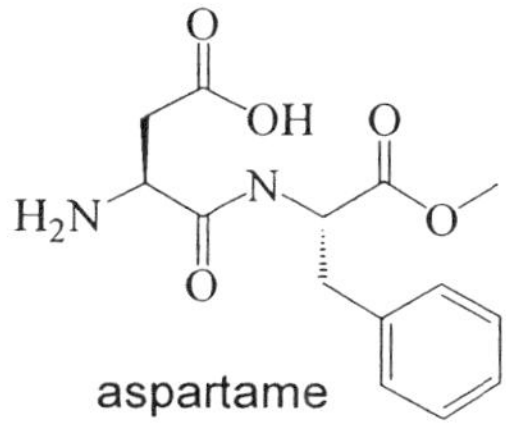

aspartame

To begin, a preliminary study established the sweetness equivalence concentration of aspartame, for a 5% (w/w) aqueous sucrose solution; the paired-comparison method was used. On average (25 panelists), this concentration was determined as 0.042% (w/w) aqueous aspartame, fully consistent with the reported values [21-25]. The panelists were asked to evaluate the flavor intensity of each chemical and the lemon flavor, in 3 different ways:

- an orthonasal way using "olfactometers" to determine, under controlled conditions, the perceived intensity as a function of the gas phase concentration,
- an orthonasal way using sniffing cups containing the 3 different media to evaluate the perceived intensity as a function of the liquid phase concentration. Prior to the panel testing, the cups were closed to attain equilibrium conditions,
- a retronasal way using tasting cups containing the 3 different media to evaluate the perceived intensity as a function of the liquid phase concentration.

We also decided to determine the gas phase concentrations of the chemicals in the different tests, so as to be able to interpret and compare the sensory results. Because olfactometry is based on a vaporization process, this concentration is readily known. However, for sniffing and tasting processes, it was necessary to apply analytical techniques to obtain such information. To determine the gas phase concentration of the volatile components during sniffing, headspace technology was used to obtain the gas phase concentrations, under equilibrium conditions, above the liquid phase in the cups. To determine the gas phase concentrations in the gas phase during tasting, Affirm® (trademark of *Firmenich SA*) technology was used under equilibrium conditions. This method allows the measurement of gas phase concentrations during exhalation

[26-28] and thereby permits an estimation of the amount of volatile components which reaches the olfactory epithelium *via* the retronasal route.

Panel and sensory evaluation

A panel of 25 *Firmenich* employees was trained to evaluate the flavor intensity of the individual chemicals and the lemon flavor. During training sessions, panelists smelled and tasted aqueous solutions (non-sweetened or sweetened with sugar or aspartame) of the chemicals. The aims of this training were to evaluate the "product space", *i.e.* to taste the whole range of concentrations, and to see reproducibility and discrimination with respect to the flavor intensity. A similar protocol was designed to evaluate orthonasal and retronasal perceptions and in which the panelists were trained to evaluate flavor intensity. Preliminary tests were organized to find the optimum range of concentrations so as to have the most complete dose-response curves between the perception threshold and the saturation level.

Test sessions

During each session, panelists smelled the same compound at different concentrations in a defined medium (gas phase (olfactometer), water, aqueous sucrose or aspartame solutions for sniffing and tasting cups). The samples were offered in a random order according to our experimental design (Latin square), in order to balance samples for 25 panelists: 9 for the olfactometers, 8 for the cups.

Scaling procedure

Using a linear scale, panelists were invited to rate the odor intensity of a given stimulus by making a mark on a horizontal line, corresponding to the size of the perceived stimulus. Two marks (anchors) are at the 2 ends: "not perceptible" and "very intense". The mark position is then converted into a number between "0" ("not perceptible") and "10" ("very intense"). For acquisition and treatment of this data, we used a specific sensory data system software (*Fizz-Biosystèmes*).

Experimental protocol

Orthonasal dose-response curves using olfactometers

The apparatus (Figure 1) utilized is a dynamic olfactometer which we designed to systematically study the perceived intensity of raw materials as a function of concentration. A syringe is loaded with the sample (neat, or in solution in an odorless solvent). It is then inserted into a press-syringe, which delivers a steady flow rate into the lower part of the olfactometer, maintained at a constant temperature (150°C for these experiments). The sample is evaporated, and a flow of nitrogen flushes the vaporized product into the upper part of the system, maintained at 22°C. Simultaneously, humidified air (relative humidity level *ca.* 50%) is introduced to dilute the product before it reaches the outlet of the olfactometer; the temperature of the sniff port is *ca.* 26°C. The gaseous concentrations of the odorants can be calculated from the delivery rates from the syringe and the flow rate of the gases. It is possible to cross-check this concentration by headspace sampling a known volume of gas on a cartridge filled with an adsorbent. In addition, this control guarantees that the compound under examination does not decompose during the process [29]. For these measurements, a set of 9 olfactometers was used.

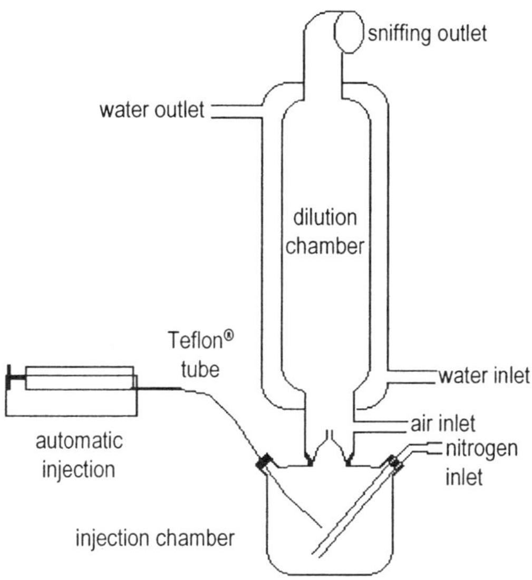

Figure 1: *Scheme of an olfactometer. (Reproduced with permission from ref. 20. Copyright 1995 Firmenich.)*

Table I: Gaseous concentrations in finalized olfactometric tests (in μg/l air)

(±)-linalool	*(+)-limonene*	*citral*
115	562	119
11.5	112	10.7
5.26	11.2	5.92
2.46	3.90	2.83
1.15	1.12	1.36
0.345	0.390	0.397
0.115	0.112 (repeated)	0.136
0.0115	0.0112	0.0136
0.00115	--	0.00136

This series of tests, using the same protocol, was effected 3 times, each time presenting the samples in a random order with respect to their concentrations. To effect the same study for the lemon flavor, we determined by headspace the liquid-gas partition of the mixture in the 3 aforementioned media. We thus loaded 0.01% w/w of the lemon flavor in water, 5% aqueous sucrose solution, and 0.042% aqueous aspartame solution in 100 ml syringes. After waiting so as to assure equilibrium between the liquid and gas phases, we then sampled a measured volume of the gas phase onto an absorbent cartridge. In this manner, we were able to quantify the amount of each of the ingredients (Figure 2).

The gas phase concentrations, averaged values of 2 consecutive measurements, are presented in Table II.

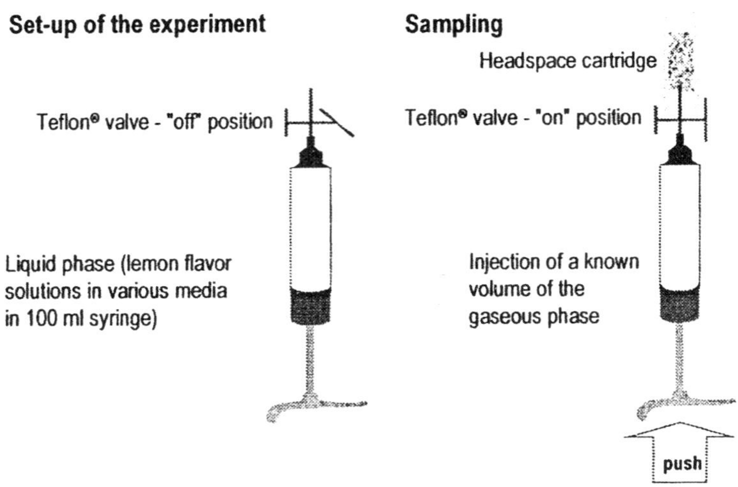

***Figure 2**: Experimental device to measure the liquid-gas partition coefficients*

Table II: Headspace (HS) concentration (%) of the lemon

	Pure lemon flavor	*HS over water*	*HS over sucrose solution*	*HS over aspartame solution*
citral	5.00	1.01	0.33	0.72
(+)-limonene	94.70	98.87	99.65	99.22
(±)-linalool	0.30	0.12	0.02	0.06

We have studied the lemon flavor (gas phase concentrations as indicated for the different media in Table II) by employing the same protocol as used above for the olfactometric tests and by selecting a concentration range taking into account the composition of the different ingredients.

Dose-response curves by sniffing and tasting solutions (orthonasal and retronasal)

Citral, (+)-limonene, and (±)-linalool were placed in closed plastic cups (30 ml) and then individually smelled and tasted in the 3 different media: water (*Henniez* "Naturelle"), 5% aqueous sucrose solution (D-(+)-sucrose: *Fluka* ref. 84100, purity >99%), and 0.042% aqueous aspartame (*Fluka* ref. 11300, purity >99%) solution. The samples were prepared 1 h preceding the tests. The lemon flavor was also smelled and tasted under the same conditions. In order to cover the maximum range of perceived intensity, the concentrations used were defined during preliminary sessions with panelists. During each session, panelists were invited to smell and taste 8 concentrations of the same compound or mixture in each medium at room temperature (Table III).

Table III: Test concentrations (mg/l) for chemicals and lemon flavor tested

citral	*(+)-limonene*	*(±)-linalool*	*lemon flavor*
0	0	0	0
2.5	50	5	25
5	100	10	50
10	200	20	100
20	400	40	200
40	800	80	400
80	1600	160	800
160	3200	320	1600

(all compounds, including lemon flavor, are prediluted in a 10% ethanolic solution)

The panelists rated the flavor intensity; the results were either presented by plotting intensity versus liquid phase concentration in the cups, or by converting this concentration into a gas phase concentration. For sniffing, this conversion depends on the partition coefficients, estimated in the closed cups after attainment of equilibrium from headspace measurements (Table IV).

For tasting, the conversion is consistent with the aforementioned Affirm® data. One expert is asked to taste 10 times the same concentration of each chemical and the lemon flavor, in each of the 3 media. The obtained results are reproducible. The partition coefficient measurements are presented in Table V.

Table IV: Air-liquid phase partition coefficients

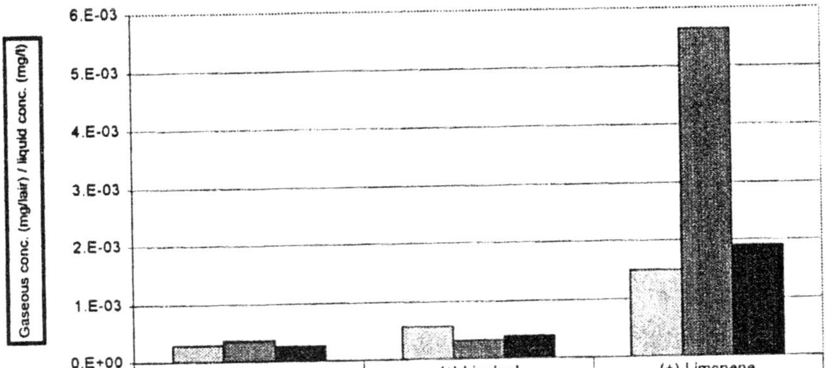

	Citral	(±)-Linalool	(+)-Limonene
water solution	2.90E-04	5.65E-04	1.50E-03
sucrose solution	3.59E-04	3.19E-04	5.65E-03
aspartame solution	2.61E-04	3.77E-04	1.91E-03

Table V: Gas phase-liquid phase partition coefficient in the nose

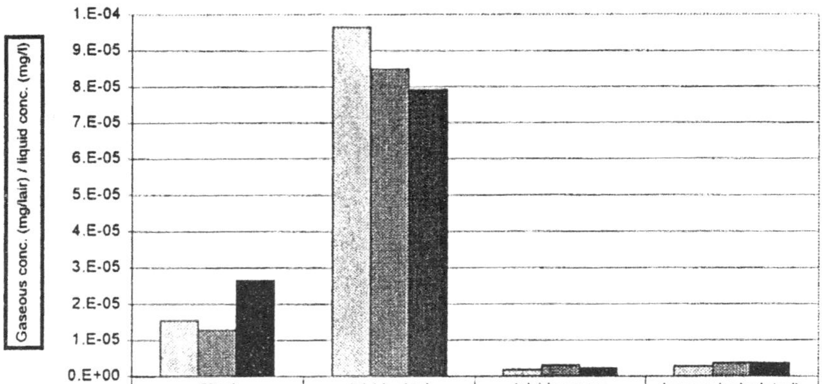

	Citral	(±)-Linalool	(+)-Limonene	Lemon (calculated)
water solution	1.55E-05	9.63E-05	1.84E-06	2.81E-06
sucrose solution	1.28E-05	8.50E-05	3.00E-06	3.74E-06
aspartame solution	2.65E-05	7.92E-05	2.30E-06	3.74E-06

Data treatment

Data treatment enables us to verify whether or not the different concentrations for a given dose-response curve are well ordered and perceived differently by the panelists (by *Analysis of variance*). For each product in each medium, the dose-response curves are determined and their parameters then calculated to compare the behavior of the products in the different media.

Intensity curves versus concentrations

The averaged intensities and confidence limits of the measurements were calculated for all the experiments, finally giving rise to a plot of intensity as a function of concentration and allowing comparisons to be made.

Analysis of variance (*Anova* approach)

Analysis of variance of products and panelists shows that, for each product in each medium, the product effect is very significant and much larger than the panelist effect. The perceived intensity rises as a function of the increasing concentration of the odorant. This approach is used to determine whether or not significantly different results in averaged perceived intensity are obtained for a defined set of concentrations. It can also be used to check whether one obtains significantly different results when comparing the same compound and concentration in 2 different media or with orthonasal versus retronasal mode.

Calculation of dose-response curve parameters

To compare the different intensity curves, we calculated the detection threshold, the intensity slope and the saturation level. We determined these 3 parameters from the intensity curves as shown in Figure 3. The threshold level is not determined as a classical detection threshold according to ASTM standard practices [30-31] but is calculated from the panel-averaged intensity dose-response curve showing how intensity varies with changes in concentration.

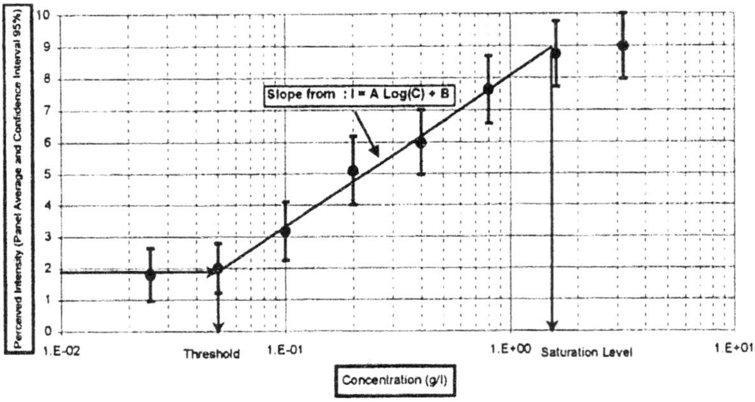

***Figure 3**: Determination of parameters for dose-response curve*

Thus the threshold level (called simply threshold in the text and graphs) is considered as being equivalent to the concentration corresponding to the perceived intensity of 2 out of 10 on our linear scale. The intensity slope is defined as "A", the director coefficient of the regression "I = A Log(C) + B"; the regression is calculated from results measured in the supraliminal range. The saturation level represents the concentration which corresponds to the maximum intensity. These 3 parameters characterize the perception of each compound in the 3 different media along the whole intensity range for both orthonasal and retronasal modes.

Experimental results and interpretation

We began experimental work by testing the reproducibility of the various evaluation methods. Figures 4, 5 and 6 present the results from the olfactometric, sniffing and tasting experiments in which a good reproducibility for the dose-response curves is generally observed.

Because of the excellent reproducibility of these olfactometric measurements, we only tested the reproducibility of sniffing and tasting for (±)-linalool in a 0.042% aqueous aspartame solution (Figure 6).

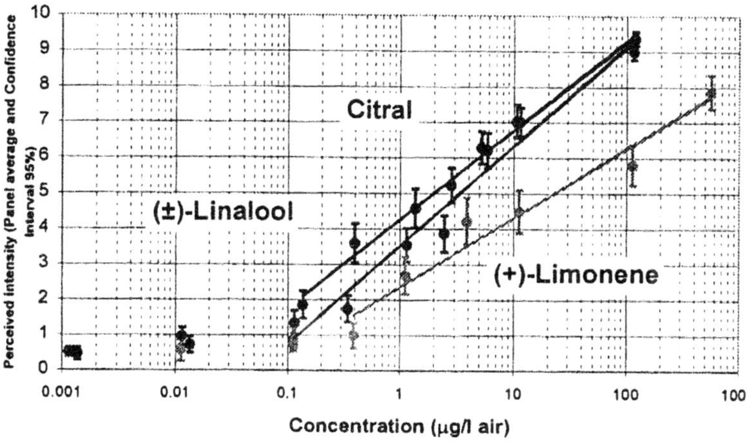

***Figure 4**: Results for orthonasal dose-response curves by olfactometry*

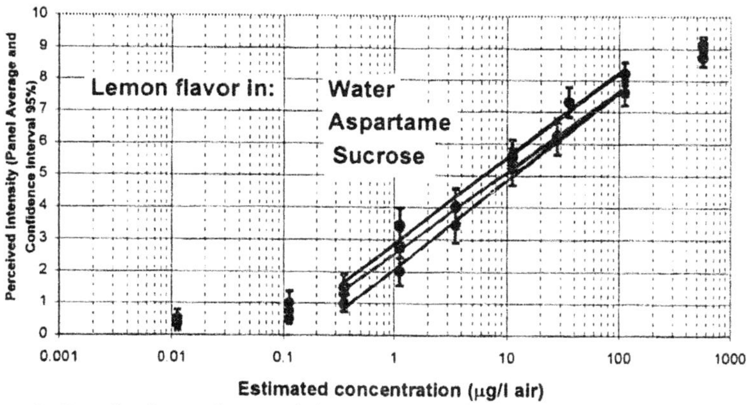

***Figure 5**: Results for orthonasal dose-response curves for the lemon flavor by olfactometry (gas phase concentrations are determined from the measured partition coefficients in closed cups)*

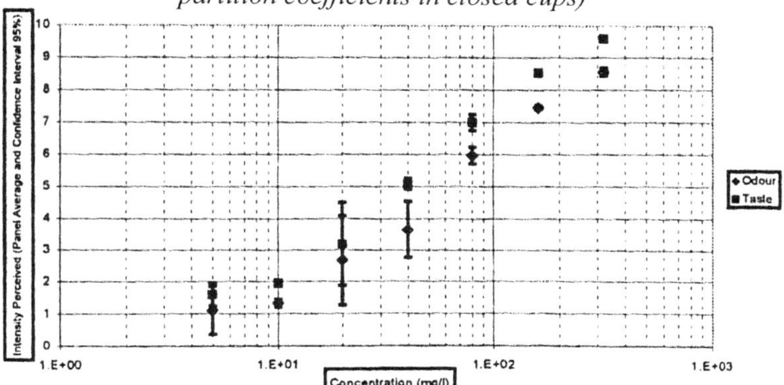

***Figure 6**: Reproducibility*

To more easily compare the results, we decided that data mapping would be appropriate. This is illustrated in Figure 7, where the averaged data (based on 3 olfactometric measurements) for the 3 chemicals and the lemon flavor are presented.

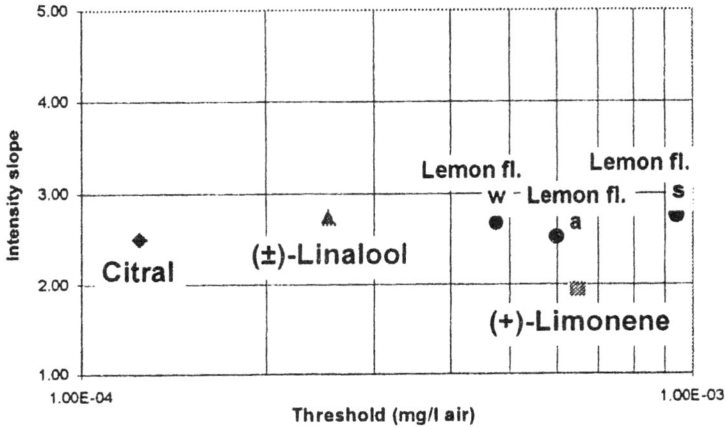

(w = water, s = sucrose and a = aspartame)

***Figure** 7: Mapping of olfactometric data*

The horizontal axis represents the above defined threshold value, and the intensity slope is represented by the vertical axis. We can thus rapidly visualize the position of the different compounds according to these 2 criteria. It is apparent that the threshold values all lie within the same order of magnitude. In addition, the slopes are extremely similar, signifying that, in the supraliminal range, each increase in concentration is accompanied by a comparable increase in intensity. It is also worth noting the similarity of the lemon flavor and (+)-limonene (its principal constituent) in the 3 different media. If we present a mapping of the sniffing measurements (Figure 8), using threshold values expressed either in liquid phase (left-hand graph) or gas phase (right-hand graph) concentrations, the threshold differences are readily apparent.

In fact, the graphs are interrelated by the air-liquid partition coefficients described earlier. It is thus evident that the interpretation of the results is critically dependent on the way they are presented. If this method of presentation is applied to the olfactometric and sniffing results (expressed in gas phase concentrations, Figure 9), there are notable differences.

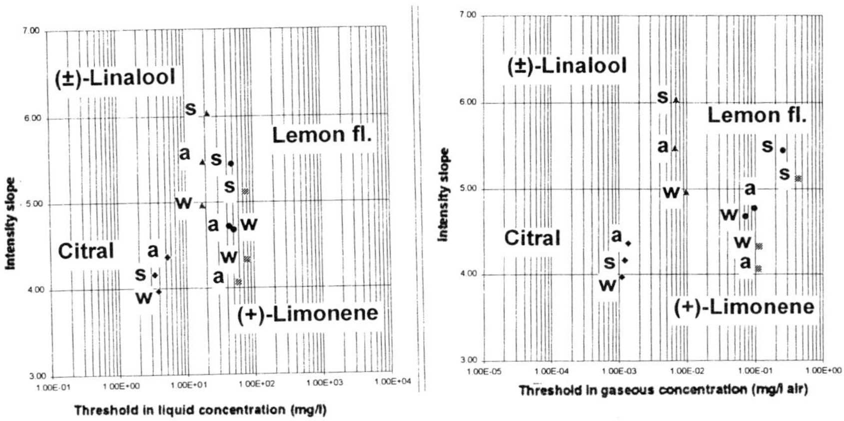

(w = water, s = sucrose and a = aspartame)

Figure 8: Sniffing mapping

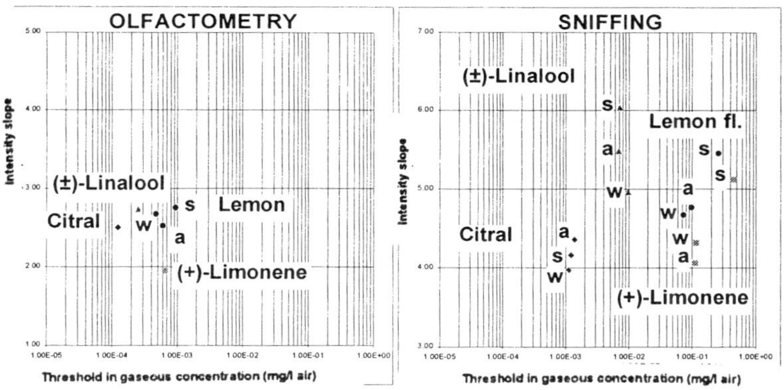

(w = water, s = sucrose and a = aspartame)

Figure 9: Olfactometric and sniffing mapping in mg/l air

At first sight this is surprising, given that, in both cases, the process is orthonasal and the results are expressed in the same units. These differences are apparently the consequence of the difficulty of estimating the gas phase concentrations in the experiments using the sniffing cups. The assumption is made that the fact of sniffing, after opening the closed cup (an essentially dynamic process), does not modify the gas phase concentration of the compound in the cup when closed. However, it is very likely that this assumption is false. In addition, the differences in the slopes are probably due to the fact that the determination of the gas phase concentration may well vary differently depending on the liquid phase concentration. It is also apparent that the differences in the slopes may have other origins. Indeed, the sniffing, tasting and olfactometric conditions are not strictly identical regarding several physical parameters. For example, both the temperature and relative humidity vary considerably. In olfactometry, a temperature of 25°C is maintained at the exit port (a value which is very similar to the natural temperature of the nostrils) with a relative humidity of 50% (similar to ambient conditions, though it should be borne in mind that the relative humidity of the nostrils is *ca.* 90%). In contrast, during tasting, the mouth temperature is approximately 35°C and the relative humidity 100%. In addition, whereas the perception of smelling involves inhalation, the process of tasting is completely different, consisting of a sequence of inhalations and exhalations. It is also worth noting that biases due to the medium do not notably affect the thresholds and the intensity slopes of the lemon flavor. Next we shall consider the olfactometric results for comparisons in the gas phase because of a better control of the experimental conditions. If one compares the olfactometric and tasting results using gas phase concentrations (Figure 10), it is evident that once again the results are not greatly influenced by the medium employed. However, there is a good correlation between the threshold values for the olfactometric and tasting data, but it is remarkable that, in the latter case, the intensity slope values are substantially higher; (±)-linalool stands out by a slope noticeably higher than the others, in contrast to its threshold value, which is only slightly different. This phenomenon may be due to the bitter character of (±)-linalool at high concentrations, a factor which could influence the evaluation of the panelists in this sense.

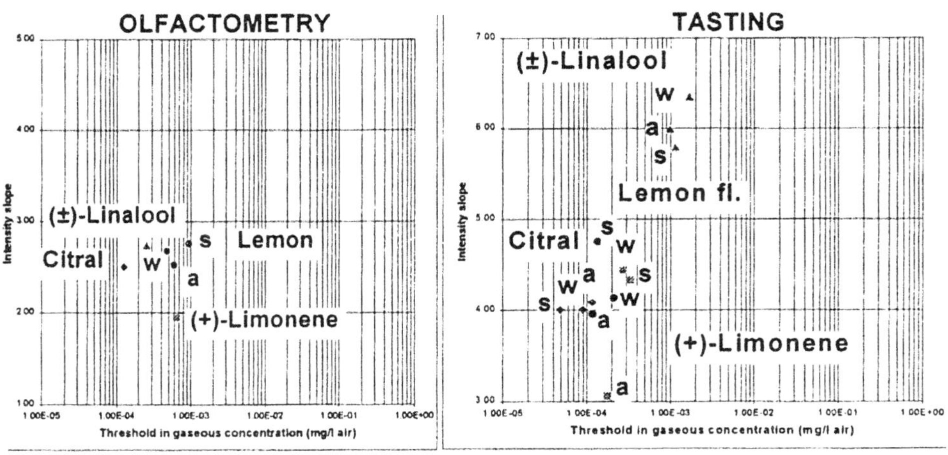

(w = water, s = sucrose and a = aspartame)

***Figure 10**: Olfactometric and tasting mappings (mg/l air)*

Conclusion

Our aim has been to compare the dose-response curves of volatile components measured in different contexts, *viz.* sniffing a gaseous emission from an olfactometer, sniffing the vapour phase above a solution, and tasting the same solution. In this manner, we have thus been able to determine the variation of intensity as a function of the concentration of the volatile products in diverse media. For a strict comparison of these curves, it is necessary to quantify the gas phase concentrations which are generated in the different media and susceptible to stimulate the olfactive epithelium. By definition, olfactometry allows us to control this gas phase concentration, and is thus considered to be our point of reference. To estimate the gas phase concentrations in the other systems, we have used partition coefficients obtained from either static headspace analysis (for solutions in closed cups) or Affirm® technology (for tasting experiments). From the sensory analysis results, these theoretical concentrations do not appear to correspond to the actual concentrations perceived when sniffing a freshly opened cup and, to a lesser extent, those perceived when tasting. For orthonasal perception, it would be a question of integrating a dynamic evaluation of the gas phase concentration above the cup during sniffing. Specifically, the airflow contributing to the dilution of the headspace in the cup when one removes the lid, and the resultant time dependent decrease in concentration. For the tasting experiments, the quantification of the process already takes these dynamics into consideration, but it is evident that other phenomena take place in the mouth, to give rise to a variety of poorly understood interactions. However, it is interesting to recall that the effect of added sugar and aspartame (in comparison with water) has a negligible effect on both sniffing and tasting. Future work will attempt to investigate more realistic conditions for the study of products in solution (drinks in particular). For this, it will be necessary to establish a correlation between the intensity measurements recorded under "normal" conditions, *i.e.* without closure of the cup prior to tasting, and the olfactometric data already in hand. All this, in accepting that knowledge of the equilibrium air-liquid partition coefficients in a cup or in the mouth is necessary information but not sufficient to describe the real behavior of products during their organoleptic evaluation.

References

1. Larson-Powers, N; Pangborn, R.M. Paired comparison and time-intensity measurements of the sensory properties of beverages and gelatins containing sucrose or synthetic sweeteners. *Journal of Food Science*, **1978**, Vol. 43.

2. Matysiak, N.L.; Noble, A.C. Comparison of temporal perception of fruitiness in model systems sweetened with aspartame, an aspartame + acesulfamK blend, or sucrose. *Journal of Food Science*, **1991**, Vol. 56, N° 3.
3. Wiseman, J.J.; McDaniel, M.R. Modification of fruit flavors by aspartame and sucrose. *Journal of Food Science*, **1991**, Vol. 56, N° 6.
4. Baron, R.F.; Hanger, L.Y. Using acid level, acesulfamK/aspartame blend ratio and flavor type to determine optimum flavor profiles of fruit flavored beverages. *Journal of Sensory Studies*, **1998**, Vol. 13, pp 269-283.
5. Muir, D.D.; Hunter, E.A.; Willliams, S.A.R.; Brennan, R.M. Sensory Profiles of commercial fruit juice drinks: influence of sweetener type. *Journal of Sci. Food Agric.*, **1998**, Vol. 77, pp 559-565.
6. Nahon, D.F.; Roozen, J.P.; De Graaf, C. Sweetness flavour interactions in soft drinks. *Food Chemistry*, **1996**, Vol. 56, N° 3, pp 283-289.
7. Nahon, D.F.; Roozen, J.P.; De Graaf, C. Sensory evaluation of mixtures of maltitol or aspartame sucrose and an orange aroma. *Chemical Senses*, **1998**, Vol. 23, pp 59-66.
8. Nahon, D.F.; Navarro y Koren, P.A.; Roozen, J.P.; Maarten, A. Posthumus. Flavor release from mixtures of sodium cyclamate, sucrose, and an orange aroma. *Journal of Agricultural Food Chemistry*, **1998**, Vol. 46, pp 4963-4968.
9. Baldwin, R.E.; Korschgen, B.M. Intensification of fruit flavors by aspartame. *Journal of Food Science*, **1979**, Vol. 44, pp 938-939.
10. Bonnans, S.; Noble, A.C. Effect of sweetener type and of sweetener and acid levels on temporal perception of sweetness, sourness and fruitiness. *Chemical Senses*, **1993**, Vol. 18, pp 273-283.
11. Hussein, M.M; D'Amelia, R.P.; Jacin, A.L.; Manz, R.P.; Chen, W.T.C. Determination of the reactivity of aspartame with flavor aldehydes by GC, HPLC and GPC. *Journal of Food Science*, **1984**, Vol. 49, pp 520-524.
12. Tateo, F.; Triangli, L.; Ciserchia, E.; Nicoletti, R.; Berte, F. Reactivity of aspartame with aldehydes in flavoring agents. *Boll. Chim. Farm.*, **1986**, Vol. 125, N° 11, pp 404-409.
13. Le Quéré, J.L.; Leschaeve, I.; Demaizières, D.; Issanchou, S.; Delache, R.; Chemical and sensory effects of intense sweeteners on the flavour of diet orange soft drinks. *In Trends in Flavour Research*, **1994**, pp 287-291, Edited by H. Maarse and D.G. van der Heij, Elsevier Science, Amsterdam.
14. Von Sydow, E.; Moskowitz, H.; Jacobs H.; Meiselman H. Odor-taste interaction in fruit juices. *Lebenssm-Wiss.Technol.*, **1974**, Vol. 7, pp 18-24.
15. Deibler, K.; Acree, T. Effects of soft drink base composition on flavor release. Internal communication from Cornell University, 1999, New York Agriculture Experiment Station , Geneva, NY 14456.

16. Frank, R.A.; Byram, J. Taste-smell interactions are tastant and odorant dependent. *Chemical Senses*, **1988**, 13, 445-455.
17. Murphy, C.; Cain, W.S. Taste and olfaction: Independence vs Interaction, *Physiol. Behav.*, **1980**, Vol. 24, pp 601-605.
18. Meilgaard, M.; Civille, B.; Carr, *T. Sensory Evaluation techniques*, 3rd Edition, **1999**, CRC Press.
19. HandBook of Olfaction and Gustation, 1991, Dekker Inc., New-York.
20. Wünsche, L.F.; Vuilleumier, C.; Keller, U.; Byfield, M.P.; May, I.P.; Kearney, M.J. Scent characterisation: from human perception to electronic noses. Proceedings of 13th International Congress of Flavours, Fragrances and Essential Oils, 15-19 October 1995, Istanbul.
21. Mori, E.E. Analise sensorial de adoçantes e edelcorantes (Revisao) *Cienc. Tecnol. Aliment.*, jul/dec. **1992**, Vol. 12, N° 2, pp 101-115.
22. Hanger, L.Y.; Lotz A., Lepeniotis S. Descriptive Profiles of Selected High Intensity Sweeteners (HIS), HIS Blends, and Sucrose. *Journal of Food Science*, **1996**, Vol. 61, N° 2.
23. Schiffman, S.S.; Gatlin, C.A. Sweeteners: State of Knowledge Review. *Neuroscience and Biobehavioral Reviews*, **1993**, Vol. 17, pp 313-345.
24. Ketelsen, S.M.; Kea, C.L.; Wiet, S.G. Time-intensity parameters of selected carbohydrate and high potency sweeteners. *Journal of Food Science*, **1996**,Vol. 58, N° 6.
25. Portmann, M.-O.; Kilcast, D. Psychophysical characterization of new sweeteners of commercial importance for the EC industry. *Food Chemistry*, **1996**, Vol. 56, N° 3, pp 291-302.
26. Blake, A. Communicating with chemicals. 219th ACS National Meeting, 26-30 March 2000, San Francisco, CA.
27. Taylor, A.J. Measuring taste release *in vivo*. 219th ACS National Meeting, 26-30 March 2000, San Francisco, CA.
28. Brauss, M.S.; Linforth, R.S.T.; Cayeux, I.; Harvey, B.; Taylor, A.J. Altering the fat content affects flavor release in a model yogurt system. *Journal of Agricultural and Food Chemistry*, **1999**, Vol. 47, N° 5, pp 2055-2059.
29. Vuilleumier, C.; Flament, I.; Sauvegrain P. Headspace analysis study of evaporation rate of perfume ingredients applied onto skin. *International Journal of Cosmetic Science*, **1995**, Vol. 17, pp 61-76.
30. ASTM, Standard Practice E679, Determination of Odor and Taste Thresholds by a Forced-Choice Ascending Concentration Series Method of Limits, *American Society for testing and Materials*, Philadelphia, **1979**.
31. ASTM, Standard Practice E1432, Defining and Calculating Sensory Thresholds from Forced-Choice Data Sets of Intermediate Size, *American Society for Testing and Materials*, Philadelphia, **1990**.

Chapter 12

Communicating with Chemicals

Anthony Blake

Firmenich S.A., Rue de la Bergère 7, P.O. Box 239, CH–1217 Meyrin 2, Geneva, Switzerland

All living cells interact and communicate with their environment using receptors which respond to the chemicals within that environment. The evolution and sophistication of these processes in humans are why we are able to smell, taste and enjoy our food and drink. However, visual and spoken communication have become so important to us that we tend to relegate our senses of taste and smell to a secondary role even thought they are powerfully emotive and especially in relation to early childhood memories. Although we fully recognize the importance of the time dimension in speech, music and reading, it is little discussed as it relates to taste and smell. It is only in recent years that techniques have become available which allow us to record the time course release of flavour molecules to the nose during eating and drinking. This paper will discuss the recent developments in quantitative measurement of flavour release kinetics and will discuss how these relate to the perception of flavour and the enjoyment of food.

It is said that we live in the age of the Information Revolution and as companies increasingly change to the status of dotcoms and the use of the internet grows more people talk about data transmission speeds and signal decoding. My talk today is also about data transmission and arguably about the oldest form of signaling between living systems, communication through chemicals. In our high speed world of digital information and from our dependence on the written and spoken word we humans tend to forget that odours, smells and tastes provide a large amount of the data that affect our daily lives. One only has to take a dog for a walk to appreciate this fact.

This presentation is essentially an introduction to the next, which will be given by Professor Taylor. Here I will try to underline why we are increasingly looking at the problems associated with the flavouring of food as issues of communication and cognitive science, topics which have traditionally been far removed from the scope of the average flavour symposium. Let me first of all briefly review the changing nature of the flavour industry over the last century. Manufacturers of flavour essences at the beginning of this century largely relied on techniques of distillation and solvent extraction for obtaining the essential oils, oleoresins and absolutes which they used as the raw materials for their products. The synthetic flavours used were limited to materials such as vanillin and simple esters and indeed there was little knowledge about the chemicals which were responsible for the flavours and odours evolved by nature. One hundred years ago it was the physical laws of fractionation and selective solubility which provided the fundamental science base of the flavour industry.

This situation changed dramatically over the first fifty years of this century and especially in the immediate post World War II era. New industries had emerged together with new scientific skills and in particular the pharmaceutical and petrochemical industries had driven the development of new techniques for separating and analyzing the compositions of complex mixtures; new detection methods allowed precise quantification of ever lower levels of such materials. Above all the emergencies of war had driven the understanding and use of electronics and radio transmission in a way that a peacetime research programme could never have done. From these new scientific skills the techniques of nuclear magnetic resonance and mass spectrometry were born. During the next twenty years the flavour industry was transformed as its scientists separated, identified and synthesized the molecules responsible for the perfume of flowers and the flavour of fruits. The industry now largely rested on a scientific base of organic chemistry and the Flavour Symposia of the 70's and 80's reflected this change.

Within the last decade, however, there has once again been a subtle change in emphasis; it is not enough to have a flavour composition in a bottle with the authentic odour of a strawberry. It is increasingly realized that it is also important to know how and why it performs in a food system in the way that it does. The mechanisms of entrapment, migration, release and perception of flavour molecules are increasingly a subject for discussion at flavour conferences and the physical chemists can once again emerge from relative obscurity. Ten years ago the Research Division of Firmenich had no more than two scientists who would admit to being physical chemists; we now have eight research teams involved in this area.

One of the most important developments within the flavour industry in recent years is the realization that by better understanding the way flavours are released from foods and how our olfactory system works we can improve the way we use flavours in food systems. As already mentioned the release of flavour from food and our sensory perception of it can be considered a system of communication. The food is the emitter of information, the mouth and nasal tract are the transmission system, finally the olfactory epithelium and the brain receive and process the data for the conscious regions of the mind. Flavour molecules carry the data but the other processes cannot be ignored in the overall communication of information and the way it is interpreted. As with all data transmission systems the time element is an essential component. To take an audio analogy a piece of music is composed of a specific sequence of sound frequencies; the order and phasing at which they arrive at the ear is essential to the quality of the music; the notes may be the right ones but if they arrive in the wrong order or with the wrong phasing then the result is not the same. The use of sound as an analogy is not inappropriate because it has been calculated that the data transmission capacity of our senses of hearing and smell are of roughly the same magnitude. The kinetics of flavour release from foods, the way that the flavour molecules are transmitted to the nose and the effects which food composition have on this are fundamentally important aspects of flavour and our perception of it. Until very recently the importance of the time dimension to flavouring was hardly recognized and no techniques existed to study it in a scientific way. It is only in the last few years that equipment has been developed which has the sensitivity and speed of response to allow the kinetics of flavour release from foods to be followed during the eating process. In particular, the work of Taylor and Linforth at the University of Nottingham, UK, has reached the stage where the flow of flavour molecules across the olfactory epithelium can be monitored in real time at levels of a few parts per billion. This pioneering work allows us to relate the input signal (the rate and intensity at which the flavour molecules pass through the nose) directly to the output from the brain (the perceptions of flavour). It is increasingly clear that, by controlling the way in which flavour molecules are released from food during eating, it is possible to change the way in which the food's flavour is perceived. The instrumental development which made this possible was the combination of a mass spectrometer with a breath sampling interface which links the nose of the volunteer taster to an ionization chamber in which any flavour molecules become electrically charged and can be separated and quantitatively measured in the mass spectrometer. The technique goes by the name Atmospheric

Pressure Chemical Ionization Mass Spectrometry but we have christened it AFFIRM®, which is a loose acronym for the **A**nalysis of **F**lavours and **F**ragrances **I**n **R**eal ti**M**e.

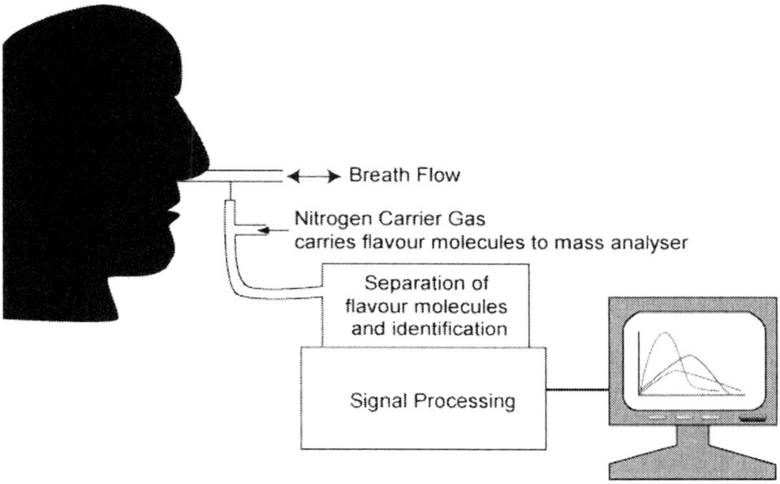

Figure 1 – Schematic representation of the AFFIRM® technique

Breath is sampled directly from the nose (nosespace) of a volunteer subject into the APCI mass spectrometer (Figure 1). The volunteer is able to breathe normally through his or her nose via a tube held gently in one nostril. A proportion of the breath is drawn by a gentle suction and is diluted with a carrier gas for APCI analysis. The connection between the volunteer and the apparatus is made to be as comfortable as possible and allows the subject to eat and drink whilst breathing normally. The continuous analysis of the gas stream can be arranged to track all volatile materials, or can be set to monitor a smaller number of molecular species of interest but at greater sensitivity and typically down to a few ppb. In this way, a time intensity plot is directly obtained of the flavour chemicals in the breath of the subject, which are responsible for the odour impression he or she is experiencing.

Figure 2 which is taken from the work of Taylor & Linforth shows the real time monitoring of three volatile components in the breath of a volunteer. The top trace shows acetone, which is a normal component of everyone's breath and is monitored to show the breathing pattern of the subject. By contrast the middle and bottom trace show ethyl butyrate and ethanol released from a

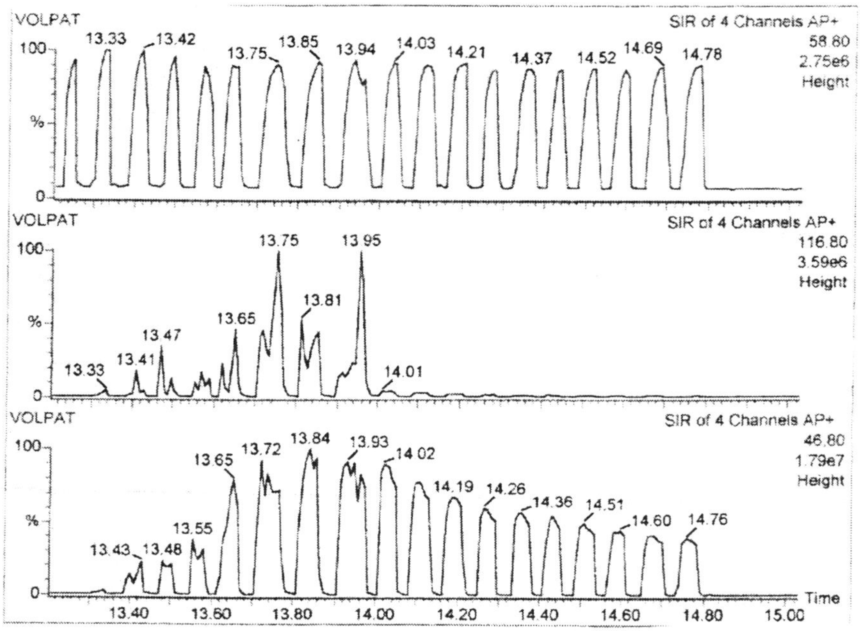

Figure 2 – AFFIRM[R] measurement of acetone, ethyl butyrate and ethanol in the breath of a volunteer eating a flavoured gelatine gel

gelatine gel. What is clear is that when the sweet is swallowed the ethyl butyrate disappears rapidly from the breath but the ethanol is much more persistent. We are now exploring exactly how different flavour chemicals behave in terms of their persistence on the breath and in relation to the composition and texture of food systems.

AFFIRM® analysis has already improved our fundamental comprehension of how flavour perception works. It is now clear that our nose responds more to the rate of change of a flavour volatile than to its absolute level in the nose. Furthermore if a flavour molecule remains in the nose for more than a few minutes the nose adapts and becomes less effective at smelling it. For these reasons a quick release of flavour from a foodstuff can be more effective at delivering a flavour sensation than a sustained release of the same component.

In hindsight, these facts give such obvious evolutionary advantages that it is surprising that they were not appreciated and understood much sooner than they were. If we take a visual analogy we are all very aware of the fact that a flashing light demands our attention much more powerfully than a light of constant intensity so it is perhaps not surprising that we have a similar reaction to smells. From the world of insects we have also known that certain moths emit pheromones in pulses rather than at a constant level and that the antennae of the recipient moth responds better to a pulsed signal. Similarly adaptation would give an evolutionary advantage in filtering new and potentially important signals from constant and less important background smells. Nevertheless the advice from the suppliers of flavours has always been "Make sure the flavour is evenly distributed", "Ensure that you do not have flavour hot-spots", well-intentioned advice but in hindsight a poor strategy for effective flavouring.

Within recent months we have also been using the AFFIRM® technology to better understand how food composition can influence flavour release and perception. To illustrate this I will show data from the product known and loved by most American children but considered a little strange by most of the rest of the world, the peanut butter and grape jelly sandwich (PBGJS). Since there are flavour components unique to peanut butter and others unique to grape jelly it is possible to quantitatively track these on the breath of a volunteer taster. Figure 3 shows a breath by breath analysis of two such aroma molecules released during the eating of either a sandwich with just peanut butter or another with only grape jelly; as can be seen there is a clear separation of the two signals.

The debate amongst enthusiasts of PBGJS is whether the peanut butter and grape jelly should be layered separately or mixed so Figure 4 shows the results on eating both types of sandwich.

You can see that there is a reasonable correlation between the thickness of the layers and the signal provided the layers are separate. However, mixing

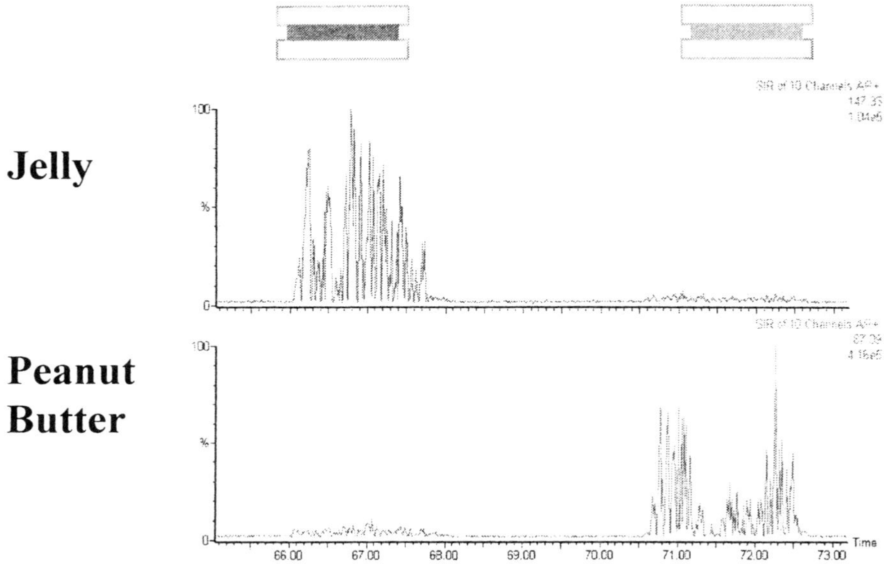

Figure 3 – Peanut butter and grape jelly analysis

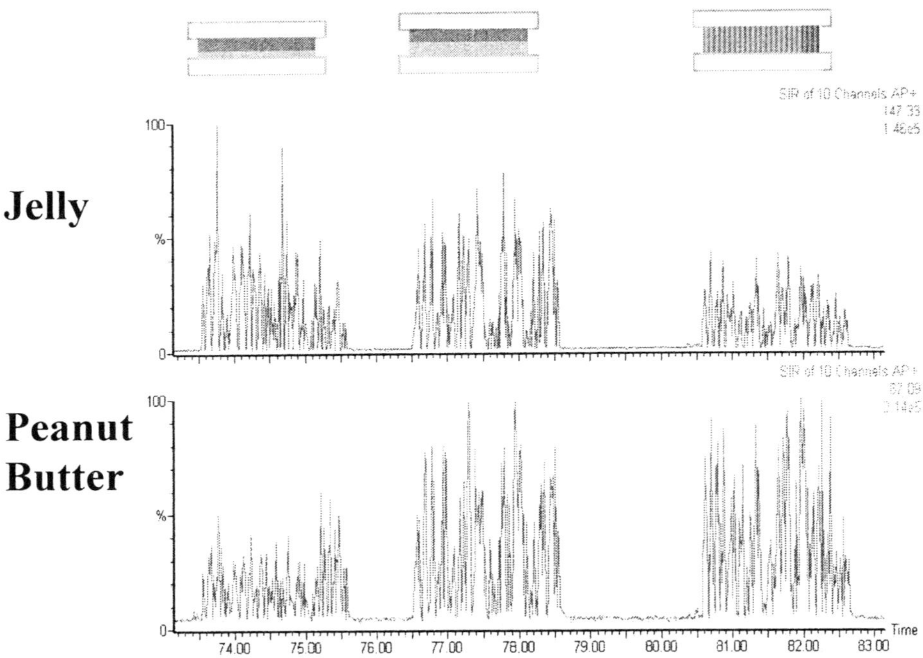

Figure 4 – Analysis of different constructions of peanut butter and grape jelly sandwich (PBGJS)

the two components leads to a loss of the grape jelly signal. This experiment was, admittedly, done with tongue in cheek but its slightly trivial nature does not detract from nor obscure the important conclusion that a relatively slight change in physical construction of a food can have a major change on the flavour perception of that food.

Within our company AFFIRM® is increasingly being used to understand how the physical make up of a food product can modify flavour perception. Even at the stage of creating a flavour system it is useful for the flavour chemist to know which flavour molecules reach the olfactory bulb when eaten in an application.

As a final comment on what is becoming a useful new analytical procedure I should point out that it need not be used only for breath analysis; it can equally be used for the continuous monitoring of aroma molecules in any gas sample. We have used the technique to monitor in real time the release or generation of flavour molecules during cooking. An unexpected result from this work was the realization that during the eating of a dry cookie the release of flavour from it is limited by the quantity of the saliva in the mouth and not simply by the amount of flavour in the cookie. Presumably at a limiting concentration there is not enough saliva in the mouth to break down the food matrix and release the flavour. Under these conditions most of the flavour is swallowed without even being functional. This explains why the dunking of cookies, particularly those with a hard glassy structure e.g. ginger biscuits does more than simply soften it. AFFIRM® offers a quick and simple way of resolving issues, which can be quite difficult to understand from a sensory analysis alone.

The measurement of the actual kinetics of flavour release and relating these to the physical properties of food systems is an area of flavour research which shows more and more surprises. It also fortunately brings us closer to an understanding of why Chefs de cuisine construct and present their creations in the way that they do. Perhaps a more accurate title for the work we are now doing would be Analytical Gastronomy.

Chapter 13

Taste Release and Its Effect on Overall Flavor Perception

Tracey A. Hollowood, Jim M. Davidson, Lucy DeGroot, Rob S. T. Linforth, and Andrew J. Taylor

Samworth Flavor Laboratory, Division of Food Sciences, University of Nottingham, Sutton Bonington Campus, Loughborough LE12 5RD, United Kingdom

Methods for monitoring the concentrations of both volatile and non-volatile flavor compounds, close to the respective flavor receptors in the nose and on the tongue during eating, have been developed. Using these techniques, a "time release" profile can be built for each of the flavor compounds in a food over the time course of eating. These release profiles can then be compared with perceived flavor, measured using conventional sensory analyses, to study the contribution of the volatile and non-volatile flavor signals to the overall perceived flavor. The methods were applied to model gel-based and viscous foods as well as a chewing gum system. From the systems tested, there was evidence that the intensity of perceived volatile flavor was not solely due to the concentration of volatile in-nose. Various explanations for this discrepancy are given along with preliminary experimental data that suggest that delivery of the non-volatile flavor component also affects the perceived intensity of volatile flavor compounds.

The study of the mechanisms and processes involved in flavor perception takes many forms, as shown by the wide range of papers in this book. For the food industry, the ultimate question is how to formulate the various components of flavor within a food, so that the consumer experiences a high quality perception when the food is consumed. Some of the physicochemical basis of flavor delivery is understood and methods like encapsulation are available to deliver flavor components in a controlled way. Flavorists use their experience to put together the appropriate blend of flavors for a particular product. However, we still lack a basic set of scientific rules to predict the exact perceived flavor of any particular flavor formulation added to a food. This paper examines existing knowledge on the link between the chemical stimulus and perceived flavor, points out areas which need further development, then goes on to show how measurement of the flavor signal *in vivo,* coupled with sensory analysis, can be used to study the phenomenon.

The release of flavor compounds from food, and their delivery to the receptors in the mouth and nose, is now acknowledged as one of the key factors determining the perceived flavor quality of many foods. With the advent of methods for following volatile release *in vivo* (*1-13*) it has become possible to monitor the concentration of several (or more) volatiles from the expired air of people eating food. The rationale for measuring the signal close to the receptors was that it was a more appropriate measure to study the correlation between the flavor signal sensed by the flavor receptors and the actual flavor perceived. The other main methods of flavor analysis provide data, either on the flavor composition of the intact food, or about the volatiles in the air above the food (the headspace). Neither of these techniques take into account entirely the changes that occur as volatiles move from food to the olfactory receptor (e.g. dilution, absorption, mastication).

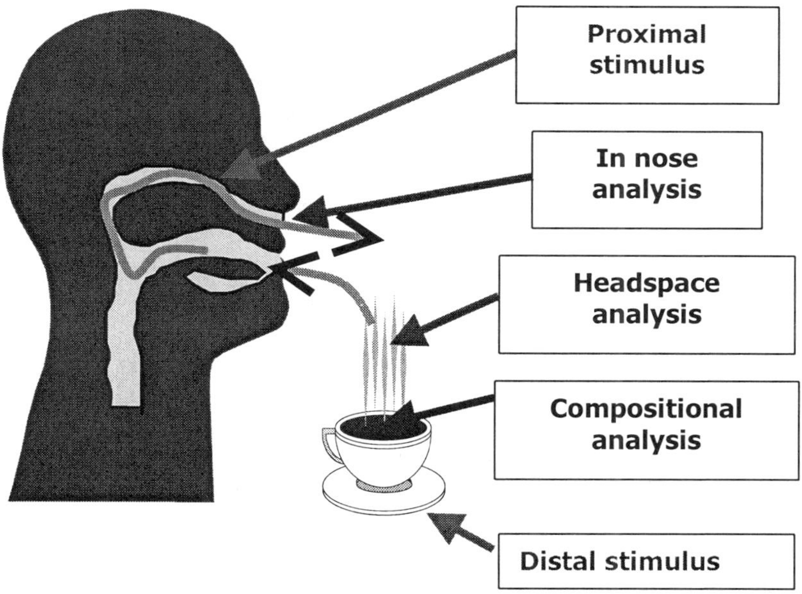

Figure 1. Relationship between analyses for volatile flavors and the distal and proximal perceptual stimuli

Measuring stimuli in different locations is a well-established principle in sensation and perception and can be illustrated by analogy with the visual perceptual process. We are aware that the signal that reaches the receptors in the eye (the proximal stimulus) is not always fully representative of the whole object which is being perceived (the distal stimulus). For instance, when we look at a person, we only receive signals from those parts of the person who is facing us. We receive no signals from the person's back (*14*). Applying this principle to odor requires some modifications as, with the visual example given above, the stimulus is either seen or not seen. With odor compounds, this simple on/off signal has to be replaced by the concept that signal strength can vary between full and zero transmission due to the physicochemical factors that differentiate the distal and proximal odor stimuli (*15, 16*). Figure 1 shows how the various methods of volatile flavor analysis relate to the distal and proximal signals.

Each analysis provides different information and, although in-nose sampling approaches the true proximal signal, it is still several levels away. For instance, collecting the air at the nostril, after it has passed over the nasal mucosa and the olfactory receptors, means the concentration measured is certainly lower than that arriving at the start of the olfactory receptor area. Secondly, the role of volatile transport through the mucus to the receptors (via odor binding protein etc.) is not considered. It is not clear what effects these processes exert on the volatile profile.

Mathematical relationships between the flavor composition of tea and its perceived flavor qualities have been published (*17*) but this approach does not seem popular and experience in our laboratory is that the approach does not produce a reliable predictive model (unpublished data). Headspace analyses have been compared with perceived flavor intensity and, for foods where much of the volatile compounds are delivered orthonasally, this may be an appropriate comparison. However, most foods deliver a significant proportion of the volatile flavor signal through the retronasal route and this can best be followed using *in vivo* monitoring, as it is difficult to mimic effectively the mastication processes in model mouth systems (*3, 18*).

Models linking volatile stimuli and perception of odor

As discussed above, the data obtained from the in-nose monitoring system should be more representative of the signals arriving at the olfactory receptors than headspace analysis and might assist in elucidating the link between stimulus and perception. The relationship between a stimulus and the perception it evokes is given by a power law function, the best known form of which is Stevens Law (*19*) $P = kS^n$ where P is the perceptual intensity, k is a constant, S is the stimulus intensity and n is the exponent. The equation is often written as $\text{Log } P = \text{log}k + n\text{Log}S$ which delivers a linear plot if LogP is plotted against LogS, with the slope equal to the exponent (n) and the intercept equal to k. Exponents for flavor compounds lie in the range 0.2 to around 1.0 (*20*). The limitations of Stevens Law are well-documented (*21*) and it is recognized that it does not adequately describe perception over the whole range of stimuli and that the values of the exponents obtained can depend on the procedures used to measure them. A review of models for the human olfactory-stimulus function has been published (*20*) and the relative merits of each model explained. However, all these models are designed to consider single compounds only; there is no term to describe any interactions between volatiles either at the receptor level or later during signal processing. Since it is known that different volatiles bind to the same receptor (albeit with different affinities) then a multi-receptor model (*22*) seems better suited to the task (*23*) but the former paper seems to have attracted few citations since publication. The possibility that two or more volatiles can produce an inhibitory or enhancing effect, through some form of interaction, has been proposed (*24-26*). Although there is a theoretical basis for the interactions (*24*), there are few published examples of volatile interactions.

Relating volatile flavor perception to volatile release *in vivo*

For the reasons described above, experimental studies in our laboratory have used either a single flavor compound or a flavor mixture with a distinctive character impact compound. This makes it easy for panelists to identify and quantify the flavor and allows us to study Stimulus-Perception relationships. Most models relating stimulus to perception apply a known, constant stimulus concentration (in mouth or in headspace) and measure the sensory intensity as an overall, discrete value. When monitoring stimulus concentration in-nose, the signal is changing with time and various methods of extracting data for comparison with sensory evaluations have been proposed and tested. The simplest method takes the maximum stimulus intensity (SImax) but investigations using gelatin-sucrose gel systems showed that, although there was a correlation between SImax and overall perceptual intensity (PI), the rate of change of volatile concentration correlated better with PI (*27, 28*).

Since the data obtained in nose has a temporal dimension, we have dubbed the traces obtained, Time Release (TR) curves and attempted to correlate TR behavior with perceived aroma intensity over time using sensory Time Intensity (TI) analysis (*29*) which can be obtained simultaneously with the TR data. Using Stevens' Law, the exponents for two compounds (benzaldehyde and isoamyl acetate) were obtained from pairs of TI and TR traces. The values obtained varied widely and no clear results were obtained (*29*). However, the power law, as expressed by Stevens, relates to single, short-lived events whereas the flavor stimuli persist over periods of 0.5 min to 20 min (chewing gum) depending on the food type. This temporal dimension to flavor perception must therefore include the effect of adaptation and, further attempts to analyze the data, used cumulative plots to smooth the data combined with damping theory to account for possible adaptation (*29*). Roozen's group has also studied this effect (*1*) and found that TI analysis of perceived chocolate flavor peaked around 30 sec whereas the oral concentration of 2-methylbutanal (a marker for chocolate aroma) increased up to 300 s. Overbosch (*30*) has modified Stevens Law to take adaptation into account in situations where the stimulus is changed on a step wise basis. In order to apply this technique to *in vivo* TR and TI data requires consistent, high quality data otherwise the mathematical analysis will be incapable of determining the adaptation constants proposed.

Interactions of volatile and non-volatile compounds

The previous section identified that there were two major problems with existing stimulus-perception models for odors; first the lack of a proven model to take the temporal (adaptation effect) into account and second, little attention has been paid to interactions between volatiles and the possible effect on odor quality. Another omission is the lack of any term to describe interaction between odors and taste molecules and/or signals. When foods are eaten, odors are normally sensed in conjunction with taste signals, whereas orthonasal signals prior to eating are exclusively odor signals. We know from experience, and from published data, that the presence of non-volatile flavor compounds can significantly affect our perception of aroma compounds (*31*) and *vice versa* (*32*). Various mechanisms can be proposed to explain how interactions between volatiles and non-volatiles could change the perception of either class. With regard to volatiles, it is known that solutes like sugar can change the headspace concentration of the volatile above a solution of the volatile compound (*33-37*). It is therefore possible that the presence of sugars in the saliva phase in mouth during eating could change the amounts of volatile released into the gas phase although the effect varies greatly from one compound to another depending on their physicochemical properties (*37*). The literature data suggest that sucrose concentrations above 10% are necessary to cause a significant effect on headspace concentration. It is interesting to consider what change in headspace concentration might produce a sensory difference. The relevant parameter is the Weber ratio which is widely quoted at 30% (a 30% increase in volatile concentration is needed to produce a sensory difference) although Goldstein (*14*) points out that this value is probably an overestimate due to poor methodology as demonstrated by Cain (*38*) who obtained values of around 10% with a reliable method of odorant presentation.

Aim of current work

Having developed methods with which the proximal stimuli can be monitored (although imperfectly) one of our current areas of investigation is to study how flavor delivery affects flavor perception. The rate of flavor delivery may be important in some foods (*27, 28*) while the delivery of both non-volatiles and volatiles may be the key to flavor quality in other products. Our aim is to obtain objective data from well-defined systems. Some preliminary information is reported here.

Experimental

Sampling of flavor compounds in vivo and their analysis

Methods for monitoring the major non-volatiles (sugars, acids, salts) which are present in-mouth at around 1 to 20g/100g have been described previously (*39-42*).

For minor non-volatiles (e.g non-sugar sweeteners, flavor enhancers) limits of sensitivity were determined using a 0.01 g/100 mL stock solution of the compounds containing 0.1% formic acid. Saccharin and cyclamate were run in negative electrospray mode on a Finnigan LCQ Mat, equipped with a nano-spray probe at 3kV while aspartame, monosodium glutamate and quinine were run in positive mode. The mobile phase was 50:50 acetonitrile water + 0.1% formic acid at 60 μL/min. A serial dilution of stock solution was run to determine the level of detection (signal:noise ratio=10). The Atmospheric Pressure Ionization Mass Spectrometry (API-MS) technique and its application to sampling *in vivo* has been reviewed (*12*).

Perception of mint flavor in presence/absence of sucrose

Subjects were screened to ensure they could reliably recognize and distinguish minty and sweet flavors in solution. They were trained to assess and record the minty flavor of a solution (using suitable reference solutions) that was fed into their mouth at 10 mL/min. This flow consisted of equal parts of a menthone solution (50mg/kg) and a make up solution that was either sucrose (5g/100g) or water. The flows were switched manually so that panelists were unaware of the composition of solutions flowing into their mouths. Ten panelists assessed the solutions in triplicate. Data for all panelists were averaged at 30 sec intervals and replotted to produce a mean response.

Flavor release and perception in viscous solutions

A basic strawberry flavor with no furaneol (Firmenich sa, Geneva, Switzerland) was added (300mg/kg) to a sucrose solution (2g/100mL) containing hydroxymethylpropyl cellulose (HPMC; E4M, Dow Chemical Co., Germany) at concentrations between 0 and 2g/100mL. The concentration at which overlap of the HPMC coils occurred (C*; (*43*)) was determined as 0.5g/100g. The static equilibrium headspace of each solution was measured using API-MS (*12*) to determine whether binding of volatiles to HPMC occurred. Portions of the solution (5 to 30mL) were then consumed by a trained sensory panel (12 people) who breathed in, introduced the portion of solution to their mouth and then swallowed it. API-MS was used to monitor ethyl butyrate in expired air from the nose (duplicate values from each panelist) and the concentration in-nose for the first exhalation after swallowing was determined after calibration of the system with standards of ethyl butyrate (*12*). Perception of sweetness and strawberry flavor intensity for these samples was carried out using magnitude estimation methods on a separate occasion using the same 12 panelists. Suitable reference solutions were provided against which all subsequent samples were scored. Each sample was assessed three times.

Results and Discussion

Previously, we have studied flavor release and flavor perception in chewing gum systems (*40, 41*). Chewing gum is ideal for these studies as it contains a simple flavor at high concentrations, forms a single bolus in mouth (allowing easy sampling of saliva) and remains in-mouth for long periods (5 to 30 min). Using in-mouth and in-nose sampling methods, coupled to Time Intensity analysis of the mint flavor (*40*), we showed that mint flavor perception was short-lived despite the fact that mint volatiles levels were high (after initial release, they remained at a high plateau level for 10 to 20 min). For both tablet and stick gums, the perception of flavor (as expressed by the TI curves) seemed to follow the changes of sugar concentration in-mouth (*41*). There are several potential explanations for this observation. One is that we are simply observing adaptation and it is just coincidence that the two traces align. Closer inspection of the traces for the tablet and stick gums, however, shows that there are differences in the TI curves, notably in the time to reach maximum sensory intensity. If adaptation were the sole explanation, both TI curves should be the same. For both samples, the sugar curves are also different and follow the same trends as the TI curves. It is therefore possible that there is some interaction occurring between the sugar sensory signal and the volatiles. There is anecdotal evidence of this from those post-war children to whom chewing gum was a scarce luxury and who revived the flavor by adding sugar to their well-chewed gum. The other potential factors include confusion by the sensory panel of mint and sugar signals and the halo dumping effect which occurs when panelists are given insufficient parameters to describe perceptual effects and ascribe the noted change incorrectly. These effects have been examined by re-testing the same sensory panel used initially. They were able to differentiate mint and sugar flavors in a series of experiments and gave the same sensory results even when offered more categories in which to assess flavor changes.

To establish the interactive hypothesis unequivocally, it is necessary to carry out other experiments and ensure the same results are obtained, irrespective of sweetener type and rate of release. Obtaining further rate of release differences with sucrose is difficult as we have already tried the two extremes, sucrose coating in the tablet (dragee) gum and incorporation of sugar in the stick gum. Using non-sugar sweeteners is an alternative but they are included at much lower levels than sugar sweeteners and so it has been necessary to develop a suitable assay procedure. Nanospray-MS was evaluated as it offers low flow rates, small sample volume and high sensitivity.

Table 1 shows the taste thresholds of some non-sugar sweeteners and related taste compounds. It is clear from the table that nanospray-MS operates at levels below the taste thresholds as shown by the column labeled “Factor”. Experiments are now underway to study release from chewing gum samples containing non-sugar sweeteners to study the relationship between mint perception, mint volatile release and sweetener release.

An alternative strategy is to present the various combinations of sucrose and mint flavor to panelists through a simple aqueous solution, rather than through the chewing gum matrix. The idea is to measure their sensory response to mint flavor when presented in mixtures with sucrose absent and present. Some preliminary data from such experiments has been obtained and a typical trace is shown in Figure 2.

Table 1. Sensory and analytical thresholds (nanospray-MS) for some taste compounds found at low levels in foods.

Compound	*Sensory threshold (mg/100mL)*	*Analytical Threshold (mg/100mL)*	*Factor Sensory/analytical threshold*
Aspartame	1.3	0.01	130
Cyclamate	53	1.0	53
Saccharin	1.2	0.01	120
Mono sodium glutamate	14	1.0	14
Quinine	0.07	0.001	72

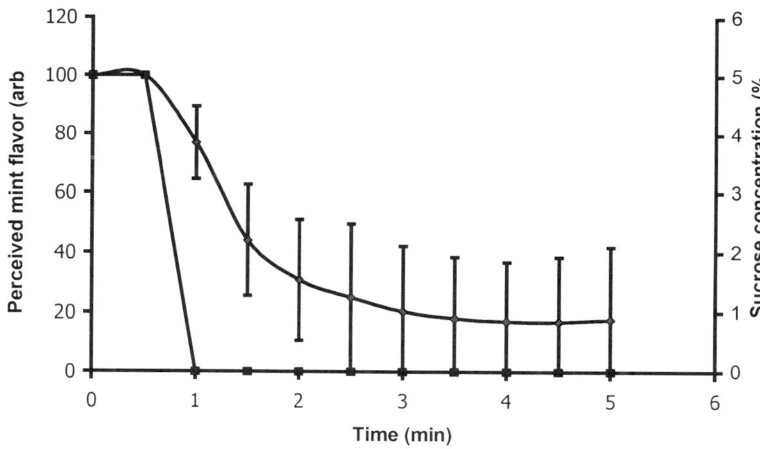

Figure 2. Mean Time Intensity curve from 10 panelists given mint solution plus 5% sucrose solution for 0.5 min then water thereafter

To study further the potential interaction of sweeteners on volatile perception, another approach is to scan the published literature for food systems where changes in perception are noted and re-investigate them using *in vivo* analyses of volatile and non-volatile compounds, combined with sensory analysis. We have previously used gelatin-sucrose gel systems as there are reports of sensory changes when the gel composition is altered (e.g. *44, 45*). Similar effects were reported for viscous solutions (*43, 45*) and Morris (*43*) put forward a hypothesis which linked flavor perception to the state of hydrocolloid interaction as a function of concentration. Figure 3 plots Morris' data for sensory perception against hydrocolloid concentration on a log-log scale which produces a clear inflection in perceived viscosity at the C* value for this system (C* is the concentration at which coil overlap and entanglement begins (*46*)). While the inflection coincides exactly with C* for perceived thickness of the solution, the fit is not so exact for perceived sweetness and flavor.

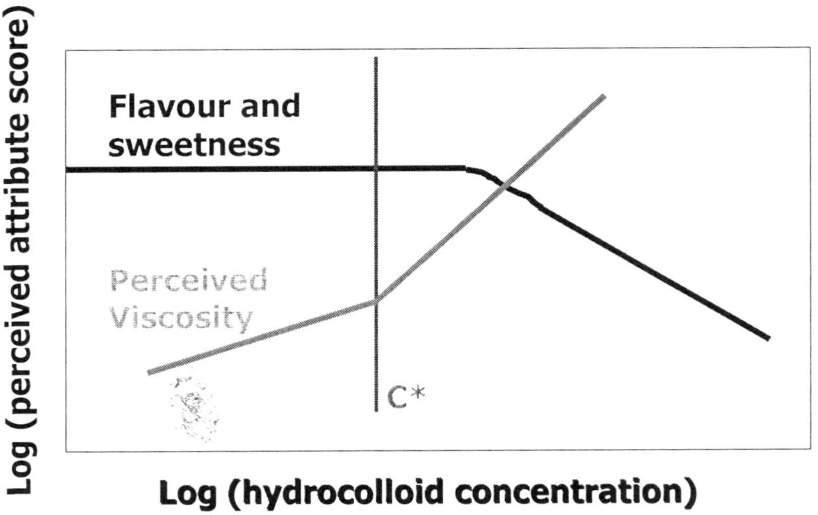

Figure 3. Relationship between hydrocolloid concentration and sensory properties of solutions containing hydrocolloid, sugar and flavor (redrawn after Morris (43)).

In our work, solutions of hydroxymethylcellulose (HPMC) were used. Initially, the static equilibrium headspace of solutions containing no HPMC and HPMC in the range 0.06 to 2% HPMC were measured and no significant differences were found between the headspace volatile concentrations, indicating that no significant binding of volatiles to the hydrocolloid was occurring. C* for HPMC was determined as 0.5%. Portions of solutions were presented to panelists who assessed them following the protocol described above. There was some concern that small portions of sample might become diluted with saliva, lowering the hydrocolloid concentration below C* in mouth. However, experiments with samples ranging from 5 to 30mL gave consistent results and a sample size of 10mL was used for convenience thereafter. Having removed potential problems that might invalidate the experiment, in-nose concentrations of ethyl butyrate were monitored as this volatile was representative of the other flavor compounds present in the strawberry flavor. Sensory analysis was obtained using conventional magnitude estimation and the data plotted in the Morris format although a semi log plot was used as the inflection could be more easily seen.

Perceived flavor (Figure 4) agreed with the Morris data, showing a clear inflection around C*. These sensory data were free of noise as reference solutions were used to calibrate the scale and the panelists performed well. The in-nose concentrations showed more "noise" and there were substantial variations from panelist to panelist. However, there was no clear inflection for the in-nose volatile concentrations and there was no significant difference between any of the in-nose concentrations across the range of HPMC solutions (0.06 to 2%).

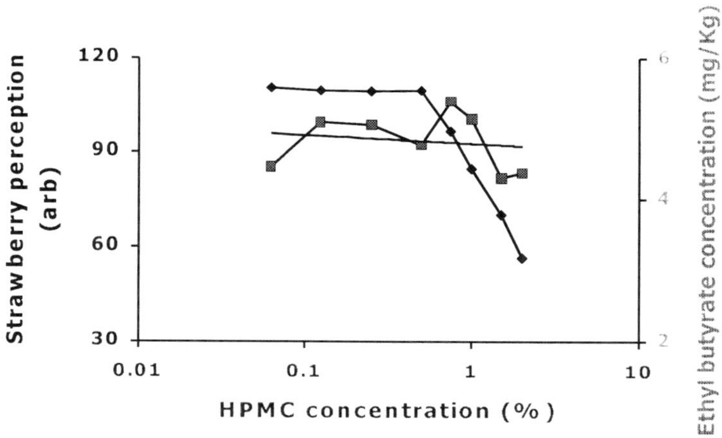

Figure 4. Plot of sensory perception of strawberry flavor (diamonds; left hand axis) against maximum volatile concentration in-nose (as represented by ethyl butyrate, squares; right hand axis). Values are the mean of 3 replicates (perception) or 2 replicates (in-nose) from 12 panelists

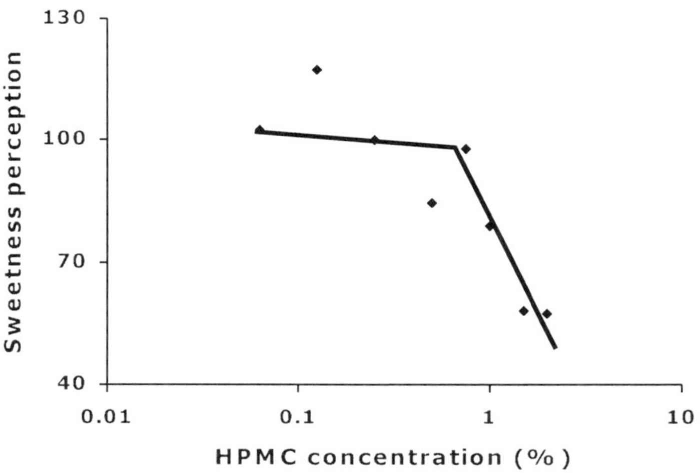

Figure 5. Effect of HPMC concentration on sweetness perception as obtained by magnitude estimation. Values are the mean of three replicates from 12 panelists.

When sweetness perception was plotted against HPMC concentration, an inflection was again observed (Figure 5) although the data showed more variation than the flavor perception. This could be due to the relatively low concentration of sucrose used (2%). Experiments with 5 and 8% sucrose are underway.

It is attractive to postulate that volatile release is largely unaffected by viscosity and that sugar transport to the receptors is somehow hindered, so that the sugar signal in viscous solution is slowed, with the result that the intensity of volatile perception is decreased. However, previous work on volatile release from viscous solutions showed that some compounds were affected by viscosity, others were not (*45*) and it could be that ethyl butyrate is one of those unaffected. A wider range of volatile compounds needs to be tested. It seems unlikely that diffusion will be hindered in viscous solutions like these, and so, it is mass transport which is thought to be the factor affected by viscosity. Unfortunately, our current methods for measuring sugar concentrations in-mouth only give values for the saliva phase which must be at the same concentration as the viscous solutions. Ideally, we need to develop a method that mimics the transport of sugar from the saliva to the taste bud but there is little information about the mechanisms by which transport is effected. It is not clear what form a model system should take (e.g. semipermeable membrane, molecular sieve). Consideration should also be given to the state of the sugar in solution. Birch has shown that the interaction of water with sugar molecules (measured as apparent specific volume) affects the intensity of sweetness perception (*47*) and there may be some effect of viscous solutions on apparent specific volume.

Conclusions

The data obtained so far cannot be interpreted as solid evidence that there is an interaction between sugar and volatiles which affects volatile perception. However, the every day evidence (that addition of non-volatiles like sugar and salt to food enhances overall flavor) is so powerful that further work to elucidate the mechanisms behind this phenomenon will no doubt continue.

Acknowledgements

The authors acknowledge financial help from Firmenich SA (Geneva), the UK Ministry of Agriculture, Fisheries and Food and the UK Biotechnology and Biological Science Research Council.

References

1. Roozen, J.P.; Legger-Huysman, A. In *Aroma: Perception, formation, evaluation*; Rothe, M.; Kruse, H.-P., Eds.; Eigenverlag Universitat Potsdam: Potsdam, 1994; pp 627-632.
2. vanRuth, S.M.; Roozen, J.P.; Cozijnsen, J.L. *Food Chem.* **1995,** *53,* 15-22.
3. vanRuth, S.M.; Roozen, J.P.; Cozijnsen, J.L. *Chem. Senses* **1995,** *20,* 146.
4. DeKok, P.M.T.; Smorenburg, H.E. In *Flavor Chemistry: Thirty years of progress*; Teranishi, R.; Wick, E.L.; Hornstein, I., Eds.; Kluwer Academic: New York, 1999; pp 397-408.
5. Delahunty, C.M.; Piggott, J.R.; Conner, J.M.; Paterson, A. *J. Sci. Food Agric.* **1996,** *71,* 273-281.
6. Lindinger, W.; Hansel, A. *Plasma Sources Sci. Tech.* **1997,** *6,* 111-117.
7. Soeting, W.J.; Heidema, J. *Chem. Senses* **1988,** *13,* 607-617.
8. Spanel, P.; Smith, D. *Rapid Comm. Mass Spec.* **1999,** *13,* 585-596.
9. Ingham, K.E.; Linforth, R.S.T.; Taylor, A.J. *Flav. Frag. J.* **1995,** *10,* 15-24.
10. Taylor, A.J.; Linforth, R.S.T. In *Trends in Flavour Research*; Maarse, H.; van der Heij, D.G., Eds.; Elsevier: Amsterdam, 1994; pp 3-14.
11. Taylor, A.J. *Crit. Rev. Food Sci. Nutr.* **1996,** *36,* 765-784.
12. Taylor, A.J.; Linforth, R.S.T.; Harvey, B.A.; Blake, A. *Food Chem.* **2000,** submitted.
13. Taylor, A.J.; Linforth, R.S.T. In *Flavor release: Linking experiments, theory and reality*; Roberts, D.D.; Taylor, A.J., Eds.; American Chemical Society: Washington D.C., 2000; in press.
14. Goldstein, E.B. *Sensation and Perception*; Brooks/Cole: Pacific Grove, 1999.
15. Taylor, A.J. In *Current Topics in Flavours and Fragrances: Towards a new millenium of discovery*; Swift, K.A.D., Ed.; Kluwer Academic: Dordrecht, 1999; pp 123-138.
16. Taylor, A.J. *Int. J. Food Sci. Tech.* **1998,** *33,* 53-62.
17. Togari, N.; Kobayashi, A.; Aishima, T. *Food Res. Int.* **1995,** *28,* 485-493.
18. Roberts, D.D.; Acree, T.E. *J. Agric. Food Chem.* **1995,** *43,* 2179-2186.
19. Stevens, S.S. *Percept. Psychophys.* **1969,** *6,* 302-308.
20. Chastrette, M.; ThomasDanguin, T.; Rallet, E. *Chem. Senses* **1998,** *23,* 181-196.
21. Hoppe, K. In *Flavor perception*; Kruse, H.-P.; Rothe, M., Eds.; Eigenverlag Universitaet Potsdam: Potsdam, 1997; pp 27-34.
22. Ennis, D.M. *Food Chem.* **1996,** *56,* 329-335.
23. Ennis, D.M. *Food Technol.* **1998,** *52,* 78-89.
24. Fritjers, J.E.R. In *Synergy*; Birch, G.G.; Campbell-Platt, G., Eds.; Intercept Limited: Andover, 1994; pp 1-26.
25. Lawless, H.T. *Food Qual. Pref.* **1999,** *10,* 325-332.
26. Lawless, H.T. In *Tasting and smelling*; Beauchamp, G.K.; Bartoshuk, L., Eds.; Academic Press: San Diego, 1997; pp 142-150.

27. Baek, I.; Linforth, R.S.T.; Blake, A.; Taylor, A.J. *Chem. Senses* **1999,** *24,* 155-160.
28. Linforth, R.S.T.; Baek, I.; Taylor, A.J. *Food Chem.* **1999,** *65,* 77-83.
29. Taylor, A.J.; Linforth, R.S.T. In *Flavours and Fragrances*; Swift, K.A.D., Ed.; Royal Society Chemistry: Cambridge, 1997; pp 171-182.
30. Overbosch, P. *Chem. Senses* **1986,** *11,* 315-329.
31. Noble, A.C. *Trends Food Sci. Tech.* **1996,** *7,* 439-443.
32. Stevenson, R.J.; Prescott, J.; Boakes, R.A. *Chem. Senses* **1999,** *24,* 627-635.
33. Voilley, A.; Simatos, D.; Loncin, M. *Lebensmitt. Wiss. Technol.* **1977,** *10,* 45-49.
34. Nahon, D.F.; Roozen, J.P.; DeGraaf, C. *Food Chem.* **1996,** *56,* 283-289.
35. Nahon, D.F.; Koren, P.; Roozen, J.P.; Posthumus, M.A. *J. Agric. Food Chem.* **1998,** *46,* 4963-4968.
36. Nahon, D.F.; Roozen, J.P.; deGraaf, C. *J. Agric. Food Chem.* **1998,** *46,* 3426-3430.
37. Friel, E.N.; Linforth, R.S.T.; Taylor, A.J. *Food Chem.* **2000,** in press.
38. Cain, W.S. *Science* **1977,** *195,* 796-798.
39. Davidson, J.M.; Linforth, R.S.T.; Taylor, A.J. *J. Agric. Food Chem.* **1998,** *46,* 5210-5214.
40. Davidson, J.M.; Hollowood, T.A.; Linforth, R.S.T.; Taylor, A.J. *J. Agric. Food Chem.* **1999,** *47,* 4336-4340.
41. Davidson, J.M.; Linforth, R.S.T.; Hollowood, T.A.; Taylor, A.J. In *Flavor release:Linking experiments, theory and reality*; Roberts, D.D.; Taylor, A.J., Eds.; American Chemical Society: Washington D.C, 2000; in press.
42. Jack, F.R.; Piggott, J.R.; Paterson, A. *J. Food Sci.* **1993,** *58,* 1313-1317.
43. Morris, E.R. In *Food hydrocolloids: Structure, properties and functions*; Nishinari, K.; Doi, E., Eds.; Plenum Prss: New York, 1994; pp 201-208.
44. Guinard, J.X.; Marty, C. *J. Food Sci.* **1995,** *60,* 727-730.
45. Bakker, J.; Brown, W.E.; Hills, B.P.; Boudaud, N.; Wilson, C.; Harrison, M. In *Flavour science: Recent developments*; Taylor, A.J.; Mottram, D.S., Eds.; Royal Society of Chemistry: Cambridge, 1996; pp 369-374.
46. Morris, E.R.; Cutler, A.N.; Ross-Murphy, S.B.; Rees, D.A.; Price, J. *Carbohydr. Polym.* **1981,** *1,* 5-21.
47. Birch, G.G.; Parke, S.; Siertsema, R.; Westwell, J.M. *Pure Appl. Chem.* **1997,** *69,* 685-692.

Taste, Smell, and Trigeminal Interactions

Chapter 14

Chemoreception of Fat

Timothy A. Gilbertson[1,2] and Insook Kim[2]

[1]**Department of Biology, Utah State University, 5305 Old Main Hill, Logan, UT 84322-5305**
[2]**Pennington Biomedical Research Center, Louisiana State University, Baton Rouge, LA 70808**

Fat has long been assumed to present only textural cues to receptors in the oral cavity. Recent evidence has demonstrated that components in fat may lead directly to taste cell activation. Our recent work has demonstrated that essential (*cis*-polyunsaturated) fatty acids activate taste cells via an inhibition of Shaker Kv1.5-like delayed rectifying K (DRK) channels. This activation may represent a gustatory cue for fat as well as have profound implications for the ability of fats to modify the taste response to other sapid molecules. Moreover, we have shown that the peripheral responsiveness to fatty acids is correlated with dietary fat preference in rats. Recordings from other fat-responsive tissues (pancreas, duodenum) reveals that the chemoreceptive cells in these organs contain Shaker Kv1.5-like channels and respond to fatty acids in a similar fashion. We hypothesize that fatty acid mediated inhibition of DRK channels may be a universal mechanism for the chemoreception of fat.

The peripheral gustatory system plays two major roles. The first is the ability to detect essential nutrients in the environment. That is, those compounds that we need in order to survive typically activate receptive mechanisms (salty, sweet, umami) in the oral cavity related to their detection. The second function of the gustatory system is a more protective role. Aversive tastants (bitter and, possibly, sour) may serve to warn about the potential ingestion of toxic or harmful compounds. This functional duality of the taste system is reflected in a wide range of different compounds that are capable of activating taste receptor cells.

The ability of the taste system to detect nutrients is reflected in a variety of transduction mechanisms involved in the detection of sodium salts and those of other mineral ions ("salty"), of carbohydrates ("sweet") and of amino acids ("umami") (1,2). Indeed, the general focus of investigation into the peripheral and central gustatory mechanisms is centered almost exclusively on the 4 or 5 basic tastes. Given the importance of the taste system in nutrient detection, it is surprising that there has been comparatively little interest in looking at the effects of fat in the gustatory system. Fat is the most energy dense nutrient (9 kcal/gm vs. 4 kcal/gm for carbohydrate or protein) and thus mechanisms for the detection of fat seem a reasonable expectation.

Until very recently, the only sensory cue for dietary fat was assumed to be its texture or "mouth-feel" (3,4). It is generally believed that these textural cues for fat are mediated through an unknown mechanism involving activation of the somatosensory containing a textural component, more recent evidence has implicated a more direct effect of fats in the activation of taste receptor cells. The essential (*cis*-polyunsaturated) fatty acids activate taste cells via direct effects on one (or more) classes of delayed rectifying K^+ (DRK) channels. The recent identification of fatty acid transporters in the apical membranes of taste cells (5,6) is indicative of other potential mechanisms that may allow the identification of fat by the gustatory system. Thus, there appear to be direct gustatory mechanisms that may enable the identification of fat in much the same manner that other more conventional taste stimuli are detected (for review see 7).

Recently, contributions from gustatory mechanisms have been garnering more interest as playing a more active role in the overall control of food intake that was previously thought (8). That is, its location at the most peripheral site in the ingestive pathway coupled with its flexibility to respond to nutritional challenges puts the taste system in a prime position to help contribute to food intake. Concomitant with this idea these taste mechanisms involving responses to fatty acids appear to be correlated with dietary fat preference. That is, taste cells from obesity prone, fat preferring rats are significantly less responsive to fatty acids than are obesity resistant, fat avoiding rats.

The detection of nutrients is not limited to the oral cavity. Indeed, all throughout the digestive system are tissues and organs that are responsive to circulating nutrients and may be considered nutrient receptors in much the same manner as taste buds. In the stomach and intestine, for example, are collections of cells structurally and functionally similar to taste buds (9,10). Furthermore, a number of specific organs (e.g. pancreas, small intestine, liver) respond to circulating levels of fat with well-documented physiological responses related to nutrient partitioning. Though the physiological effects of fat intake have been examined in great detail, comparatively little is known about the mechanisms by which nutrients, especially fats, are detected in these fat-responsive tissues.

Because our data in the gustatory system would suggest that the essential fatty acids may be a salient cue for dietary fat, we looked for the presence of the fatty acid mediated inhibition of delayed rectifying K^+ channels in these fat-responsive post-ingestive organs. Similar to the taste cells, cells from the pancreas, small intestine and liver all contain the fatty acid sensitive Shaker Kv1.5 K^+ channels. A pancreatic beta cell line and one from cholecystokinin (CCK)-secreting duodenal enterocytes both respond to the essential fatty acids in a qualitatively and quantitatively similar fashion. Moreover, these fatty acids are able to cause release of insulin and CCK from the beta cells and enterocytes, respectively, in the concentration range that affects the delayed rectifying K^+ channels in these cells. Thus, the fatty acid mediated activation of cells involving inhibition of delayed rectifying K^+ channels, may represent a common mechanism for the detection of fat pre- and post-ingestively.

This chapter will summarize some of the recent work demonstrating that activation of fat-responsive cells by free fatty acids may play a role in the recognition of fat that subsequently may have implications for the control of fat intake.

Fatty Acids as Primary Signaling Molecules

Triglycerides are the predominant form of fat that is ingested, accounting for more than 98% of the total fat content (11). Triglycerides, which consist of three fatty acids bound to a glycerol backbone, are broken down during digestion into free fatty acids and monoglycerides and diglycerides by a generic class of enzymes, known as lipases (12). Lipases, which occur in various subtypes in the oral cavity and digestive tract, are responsible for the breakdown of the triglycerides prior to their absorption. There is a great deal of literature surrounding the role of the gastric lipases and the breakdown of ingested fat.

Less well known is the role of lipases in the oral cavity. It has been assumed that these enzymes merely aid in the predigestion of fat before it enters the stomach (13). Lingual lipase, which exists in the oral cavity of a variety of species in varying degrees, is released from the von Ebner's gland at the base of the circumvallate papillae (12,13) that contains hundreds of taste buds. Thus, one would predict that this lingual lipase would generate free fatty acids in the vicinity of the taste receptor cells during fat ingestion. The highly lipophilic nature of free fatty acids, and fats in general, may implicate the need for a lipophilic carrier molecule to transport these molecules in the aqueous environment of the oral cavity. The von Ebner's gland proteins (VEG-P), which like lingual lipase are released from the serous von Ebner's glands, are closely related to the lipocalins, a class of lipophilic carrier molecules (14-16). Binding studies of the VEG-P revealed that they were able to bind free fatty acids with high affinity (17). Thus, this ability to release and transport free fatty acids in the area of the taste buds is consistent with the hypothesis that these may be important gustatory cues for fat.

The idea that free fatty acids can themselves act as signaling molecules is not a new one. In addition to their well-documented roles as second messengers intracellularly, a significant number of studies have demonstrated that free fatty acids may be very potent *primary* messengers in a number of systems (18,19). Fatty acids have been shown to directly affect a number of different types of ion channels in a variety of tissues. Voltage-activated K^+ channels, Na^+ channels and Ca^{2+} channels have all been shown to be modulated through a direct interaction with fatty acids (for review see 19). Thus, there seems precedence to the theory that free fatty acids, released in the oral cavity and gastrointestinal tract by the action of lipases, may be important messengers for signaling the presence of dietary fat.

Fatty Acids Activate Taste Cells via Inhibition of Delayed Rectifying K^+ Channels

Given the role of the taste system in nutrient detection, it is surprising how little attention has been paid to looking directly at the effects of fat and its components (triglycerides, fatty acids) on the activity of the chemoreceptive taste cells (20). Because the cellular machinery is present to release and transport fatty acids in the oral cavity (see above) and the reports demonstrating that fatty acids directly affected ion channels in a number of systems, we looked for the effects of free fatty acids on taste cells from rat tongue. To determine if free fatty acids modulated ion channels in taste cells in a manner similar to that found in other systems, we have used whole-cell patch clamp recording to monitor changes in ion channel activity in isolated rat receptor cells.

Taste cells contain a variety of voltage-activated ion channels that act both in the transduction pathway and as targets for taste stimuli (1,2). These channels include the tetrodotoxin-sensitive Na^+ channels, delayed rectifying K^+ channels, Ca^{2+}-activated K^+ channels, inwardly rectifying K^+ channels, ATP-sensitive K^+ channels, Cl^- channels and Ca^{2+} channels. Free fatty acids, like arachidonic acid when applied in the low micromolar range ($EC_{50} \sim 1$ μM), act predominately by inhibiting the delayed rectifying K^+ (DRK) channels in taste cells (Figure 1A). These channels are present in most electrically active cells, including taste cells, and play a role in the repolarization of cells following activity. Moreover, in taste cells specifically, DRK channels are apparently the target for inhibition by a number of taste stimuli, either directly or indirectly. Sweet compounds, bitter taste stimuli and acidic (sour) tastants have all been shown in a variety of species to be transduced, at least in part, via an inhibition of DRK channels (2,7). This inhibition would be predicted to lead to a depolarization of the taste receptor cell. Thus, it appears that fatty acids may act similarly to a variety of other taste stimuli to lead to taste cell depolarization.

Interestingly, the fatty acid mediated activation of taste cells via interactions with DRK channels was limited to the essential, or *cis*-polyunsaturated, fatty acids (Figure 1B). As mentioned above, one of the major roles of the taste system is nutrient detection. In concert with this idea, the peripheral gustatory system is apparently tuned to detect those fatty acids required in the diet for survival (20,21). Triglycerides, the major component in dietary fat (11), were ineffective in altering DRK channel activity. The effect on DRK channels appears unrelated to the behavioral results reported by Ramirez (22) who proposed that the impurities or oxidative by-products in oils, rather than the triglycerides themselves, may be recognized by the taste system. Like our results (21), however, the only effective compounds found were those containing unsaturated fatty acids. Fatty acid methyl esters were also ineffective in modulating the activity of DRK channels (Figure 1B; 21). This implies that the charged carboxyl carbon is important for the effects reported. Moreover, it may also suggest that many of the fat substitutes on the market, which are based upon methyl esters and polyester derivatives of fatty acids (23), would be ineffective in activating this taste mechanism. Finally, conjugated linoleic acid (CLA), which is a mixture of *cis, trans*-linoleic acid and *trans, cis*-linoleic acid (24), was moderately effective, implying that to be effective the polyunsaturated fatty acids requires a minimum of one double bond in the *cis* configuration. The effects of fatty acids were concentration-dependent, reversible and via a direct interaction with the DRK channel (or a closely associated protein) on its extracellular face (*cf.* 21).

To date, there has been cloned at least 9 different subtypes of delayed rectifying K^+ channels in a variety of tissues. The particular subtype of DRK channel that is sensitive to inhibition by fatty acids in taste cells has not been unequivocally identified. However, our recent work using a variety of techniques has implicated a Shaker Kv1.5-like channel as a likely candidate (25). Similarly, the Shaker Kv1.5 channel, which is the major delayed rectifying K^+ channel in the heart (26), is inhibited by *cis*-polyunsaturated fatty acids in a similar manner, though at somewhat higher concentrations (27). Other Shaker family channels, including Kv1.1 and Kv1.2 (28) are inhibited by fatty acids, suggesting that this may be common to a variety of channels in this subfamily. We have identified by polymerase chain reaction, immunocytochemistry and Western blotting the presence of two other DRK channels that appear to be highly expressed in taste buds, the Shab Kv2.1 and the Shaw Kv3.2 (L. Liu, I. Kim, L. Nikonova, T. Gilbertson, unpublished observations). Our data using pharmacological approaches and molecular biological techniques are consistent with the interpretation that the predominant DRK channel in rat taste buds is the Shaker Kv1.5 channel. However, we cannot rule out that other DRK channels may also contribute to the fatty acid

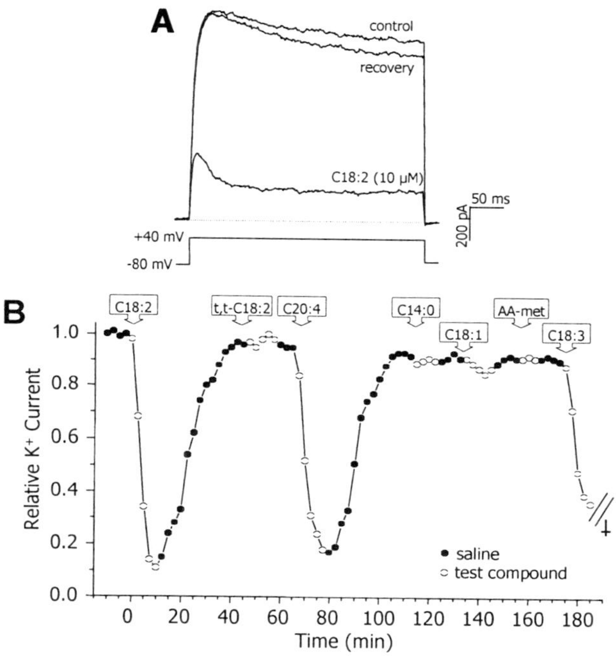

*Figure 1. Inhibition of DRK channels by fatty acids in a vallate taste receptor cell. **A**, DRK currents elicited from a voltage step to +40 mV (holding potential = -80 mV) were reversible inhibited by 10 μM linoleic acid (C18:2). Recovery is shown 15 minutes after washing with saline. **B**, effect of various fatty acids on DRK currents in a taste cell. Only the* cis-*polyunsaturated fatty acids were effective in inhibiting these currents. All compounds were tested at 10 μM. C18:2, linoleic acid; t,t-C18:2, linolenelaidic acid; C20:4, arachidonic acid; C14:0, myristic acid; C18:1, oleic acid; AA-met, arachidonic acid methyl ester; C18:3, linolenic acid. See text for additional details. Reproduced from (20, Copyright 1998) with permission.*

response seen. Since fatty acids may inhibit as much as 90% of the total outward K current, this may imply that two (or more) DRK channels in taste cells may be responsive to fatty acids. This will require experiments designed to directly test this hypothesis.

The fatty acid mediated inhibition of DRK channels may not be the sole mechanism that taste cells use to recognize components in dietary fat. Fatty acids have also been implicated in another pathway in the activation of taste cells. Fushiki and colleagues have identified the presence of a fatty acid transporter (FAT) in vallate taste cells by immunocytochemistry using antibodies against the CD-36 fatty acid binding protein and polymerase chain reaction (5). This transporter, which leads to the uptake of fatty acids into cells, appears apically localized and thus is in close proximity to the regions where the lingual lipase and VEG-P act in the vallate trench. Though the downstream effect of the FAT in taste cells is unclear presently, a preliminary report has suggested that it may lead to increases in intracellular free Ca^{2+} (6) and, presumably, to neurotransmitter release. Therefore, there may be multiple mechanisms that allow the gustatory system to respond to dietary fat in the form of free fatty acids.

The elucidation of specific fatty acid-activated pathways in taste cells may, at least in part, be representative of a transduction mechanism for the detection of dietary fat. Recent data are consistent with the idea that there may be a gustatory cue for dietary fat. Experiments demonstrating that oral fat exposure could lead to postprandial changes in lipid metabolism are supportive of there being a "taste" for fat (29). Thus, there must have been some sensory system activated in these experiments that allowed the recognition of fat in the oral cavity that led to these subsequent changes. We hypothesize that the fatty acid inhibition of DRK channels in taste cells or activation of taste cells via FAT activity would fulfill this role.

The ability of fatty acids to alter the activity of taste receptor cells may have other implications, as well. As reported (21), greater that 95% of taste cells responded to the essential fatty acids. Furthermore, as mentioned above, DRK channels may be the targets of inhibition by a number of other classes of taste stimuli. One might predict, then, that the overlap among these transduction pathways might affect the gustatory response to the complex mixtures of stimuli found commonly in foods. That is, the presence of fat (*cis*-polyunsaturated fatty acids) in food may profoundly influence the gustatory response to other taste stimuli. Numerous psychophysical studies have demonstrated that fats alter the hedonic response to other taste stimuli, particularly sweet tastants (30-32). Therefore, there may be two effects of fatty acid inhibition of DRK channels in taste cells. *One*, it may serve as the transduction mechanism for dietary fat. *Two*, it may affect or modulate the gustatory responsiveness to other taste stimuli. Additional experiments will be required to determine the precise role this transduction pathway plays in taste.

Correlation of Fatty Acid Responsiveness and Dietary Fat Preference

The peripheral gustatory system has long been thought of a passive transducer of chemical signals, one that is poorly responsive to the nutritional needs of an organism (33). That is, there has been relatively little importance given to the taste system in the *control* of ingestive behavior. However, recent evidence has begun to challenge this limited view of the taste system and suggests instead that it may be responsive to a number of extrinsic signals (hormones, peptides, neurotransmitters) and may be a more important contributor to nutrient intake than it was once thought (for review see 8). To determine the role, if any, that these fatty acid responsive pathways in taste cells may play in dietary fat intake, we have compared fatty acid responsiveness using patch clamp recording in two rat strains that differ in their dietary preference for fat.

The two rats strains examined differ in their macronutrient preferences. When placed on a three-choice diet of protein, carbohydrate and fat, Osborne-Mendel (O-M) rats prefer fat, while the S5B/Pl rats prefer carbohydrate and eat comparatively little fat (34). Thus, we may broadly classify these rats as fat-preferring and fat-avoiding, respectively. In addition, when these two strains are placed on high fat diets exclusively, the O-M rats get obese due to hyperphagia while the S5B/Pl is able to compensate for the high fat diet, reducing its intake and remaining lean. One hypothesis to account for this difference is that there may be some signal in the fat that the S5B/Pl rats respond to that allows them to reduce their intake and remain normal weight and that this same signal is reduced or absent in the O-M rat.

Consistent with this hypothesis, DRK channels in taste cells are inhibited to a *greater* degree by *cis*-polyunsaturated fatty acids than are DRK channels from O-M rats (35). There is, it appears, an inverse correlation between peripheral fatty acid responsiveness and the preference for dietary fat. The inhibition constants for the effective fatty acids were identical in the two strains (EC_{50} ~1 μM for all *cis*-polyunsaturated fatty acids). This suggests that there is not an affinity difference but rather a difference in expression of DRK channels in the taste cells of these rat strains (35). The result of this is that one would predict a greater activity per unit concentration in the taste cells of the S5B/Pl rat during fatty acid stimulation than in the O-M rat. Though the contribution of this gustatory pathway to ingestion of fat is not known, it may provide the first insights into the sensory mechanisms for the detection of dietary fat that contributes to or is correlated with dietary fat preference. Clearly, this will be a fruitful area of continued research.

Fatty Acids Directly Activate other Fat Sensitive Cell Types

There are a number of organs that may be broadly classified as fat-responsive in terms of their ability to sense dietary fat and activate systems related to the chemoreception of fat and its eventual utilization and partitioning. To determine if fatty acid mediated activation of cells via their interaction with DRK channels was a common mechanism for the detection of dietary fat, we explored these mechanisms in a variety of cell types that show fat responsiveness. Since our data in rat taste buds suggested that Shaker Kv1.5 channels were likely the major fatty acid sensitive channel, we looked for the presence of this channel in cells from the pancreas and small intestine. Western blotting using an antibody against Shaker Kv1.5 revealed the presence of a Kv1.5-like protein (66-70 kD) in pancreas and duodenum, as well as in cell lines from these two tissues (Figure 2). Both the liver and adipose tissue also

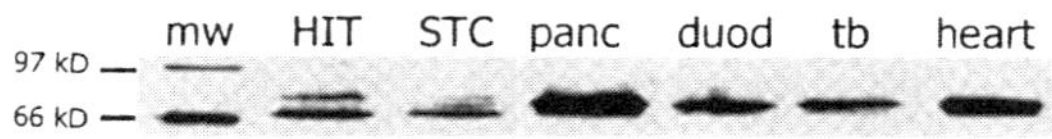

Figure 2. A Kv1.5-like protein is found in various peripheral tissues. Western blotting using an antibody against the Shaker Kv1.5 channel (Alomone Labs, Jerusalem Israel) reveals the presence of a 66~70 kD Kv1.5-like protein in HIT-T15 cells, STC-1 cells, in the pancreas (panc), duodenum (duod), taste buds (tb) and heart. All tissues were from rat. mw, molecular weight markers.

contained this protein (data not shown). RT-PCR revealed the presence of Kv1.5 messenger RNA in these areas (L. Liu, I. Kim, L. Nikonova and T.A. Gilbertson, unpublished observations).

Patch clamp recording on a pancreatic ß cell line (HIT-T15) and a cell line of cholecystokinin (CCK)-secreting duodenal enterocytes (STC-1) reveals that there is apparently a common effect of the essential fatty acids on DRK currents in these cells. Like the taste cells, *cis*-polyunsaturated fatty acids reversibly inhibit DRK currents in the pancreatic ß cell line, HIT-T15 (Figure 3) and in the enterocyte cell line, STC-1 (Figure 4) in a concentration-dependent manner that would lead to activation of these cells. Moreover, consistent with the known effects of fat intake in the pancreas and duodenum, our preliminary data show that the fatty acids that act on the DRK channels cause release on insulin and CCK from the HIT and STC-1 cell lines, respectively (36). Thus, the inhibition of DRK channels by fatty acids may represent a common mechanism that allows the recognition of dietary fat in a number of chemosensory and fat-responsive tissues.

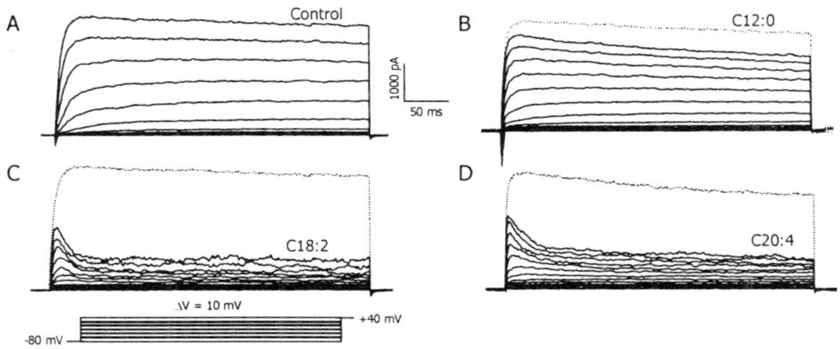

Figure 3. Effect of fatty acids on DRK currents in an insulin-secreting pancreatic β cell line, HIT-T15. ***A****, control DRK currents in saline.* ***B-D****, DRK currents in the presence of linoleic acid (C18:2), lauric acid (C12:0) and arachidonic acid (C20:4), respectively. All fatty acids were tested at 10 μM. Dotted line in B-D represent the DRK current at +40 mV in saline immediately preceding fatty acid treatment.*

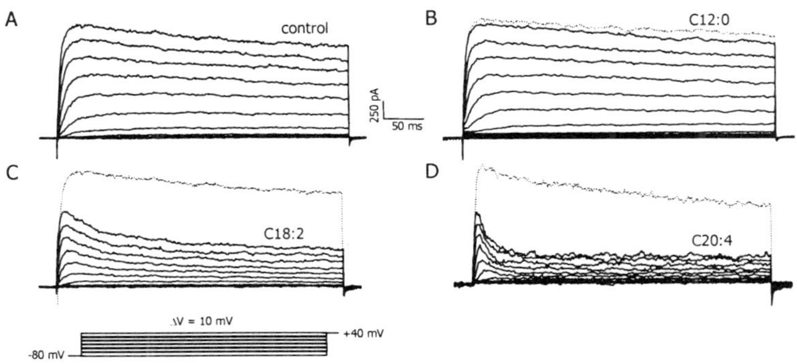

Figure 4. Effect of fatty acids on DRK currents in an CCK-secreting duodenal enterocyte cell line, STC-1. ***A****, control DRK currents in saline.* ***B-D****, DRK currents in the presence of linoleic acid (C18:2), lauric acid (C12:0) and arachidonic acid (C20:4), respectively. All fatty acids were tested at 10 μM. Dotted line in B-D represent the DRK current at +40 mV in saline immediately preceding fatty acid treatment.*

Conclusions

Though we have identified potential gustatory cues for fat, we have yet to determine its role in the ability of the taste system to respond to this important nutrient. Though based upon the cellular effects of fatty acids one can predict the effects downstream, it will require experiments aimed at identifying the effects of fat at the level of the gustatory afferent nerve, taste nuclei in the central nervous system and, ultimately, carefully-conducted behavioral studies to determine the importance of these proposed pathways for the detection of dietary fat. It is nonetheless intriguing that the responsiveness to fatty acids in the taste system is correlated with underlying dietary preferences for this nutrient. Moreover, we found that the fatty acid mediated inhibition of DRK channels is apparently common in fat responsive tissues. Thus, this signaling cascade may be important in understanding mechanisms contributing to the overall control of fat intake. Future studies should help explore this relationship in greater detail and may also shed some light into the contributions of the both the pre- and post-ingestive sensing systems to the control of ingestive behavior.

References

1. Lindemann, B. *Physiol. Rev.* 1996, 76:719-766.
2. Herness, M. S.; Gilbertson, T. A. *Ann. Rev. Physiol.* 1999, 61, 873-900.
3. Mela, D. J.; Marshall, R. J. In *Dietary fats: determinants of preference, selection and consumption.* Mela, D. J., Ed. Elsevier Applied Science: New York, NY, 1991; pp. 43-57.
4. Drewnowski, A. 1997, *Ann. Rev. Nutr.* 17, 237-253.
5. Fukuwatari, T.; Kawada, T.; Tsuruta, M.; Hiraoka, T.; Iwanaga, T.; Sugimoto, E.; Fushiki, T. *FEBS Lett.* 1997, 414, 461-464.
6. Fushiki, T.; Kawada, T.; Fukuwatari, T. *Chem. Senses* 1997, 22, 332.
7. Gilbertson, T. A. *Front. Oral. Biol.* 1998, 9, 1-28.
8. Gilbertson, T. A. *Ann. NY Acad. Sci.* 1998, 855, 860-867.
9. Höfer, D.; Asan, E.; Drenckhahn, D. *News Physiological Sci* 1999, 14, 18-23.
10. Raybould, H. E. *News Physiological Sci.* 1998, 13,275-280.
11. Gunstone, F. D.; Norris, F. A. *Lipids in Foods: Chemistry, Biochemistry and Technology.* Pergamon Press: Oxford, 1983.
12. Lohse, P.; Lohse, P.; Chahrokh-Zadeh, S.; Seidel, D. *J. Lipid Res.* 1997, 38, 880-891

13. Hamosh, M.; Scow, R. O. *J. Clin. Invest.* 1973, 52, 88-95.
14. Kock, K.; Ahlers, C.; Schmale, H. *Eur. J. Biochem.* 1994, 221, 905-916.
15. Kock, K.; Bläker, M.; Schmale, H. *Cell Tiss. Res.* 1992, 267, 313-320.
16. Schmale, H.; Holtgreve-Grez, H.; Christiansen, H. *Nature* 1990, 343, 366-369.
17. Spielman, A. I.; Huque, T.; Whitney, G.; Brand, J. G. In *Sensory Transduction*; Corey, D. P.; Roper, S. D. Eds. Rockefeller University Press: New York, NY, 1992; pp. 307-324.
18. Ordway, R. W.; Singer, J. J.; Walsh, J. V. Jr. *Trends Neurosci* 1991, 14, 96-100.
19. Petrou, S.; Ordway, R. W.; Kirber, M. T.; Dopico, A. M.; Hamilton, J. A; Walsh, J. V. Jr.; Singer, J. J. *Prostaglandins, Leukotrienes and Essential Fatty Acids* 1995, 52, 173-178.
20. Gilbertson, T. A. *Curr. Opin. Neurobiol.* 1998, 8, 447-452.
21. Gilbertson, T. A.; Fontenot, D.T.; Liu, L.; Zhang, H.; Monroe, W.T. *Am. J. Physiol.* 1997, 272, C1203-C1210.
22. Ramirez, I. *Appetite* 1992, 18, 193-206.
23. Akoh, C. C. *Crit. Rev. Food Sci. Nutr.* 1995, 35, 405-430.
24. Belury, M. A. *Nutr. Rev.* 1995, 53, 83-89.
25. Liu, L.; Kim, I.; Hu, S.; Wang, S.; Zhang, H.; Gilbertson, T. A. *Chem. Senses* 1998; 23, 612-613.
26. Takimoto, K.; Levitan, E. S. *Circulation Res.* 1994, 75, 1006-1013.
27. Honoré, E.; Barhanin, J.; Attali, B.; Lesage, F.; Lazdunski, M. *Proc. Natl. Acad. Sci. USA* 1994, 91, 1937-1944.
28. Garratt, J. C.; McEvoy, M. P.; Owen, D.G. *Eur. J. Pharmacol.* 1996, 314, 393-396.
29. Mattes, R. D. *Am. J. Clin. Nutr.* 1996, 63, 911-917.
30. Bacon, A. W.; Miles J. S.; Schiffman S. S. *Physiol. Behav.* 1994, 55, 603-606.
31. Drewnowski, A.; Krahn, D. D.; Demitrack, M. A.; Nairn, K.; Gosnell, B. A. *Physiol. Behav.* 1992, 51, 371-379.
32. Johnson, S. L.; McPhee, L.; Birch, L. L. *Physiol. Behav.* 1991, 50, 1245-1251.
33. Friedman, M. I.; Mattes, R. D. In *Smell and Taste in Health and Disease.* Getchell, T. V., Ed. Raven Press: New York, NY, 1991; pp. 391-404.
34. Okada, S.; York, D. A.; Bray, G. A.; Erlanson-Albertsson, C. *Am. J. Physiol.* 1992, 262, R1111-R1117.
35. Gilbertson, T. A.; Liu, L.; York, D. A.; Bray, G. A. *Ann. NY Acad. Sci.* 1998, 855, 165-168.
36. Kim, I.; Keller, B. L.; Liu, L.; Liou, S.; York, D. A.; Gilbertson, T. A. *Chem. Senses* 1999, 24, 571.

Chapter 15

Astringency and Bitterness of Flavonoid Phenols

A. C. Noble

Department of Viticulture and Enology, University of California, One Shields Avenue, Davis, CA 95616

In beverages and fruits the lingering tactile sensation of astringency, as well as the persistent taste of bitterness, are primarily elicited by flavonoid phenols. Chemically, relative astringency of a compound is defined by its effectiveness in precipitating protein. As the degree of polymerization of tannin increases, bitterness decreases, while astringency increases. The astringent flavanol polymers or condensed tannins have a strong affinity for binding with proline rich proteins, such as those found in saliva. Sensorially astringency is a drying or rough mouthfeel thought to result from decreased oral lubrication, following binding of salivary proteins by tannins. This may explain the increase in intensity of astringency over several sips or wine or tea. Also consistent with this hypothesis, subjects with low salivary flow rates perceive astringency more intensely than high-flow individuals.

affected by the identity of the monomeric units; catechin-(4→8)-catechin, was lower in astringency than catechin-(4→8)-epicatechin. The increase in astringency intensity with molecular size is consistent with the reported increase in effectiveness of tannins to precipitate protein from dimers to higher oligomers (*20*).

No studies have addressed the contribution of anthocyanins to wine flavor. Anthocyanins interact strongly with other phenolic compounds (*21,22*) and these pigments have been reported to increase tannin solubilization (*23*). However, whether these interactions interfere with the binding of tannins with proteins is unknown.

Tannins in red wines are extracted during fermentation from grape seeds and skins. Grape seed tannins are procyanidins (*24*), whereas skin tannins also contain prodelphinidins (*25*) and flavonols, such as the glycoside and glucuronide of quercetin. Very little sensory work has studied flavonols but their levels are so low that they are unlikely to contribute significantly to bitterness or astringency of red wines. Seed tannins typically have lower average degree of polymerization (DP) and larger proportions of galloylated units than skin tannins (*24,25*). Whereas the lower DP of the seed tannins would suggest that they may be less astringent than skin tannin, no sensory studies have addressed the effect of galloylation. The amount of galloylation may increase astringency since it has been shown to increase the interaction of tannin with proteins (*26,27*).

There is considerable controversy in the wine industry as to the role of skin *vs* seed tannins in affecting taste and mouthfeel of wine. To compare their sensory properties, tannin fractions were extracted from ripe Cabernet franc grapes from the Loire Valley and from varying ripeness levels of Cabernet Sauvignon in Napa Valley, CA. The average DP of the seed and skin extracts were, respectively, 6 and 40 for the Cabernet Franc (*28*) and 5.5 and 33 for the Cabernet Sauvignon (29). Despite this large difference in composition, the astringency of the Cabernet franc skin and seed tannins were virtually identical (*28,30*) as shown by the average astringency curves in Fig. 2 (left). Conflicting with these results, as shown in Fig 2 (Right), skin tannin extracted from ripe Cabernet Sauvignon grapes (26.5°B) was significantly more astringent than seed tannin (extracted from less ripe 21.5°B grapes) (*30*). The difference between these two studies may arise from the variation in ripeness of the grapes from which the extracts were isolated, in extraction procedures, in the varieties or in the location of origin.

Effect of sensory methodology

Evaluation of astringent products such as red wine or tea cannot be made in typical side by side comparisons. Intensity of astringency, as well as that of bitterness, builds up when several samples are tasted. Correspondingly, astringency increases upon taking many sips of the same sample. Single sip TI

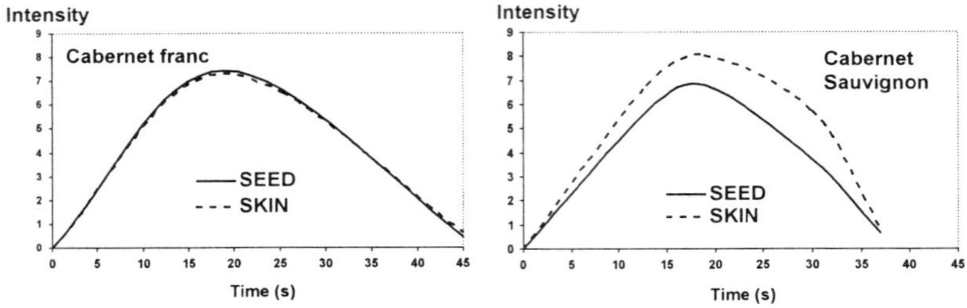

Fig. 2. Average time-intensity curves for astringency of skin (Solid line) and seed (Dashed line) tannin fractions. Left: Cabernet franc (1500 mg/L) (n =12 judges x 2 reps)(28). Right: Cabernet Sauvignon (1000 mg/L) n= 10 judges x 2 reps)(30).

protocols such as that shown in Fig. 2 minimize "carry over" effects, but do not reflect perception during normal consumption of beverages. This phenomenon has been illustrated in studies in which judges repeatedly sip astringent solutions at defined intervals, while continuously rating astringency. In Fig. 3 (Top), maximum intensity of astringency increased upon repeated sipping of red wine at 25 second intervals (*31*). At each sip, astringency increased rapidly, reaching a maximum 6-8 s after the wine was swallowed. Intensity then decreased until the next sip was taken, whereupon another rapid increase occurred. Increasing the time between sips to 30 seconds considerably reduced the enhancement of astringency of the second sip as illustrated in Fig. 3 (Bottom) (*32*). A similar trend was observed when a significant increase in astringency occurred when red wines were sipped at 20 s, whereas the increase in astringency with 40 s intervals was not significant (*33*).

Astringency can continue to build over further sips as shown for two strengths of tea (Fig. 4). Similar to the red wines, maximum intensity for each sip and the minimum value of astringency between sips increased as the number of sips increased. However, after the third sip the increase between sips was not significant suggesting a plateauing effect may have occurred by sip 3 in contrast to red wines where the increase between each of 4 sips was significant (*34*).

Effect of salivary flow rate

Saliva flow rate is increased dramatically by sour or astringent stimuli. Monitoring flow of the parotid salivary gland, while subjects sipped wine and expectorated at 10 seconds, revealed that the increase in flow-rate is almost instantaneous, reaches a maximum approximately 20 seconds later and rapidly decreases until further stimulation occurs (*35*). When subjects were partitioned into groups based on their salivary flow-rates and their data analyzed separately,

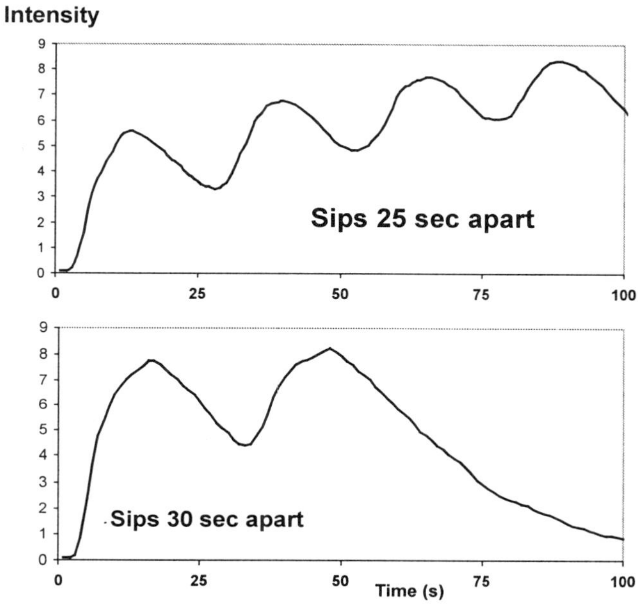

Fig. 3. Average intensity of astringency over time of red wine. Top: Sipping a Merlot wine at 25 s intervals. (n=18 judges x 2 reps) (31); Bottom: sipping another Merlot wine at 30 s intervals (n= 16 judges x 2 reps) (32).

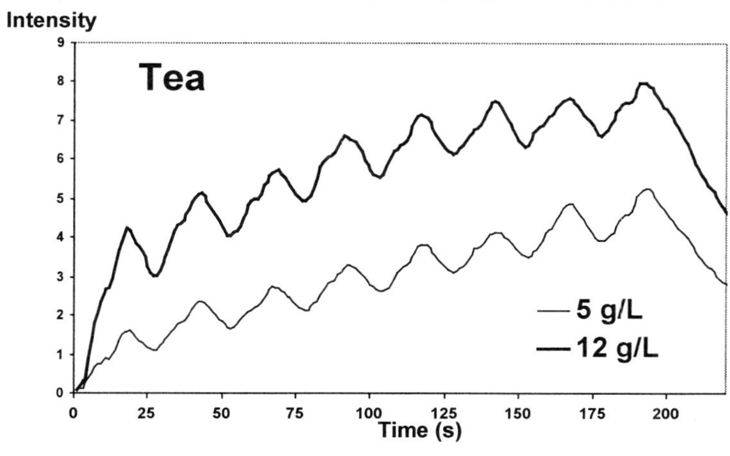

Figure 4. Average intensity of astringency of 2 concentrations of black tea sipped at 25 s intervals (n=16 judges x 3 reps) (34).

low-flow subjects perceived maximum intensity of astringency later, rated it more intensely and for a longer duration than high-flow subjects both in white (*35*) and red wine (*36*). The same results occurred over multiple sips for the astringency ratings of red wines (37) and tea (*34*). As shown for tea in Figure 5, low flow subjects rated astringency higher than the high-flow subjects over 8 successive sips (*34*). Conflicting with these results, no effect of salivary flow on perception of astringency was observed for flavanol monomers, dimers and trimers(*13*), hydroxy benzoic acids (*38*) or soymilk (*13,39*).

When the flow of saliva is stimulated, salivary pH rises and the protein concentration decreases although the total protein content remains fairly constant (*40*). Over twenty percent of saliva proteins are basic, proline-rich proteins (PRPs) (*41*) for which polyphenols have a strong affinity (*42*). The oral sensation of astringency, elicited by polyphenolic compounds, is thought to be linked to precipitation of these salivary proteins. It has been speculated that astringency is the friction perceived when oral lubrication is reduced upon binding of astringent compounds with salivary proteins (*43-46*). If this is true, then the lower astringency ratings by individuals with high salivary flow rates may reflect the greater ability of the higher volume of saliva to restore oral lubrication. Similarly, the reduced buildup of astringency observed when the interval between sips is increased by 5 sec (Fig. 3) may be due to the additional amount of secreted saliva. Correspondingly, the progressive buildup of astringency over successive sips may be the result the inability of saliva to restore lubrication during prolonged ingestion of astringent fluids.

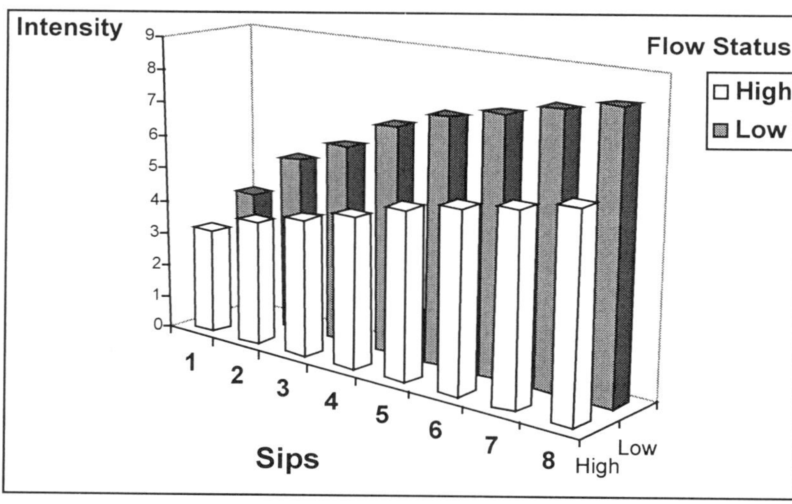

Fig. 5. Average maximum intensity of astringency for successive sips of black tea (averaged over 3 tea concentrations) as a fucntion of salivary flow-rate for 7 subjects with high salivary flow rates (Hi-Flow) and 3 subjects with low flow (34).

Summary

Astringency is evoked by compounds which precipitate or complex proteins, whereas the taste of bitterness is elicted by very diverse compounds. Flavonoid phenols are both bitter and astringent, and the intensity on an equal weight basis varies inversely with increasing molecular size. It is thought that the drying sensation of astringency is the friction resulting from decreased salivary lubrication upon binding of astringents with salivary proteins. However interactions with oral surfaces have not been studied and may play a role in the buildup of astringency upon continued ingestion.

Acknowledgement

This research was supported in part by grants from the American Vineyard Foundation.

Literature Cited

1. Piretti, M. V.; Ghedini M.; Serrazanetti, G. *Annali di Chimica*, **1976**, *66*, 429-437.
2.Czochanska, Z; Foo, L.Y.; Newman, R.H.; Porter, L.J.; Thomas, W.A.; Jones, W.T. *J. Chem Soc. Chem. Comm.* **1979**, 375-377.
3. Balentine, D. A. In *Phenolic Compounds in Food and Their Effects on Health;* Ho, C.;Lee, C.Y.; Huang, M.; Ed.; American Chemical Society, Washington, DC, 1992, Vol. 1. pp 103-117.
4. Belitz, H.-D.;Wieser, H. *Food Rev. Intl.* **1985**, *1,* 271-354
5. Thorngate, J.H. *Am. J. Enol. Vitic.* **1997**, *48*, 271-279.
6. Naim, M.; Seifert, R.; Nurnberg, B.; Grunbaun, L.; Schultz, G. *Biochem. J.* **1994**, *297,* 451-454.
7. Martin, J.H.; Jessell, T.M. In *Principles of Neural Science,* Kandel, E.R.; Schwartz, J.H.; Jessell, T.M. Eds. Appleton & Lange, Norwalk, CT, pp. 341-352.
8. Bate-Smith, E.C.; Swain, T., In *Comparative Biochemistry*; Mason, H.S.; Florkin, A.M..; Ed.; Academic Press, New York, N, 1962; pp 755-809.
9. Adams, D. O.; Harbertson, J. *Am. J. Enol. Vitic.* **1999**, *50*, 247-252.
10. Robichaud, J.L.; Noble, A.C. *J. Sci. Food Agric.* **1990,** *53,* 343-353.
11. Thorngate, J.H.; Noble, A.C. *J. Sci. Food Agric.* **1995**, *67,* 531-535.
12. Kallithraka, S.; Bakker, J.; Clifford, M.N. *J. Sens. Studies,* **1997**, *12,* 25-37.

13. Peleg, H.; Gacon, K.; Noble, A.C. *J. Sci. Food Agric.* **1999**, *79*: 1123-1128.
14. Yokotsuka, K.; Singleton, V.L. *Am. J. Enol. Vitic.* **1987**, *38*, 199-206.
15. McManus, J.P.;, Davis, K.G.; Lilley, T.H.; Haslam, E. *J.Chem. Soc. Chem. Comm.* **1981**, 309-311.
16. Arnold, R.A.;Noble, A.C., *Am. J. Enol. Vitic.* **1978,** *29,* 150-152.
17. Arnold, R.A.; Noble, A.C.; Singleton, V.L. *J. Agric. Food Chem.*. **1980** *28,* 675-678.
18. Lea, A.G.H.; Arnold, G.M. *J. Sci. Food Agric.* **1978**, *29,* 478-483.
19. Fischer, U. M.Sc. Thesis, University of California, Davis, CA, 1990.
20. Haslam, E. *Biochem. J.* **1974,** *139,* 285-288.
21. Brouillard R.; Mazza G.; Saad Z.; Albrecht-Gary A. M.; Cheminat A.. *J. Am. Chem. Soc.* **1989,** *111*, 2604-2610.
22. Brouillard, R.; Dangles, O. In *The flavonoids: Advances in Research since 1986.* Harborne, J., Ed ; Chapman & Hall, London, UK 1993.
23. Singleton, V. L.; Trousdale E. K. *Am. J. Enol. Vitic.* **1992**, *43,* 63-70.
24. Prieur, C.; Rigaud J.; Cheynier V.; Moutounet M.; *Phytochem.* **1994,** *36*, 781-784.
25. Souquet, J.-M.; Cheynier, V.; Brossaud, F.; Moutounet, M. *Phytochem.* **1996**, *43,* 509-512.
26. Ricardo-da-Silva, J. M.; Cheynier, V; Souquet, J. M.; Moutounet, M.; Cabanis, J-C.; Bourzeix, M. *J. Sci. Food Agric.* **1991**, *57*: 111-125.
27. Cheynier, V.; Prieur, C.; Goyot,S.; Rigaud, J.; Moutounet, J. A. In *A.C.S Symp. Series.*, American Chemical Society, Washington, D.C., 1997, *661*, 81-93.
28. Broussaud, F. Ph. D. Thesis. ENS Agronomique de Rennes, Rennes, France, 1999.
29. Broussaud, F.; Cheynier, V.; Noble, A.C. 2000. Unpublished..
30. Kennedy, J.A. Ph. D. Thesis. University of California, Davis, CA.,1999.
31. LeDrean, E.; Noble, A.C. **1999.** Unpublished.
32. Gillespie, D.; Noble, A.C. **2000**. Unpublished
33. Guinard, J.-X.; Pangborn, R.M.; Lewis, M.J. Am. *J. Enol. Vitic* **1986**, *37,* 184-189.
34. Noble, A.C.; Sturzenegger, K. **1999**. Unpublished.
35. Fischer, U.; Boulton, R.B.; Noble, A.C, *Food Qual. Pref.* **1994,** *5,* 55-64.
36. Ishikawa, T.; Noble, A.C. *Food Qual. Pref.* **1995**, *6,* 27-33.
37. Lange, M.; LeDrean, E.; Noble, A.C. **1999.** Unpublished.
38. Courregelongue, S.; Schlich, P.; Noble, A.C. *Food Qual. Pref.* **1999,** *10*, 273-279.
39. Peleg, H.; Noble, A.C. *Chem. Senses,* **1995**, *20,* 393-400.
40. Froehlich, D.A.; Pangborn, R.M.; Whitaker, J.R. . *Physiol. Behav.* **1987**, *41,* 209-217.
41. Bennick, A. *Molec. Cell. Biochem.* **1982**, *45,* 83-99.

42. Hagerman, A.E.; Butler, L.G. *J. Biol. Chem.* **1981, 256,** 4494-4497.
43. Lyman, B.J.; Green, B.G. *Chem. Senses,* **1990,** *15,* 151-164.
44. Naish, M.; Clifford, M.N.; Birch, G.G. *J. Sci. Food Agric.* **1993**, *61,* 57-64.
45. Smith, A.K.; June, H.; Noble, A.C. *Food Qual. Pref.* **1996**, *7,* 161-166.
46. Clifford, M. N. In : Tomas-Barberan, F.; Robins, R. *Proc. Phytochem. Soc. Europe*, **1997,** *41*, 87-108.

Chapter 16

Pungency and Tingling: Sensations and Mechanisms of Trigeminal Chemical Sensitivity

Bruce Bryant and Igor Mezine

Monell Chemical Senses Center, 3500 Market Street, Philadelphia, PA 19104

Distinct from taste and olfaction, the trigeminal nerve is the third sensory pathway in the cranial sensory system that is sensitive to chemical stimuli. Trigeminal nerve endings in the nose and mouth contribute to flavor through the sensory modalities of touch, thermal sensation and pain. The best-characterized example of chemically induced trigeminal sensation is the pungency produced by hot peppers, the result of the activation of ion channels on pain-sensitive and thermally sensitive nerve fibers. Compounds commonly found in spices, food and beverages also elicit sensations other than pain. Menthol and other related compounds stimulate a subclass of thermal nerve endings to produce cooling. Yet other compounds, stimuli as diverse as CO_2 and fatty acids as well as some unsaturated alkylamides found in non-capsicum peppers and other plants, activate cooling-sensitive and tactile nerve endings. This particular combination of modalities gives rise to the novel tingling sensations associated with these stimuli.

Introduction

The pungency of spices in Western cuisine is best known as the heat and pain of the Capsicum and Piper species of peppers as well as ginger and cinnamon. The active compounds are best exemplified by capsaicin, the pungent constituent of hot peppers. Capsaicin activates polymodal nociceptors, the pain-sensitive neurons that are sensitive to noxious thermal or mechanical stimulation and some potentially harmful chemical compounds. Initially, the existence of a specific receptive mechanism for capsaicin was inferred by structure-activity studies (1). More recently, a non-specific cation channel that is sensitive to noxious heat as well as capsaicin has been cloned and characterized (2), explaining the hot, pungent and sometimes painful sensation of peppers. Other pungent compounds such as zingerone and piperine presumably also act on this receptor/ion channel (3,4). Thus, one type of pungency is reasonably well understood.

Less well appreciated by sensory physiologists but well known in many Asian cuisines are spices that produce a tingling or buzzing sensation in the mouth that is distinct from the hot and painful sensations elicited by the Capsicum or Piper peppers. Fruits from plants of the genus Xanthoxylum are used as spices throughout Asia and are well known for their unique sensory character. The active compounds in these plants are unsaturated alkyl amides (UAA), a class of compounds that is also found in the Compositae, the Piperaceae and the Rutaceae (5). In addition to being used in Asian cuisine, plant extracts containing these compounds are also used in ethnobotanical preparations for their anesthetic properties (6).

Because the sensory qualities of UAAs and the spices that contain them differ from the pungency induced by capsaicin and the other "hot" spices, it is likely that UAAs stimulate different sensory mechanisms. While the basis of capsaicin-induced pungency is well known, little is known about the cellular and molecular mechanisms underlying oral sensations such as tingling, electrical sensations and numbness. In order to elucidate the neural basis of these unique sensations we used both physiological and human sensory measures to examine 1) which sensory neurons are activated by UAAs and 2) what molecular determinants for sensory activity are.

The Compounds

UAAs have been reported from a variety of plant sources (7,8,9). We isolated UAAs from the fruit of various Xanthoxylum species obtained from the Morris Arboretum, University of Pennsylvania, from local Asian grocery stores, as well as from oleoresin (Synthite Inc., India)) using solvent extraction, liquid-liquid extraction column chromatography and high-performance liquid chromatography. These included hydroxy-α-sanshool (**1**), N-(2-methyl 2- hydroxypropyl)-dodecateraen-2E,6Z,8E,10E amide; hydroxy-_-sanshool (**7**), N-(2-methyl 2-hydroxypropyl)-dodecateraen-2E,6E,8E,10E amide; hydroxy-_-sanshool (**3**), N-(2-methyl 2- hydroxypropyl)-dodecateraen-2E,6Z,8E,10Z amide; and _-sanshool (**2**), N-(2-methyl propyl)-dodecateraen-2E,6Z,8E,10E amide (Figure 1). Spilanthol (**4**), N-(2-methyl propyl)-decatrien-2E,6Z,8E amide, was isolated similarly from an extract provided by Takasago, Intl.(NJ). Pellitorine (**12**) was a gift from Dr. D. Rusterholz (Wisconsin). The structure of all natural compounds and the synthetic analogs described below have been confirmed by spectral methods (MS, ^{13}C-NMR, ^{1}H-NMR, COSY).

Oral sensations of unsaturated fatty acid amides.

In the literature, the ability of these compounds to produce a sensation has been noted and descriptors include "pungent" (8,10) or "twingeing" (11). Human judgments of the sensation of HO-α-sanshool (HOαS; 25-50 μg), applied directly, to the tongue, indicate that it does not elicit the same warming sensations as do capsaicin or other similar spices. Rather, after a delayed onset (30-90 sec) the tingling sensation from HOαS is more similar to a mild electric shock (5-7 V, either DC or AC) or a weakly carbonated solution applied to the tongue. Such tingling sensations are localizable to the correct side of the tongue and last from 10 - 20 minutes. At higher concentrations (> 100 μg), the sensation is painful. In addition to tingling induced by HOαS, the sensation of evaporative cooling of the tongue during inspiration is enhanced. Control applications of the vehicle, 70% ethanol, produce a brief burning sensation that disappears within 30-45 sec. In addition to stimulating the above sensations, which are all mediated by the general body sense, sometimes, the unsaturated alkylamide, HOaS also evokes a mild sour sensation. Thus, the alkylamides appear to be active at both gustatory receptors as well as somatosensory nerve endings. We will focus on the somatosensory mechanisms in this paper.

Neural Responses to UAA

In order to determine the neural mechanisms underlying the tingling sensations of UAA, we measured the response of the trigeminal nerve fibers that innervate the tongue. Extracellular nerve recordings were obtained from anesthetized (50-65 mg/kg b.w. pentobarbital plus 150 mg/kg urethane) Sprague Dawley rats (250-370 g) under an IACUC-approved protocol. The response characteristics of nerve fibers were determined by successively presenting cool H_2O(16°C), hot H_2O(43°C), tactile stimulation (camel hairbrush and pinching), and 100 mM pentanoic acid.

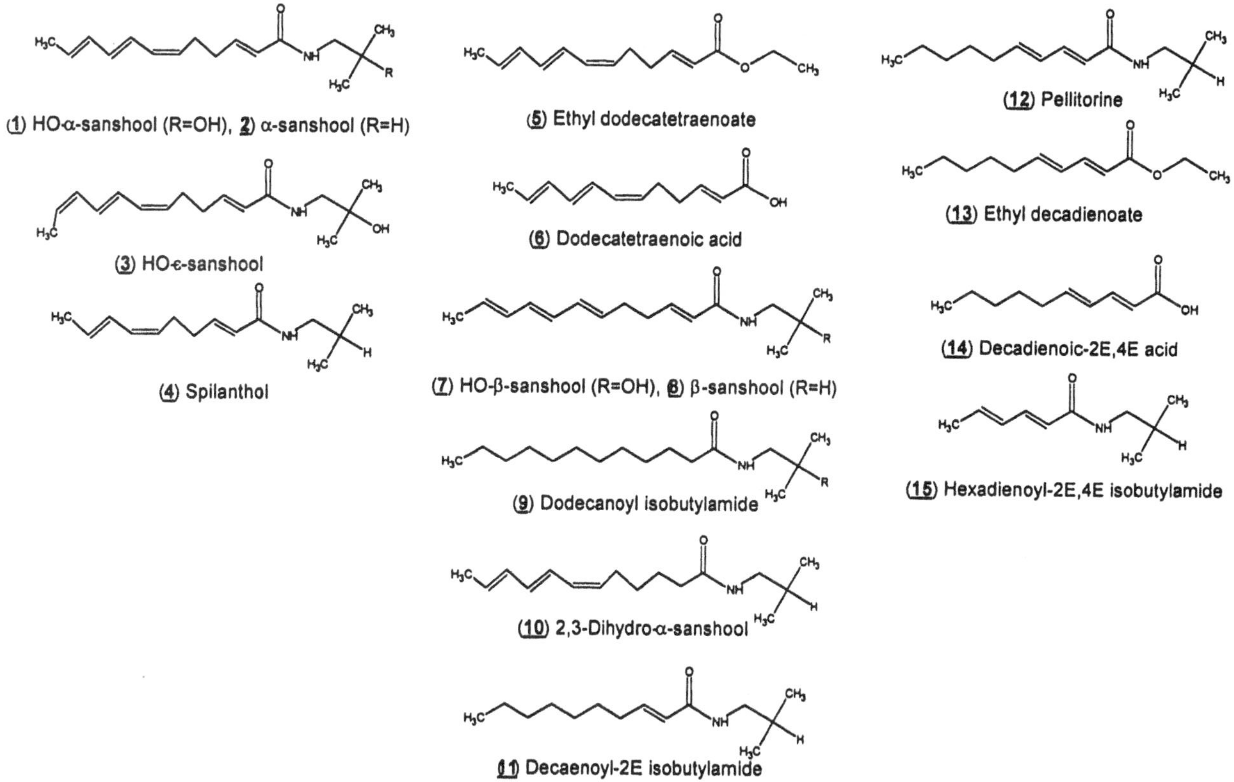

Figure 1. Naturally occurring unsaturated alkylamides and synthetic analogs.

These stimuli were flowed over the tongue in a background flow of deionized water at 31-33°C (thermally neutral). Following the series of characterizing stimuli, the effects of Xanthoxylum compounds on nerve activity was determined. The tongue was exposed by stopping the flow and allowing the test compound (200 μl of 4 mg/ml in 70% ethanol) to remain on the tongue for 5 min. Immediately after rinsing and at 20 and 40 minutes following the exposure, the characterizing tactile, chemical and thermal stimuli were again presented to determine if the modality of each neuron remained constant. Following application of HOαS to the animals' tongues, there was an initial delay of 30 sec to 4 min before a rapid increase occurred in the spontaneous firing rate of silent or spontaneously active fibers. Fig. 2 shows the delayed and sudden onset of activity in an initially silent neuron 1 minute after application. The induced activity frequently lasted up to 20-45 min after the tongue was rinsed. The activity of spontaneously firing neurons (mostly cool-sensitive fibers) frequently doubled in their rate of discharge. Both the delay in onset and the duration of excitation were comparable to the time course of the tingling sensation in humans.

Nerve recordings indicate that HOαS activated several classes of trigeminal neurons. In 26 of 31 neurons tested, the level of spontaneous activity of the neurons increased after the application of HOαS. Most tactile-sensitive (8 of 9) and cooling-sensitive (5 of 6) fibers were excited as were neurons that were sensitive to both touch and cooling (3 of 4). Cold nociceptors (2 of 2) were also stimulated. In addition, 7 neurons that were initially silent and were not driven by any of the characterizing stimuli were also excited by HOαS. The data in Fig. 3 show the rate of action potential discharge in response to the lingual application of HOαS and 70% ethanol (vehicle). Note the increase in discharge frequency in both initially silent and spontaneously active fibers. Ethanol had little or no effect on the activity of most neurons.

Following activation of trigeminal neurons by HOαS, some neurons became more sensitive to their effective stimuli. For example, the response of an acid-sensitive fiber to propanoic acid became larger after lingual application of HOαS. (Fig. 4).

In addition to inducing increased sensitivity of some nerve fibers to their preferred stimuli, HOαS also induced novel sensitivity in some trigeminal fibers. For example, following HOαS, neurons that were previously insensitive to cooling (Fig. 5a) or warming (Fig. 5b) responded to these stimuli. Similarly, tactile sensitivity was induced in 1 of 6 cool neurons and 3 of 7 silent neurons.

How do these effects correspond to the sensations that are evoked? HOαS activated several types of sensory neurons in a manner that is consistent with the quality of sensations that are evoked by the compound. Specifically, a tingling sensation, as is observed with electrical stimulation of the skin, is characteristic of

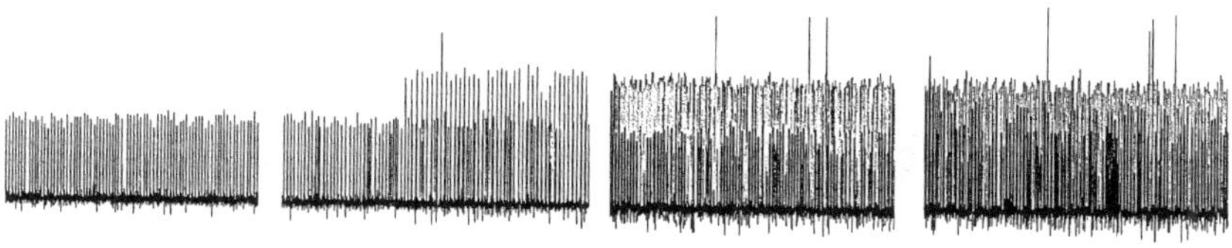

Figure 2. Trigeminal nerve fiber discharges. Note the sudden appearance of the higher amplitude action potential approximately 1 min after the application of HOαS onto the tongue. Reprinted from reference 14 with permission from Elsevier Science.

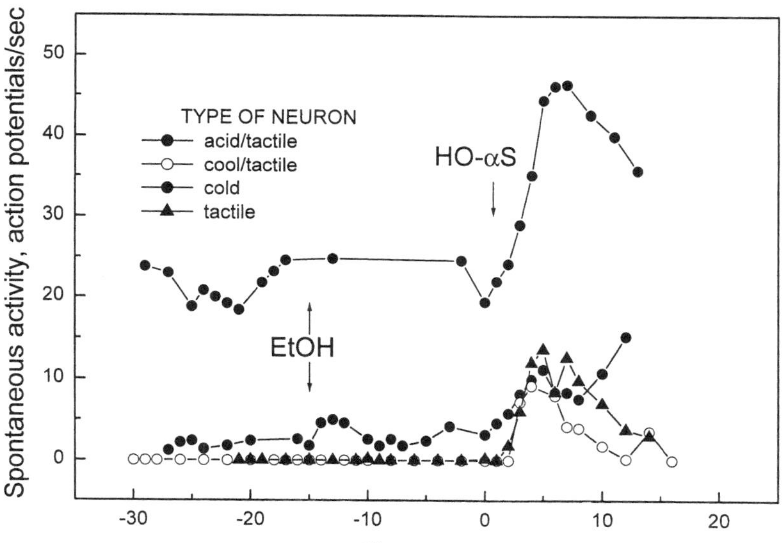

Figure 3. Spontaneous discharge rates of 4 single trigeminal fibers. Reprinted from reference 14 with permission from Elsevier Science.

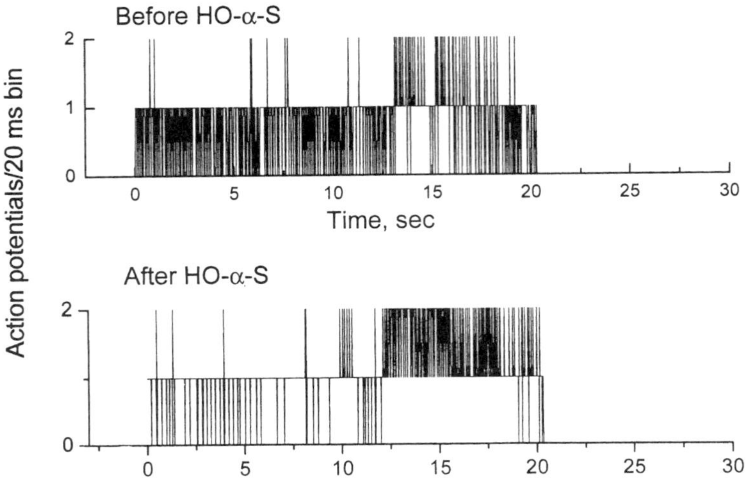

Figure 4. Discharge rate of a single acid/cool sensitive nerve fiber. n-Propanoic acid was applied at t=0. The response to the acid occurred approximately 12 sec after application. Reprinted from reference 14 with permission from Elsevier Science.

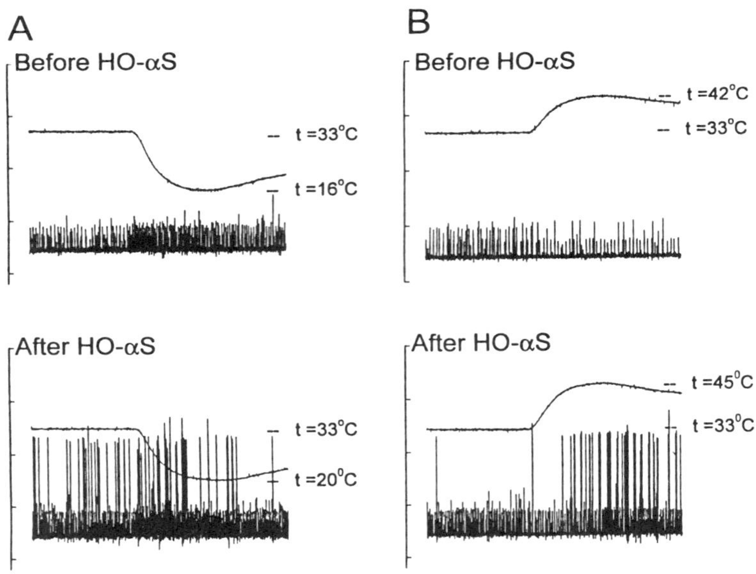

Figure 5. Nerve responses to cooling (A) and warming (B) before and after application of HOαS to the tongue. Note the appearance of responsive neurons (taller action potentials) to both cool and warm stimulation following HOαS. Reprinted from reference 14 with permission from Elsevier Science.

the excitation of multiple types of sensory fibers, especially tactile nerve fibers (12). Moreover, the activation and sensitization of cool-sensitive fibers is consistent with reported enhanced oral cooling. Finally, tingling may be due to a more complex interaction arising from a paradoxical pattern of stimulation arriving at the central nervous system. For instance, under UAA stimulation, a cool stimulus could excite some but not all tactile neurons giving rise to an incomplete or aberrant pattern of input from tactile neurons that the central nervous system does not usually receive. A related phenomenon is seen in the "thermal grill illusion" (13) in which spatially differentiated innocuous warm and cool stimuli applied simultaneously to the skin elicit a painful sensation.

Effects of HOαS on Intraneuronal Calcium

Calcium ions are an important intracellular messenger, the concentration of which affects the mechanisms that determine neuronal sensitivity. Moreover, levels of intracellular calcium $[Ca^{2+}]_i$ reflect the state of neuronal activation. We determined intracellular calcium concentrations using digital fluorescent imaging of the calcium sensitive dye, fura 2, in individual cultured rat trigeminal neurons. In addition, we were able to determine further how the sensitivity to alkylamides was distributed across different functional types of neurons.

Some neurons responded to HOαS with immediate increases in $[Ca^{2+}]_i$, suggesting that the delay in response seen with the application of HOαS to the intact tongue (human and animal nerve studies) is due to the time necessary for diffusion across the lingual epithelium.

Cultured trigeminal neurons responded to HOαS and capsaicin with increases in intracellular calcium. However, sensitivity to capsaicin and HOαS was not found to be coincident in all neurons (14). Only 17.1% of neurons responded to both HOαS and capsaicin. Neurons responding solely to HOαS comprised 38.2% of the total and those responding solely to capsaicin were 44.7% of the total. These findings imply that there are independent transduction mechanisms for these two compounds. Moreover, it means that HOαS activates some but not all capsaicin-sensitive polymodal nociceptive neurons (PMNs). PMNs are multimodal nerve endings sensitive to potentially damaging levels of thermal or mechanical stimulation. In addition they are activated by chemical conditions in the tissue that are potentially damaging as well as to some endogenous neurotransmitters and exogenous plant-derived compounds such as capsaicin. PMNs are defined as the subpopulation of nociceptors that are sensitive to capsaicin.

External rather than internal stores of calcium appear to be the source of HOαS-induced increases in $[Ca^{2+}]_i$. Removal of external calcium abolished alkylamide-induced increases in $[Ca^{2+}]_i$. Extracellular sodium also plays a role in HOαS-induced changes in intracellular calcium. Substitution of an impermeant cation, N-methyl D-gluconate, for sodium strongly attenuated the responses of intracellular calcium to HOαS. These findings suggest that HOαS-induced sodium fluxes and the concomitant depolarization of sensory endings may be responsible for changes in intracellular calcium.

Structure-Activity Relationships

Human Sensory Tests

Natural UAAs provide some insights into structure - activity relationships, however, they are insufficient for the complete elucidation of molecular features required for sensory activity. Therefore, we designed, synthesized and tested a number of analogs of the naturally occurring sensory active UAAs,(**1, 2, 3, 4, 12**) (Fig 1) to test for particular structural requirements. Testing consisted of placing 100 μg of test compound (in 50 μl of 70% ethanol) on the tongue and obtaining an intensity rating of tingling during the subsequent 10 min..

Across all UAAs, sensory activity was dependent upon the presence of a conjugated alkylamide. Removal of the 2,3 double bond yielded inactive compound (**9**, **10**). Moreover, the free acids and ethyl esters (**5**,**6**,**13**,**14**) were similarly inactive.

The presence and configuration of double bonds in the alkyl chain was also an important determinant of sensory activity. In C12 UAAs, the presence of the 6Z bond was crucial because the all-E compounds (**7**, **8**) were inactive. In C10 compounds, however, the conjugated 2E,4E system (**12**) appears to substitute for the 6Z bond present in **4**, as both were active. The terminal double bond was not important as activity was preserved regardless of its configuration (**1** and **3**) and even in its absence (**4** and **12**). While the 2,3 double bond is important, it alone did not render a C10 analog active (**11**). Finally, the conjugated 2E,4E system was not sufficient to confer activity upon an alkylamide with a 6 carbon chain (**15**).

Neural Studies

Results with cultured trigeminal neurons were generally concordant with human sensory studies; compounds that elicited tingling on the tongue induced increases in intracellular calcium *in vitro*. For instance, neurons that responded to HOαS (**1**) also all responded to pellitorine (**12**). Conversely, analogs of pellitorine that were inactive in sensory tests (**13** and **14**) were also inactive on cultured neurons, failing to cause increases in intracellular calcium. This suggests that as a class these compounds act through the same neurons. 2,3 Dihydro-αS (**10**), a compound that was inactive in sensory tests, did affect intracellular calcium, causing both increases and decreases in intracellular calcium in different neurons. This is being examined further.

Discussion

What do these physiological studies reveal about the sensory mechanisms underlying the sensation of unsaturated alkylamides? A most informative distinction is the difference in the quality of the tingling of the alkylamides as opposed to the heat and painful sensations of capsaicin and other pungent compounds such as allyl isothiocyanate, cinnamic aldehyde, zingerone and zingerol. These latter compounds excite polymodal nociceptors while the alkylamides additionally excite cool-sensitive thermal and tactile receptors. The excitation of these other populations of neurons rationalizes the enhanced cooling sensation of alkylamides as well as the tactile, tingling sensation.

The general somatosensory system is comprised of the senses of touch, temperature and pain. Nerve endings sensitive to each of these modalities are located in and below the epithelium. The somatosensory nerve that innervates the face, mouth, nose and eyes is known as the trigeminal nerve, the 5th cranial nerve, and is distinct from the other chemical senses of taste (7th, 9th and 10th nerves) and olfaction (1st nerve).

Generally speaking, activation of a given subpopulation of somatosensory neurons gives rise to a characteristic sensation. Thus, when low threshold cooling-sensitive endings are stimulated, cool is felt. With more intense cooling, to noxious levels of cold, thermal nociceptors are stimulated giving rise to pain. Sensitivity to chemically-induced sensation is best understood in terms of the response of PMNs to capsaicin. Activation of PMNs gives rise to pain and elimination of these nerve endings produces insensitivity to many forms of chemically-induced pain (15). Activation of a subset of PMNs, however, can give rise to forms of irritation other than outright pain. Histamine evokes the sensation of itch by activating such a subset of PMNs (16). Thus, not all situations that activate PMNs produce frank pain. Therefore, a potential contributing set of neurons to tingle may be a subset of PMNs, similar to the histamine-sensitive PMNs that are sensitive to UAAs. Another population of nerves that contributes to the oral sensation of alkylamides and possibly to the tingling sensation are the cool-sensitive nerve endings. This is supported by the activation of cool neurons by alkylamides and by the concomitant enhancement of the perception of cooling produced in humans. Unsaturated alkylamides may act similarly to menthol on cool-sensitive nerve endings. In addition, cooling enhances the oral perception of irritation produced by carbon dioxide (17), suggesting that the multiple effects of HOαS may be due to summation in the periphery at cool fibers or at higher levels in the central nervous system. Finally, as demonstrated in nerve recordings, UAAs act upon tactile receptors, likely contributing to the tingling aspect of the sensations.

Summary

Unsaturated alkylamides induce novel tingling sensations in the mouth as well as altering thermal sensitivity. The activity of these compounds depends upon the presence of specifically configured double bonds and a conjugated amide moiety. The activation of cool-sensitive and tactile trigeminal neurons explains the sensory attributes of unsaturated alkylamides.

References

1. Walpole, C.S.;Wrigglesworth, R.; Bevan, S.; Campbell, E.A.; Dray, A.; James, I.F.; Masdin, K.J.; Perkins, M.N.;,Winter, J. *J. Med. Chem.* **1993,**36,2381.
2. Caterina, M..J.; Schumacher, M.A.; Tominaga, M.; Rosen, T.A.; Levine,J.D.; Julius,D. *Nature* **1997,**389,816.
3. Liu, L. and Simon, S.A. *J. Neurophysiol.* **1996**,76,1858.
4. Martenson, M.E.; Arguelles, J.H.; Baumann, T.K. *Brain Res.* **1997**,761,71.
5. Hegnauer,R.,in, *The Biology and Chemistry of the Compositae*. Heywood, V.H., Harborne, J.B. and Turner, B.L., eds. .Academic Press, London. 1977.
6. Duke, J.A., *CRC Handbook of Medicinal Herbs*; CRC Press: Boca Raton, FL,1985
7. Kashiwada, Y.; Ito, C.; Katagiri, H.; Mase, I.; Komatsu, K.; Namba, T.; and Ikeshiro, Y. Phytochem. **1997**,44,1125.
8. Yasuda, I.; Takeya, K.; Itokawa, H. Chem. Pharm. Bull. **1981**,29,1791.
9. Greger, H. Planta Medica **1984**, 50,366.
10. Jacobson, M. J. Am. Chem. Soc.**1948**,70,4234.
11. *Fenaroli's Handbook of Flavor Ingredients*; CRC Press: Cleveland, OH, 1971; p 445.
12. Garnsworthy, R.K.; Gully, R,L,; Kenins, P.; Westerman, R.A. *J. Neurophysiol.* **1988**,59,1116.
13. Craig, A.D.; Bushnell, M.C.; *Science* **1994**,265,252.
14. Bryant, B.P.; Mezine, I. *Brain Res.* **1999**,842,452.
15. Holzer P. *Pharmacol. Rev.* **1991**,43,143.
16. Schmelz, M.; Schmidt, R.; Bickel, A.; Handwerker, H.O.; Torebjork, H. *J. Neurosci.* **1997,** 17,8003.
17. Green, B.G. *Chemical Senses* **1992**,17,435.

Taste Preference and Consumer Models

Chapter 17

Sensory Drivers of Liking and Sensory Preference Segmentation

Howard R. Moskowitz

Moskowitz Jacobs Inc., 1025 Westchester Avenue, White Plains, NY 10604

Abstract

People differ in what they like and dislike. Nowhere is this inter-individual variability so evident as in the chemical senses, taste and smell. This paper presents an analysis of the individual differences, focusing on food products (specifically coffee and pasta sauce). The paper shows how to measure 'drivers of liking' (sensory attributes that drive product acceptance). The relation between sensory intensity and liking is typically an inverted U shaped curve. On a person by person basis, however, this relation takes on different shapes. One can segment consumers by the specific shape of their sensory-liking function. This segmentation generates different groups of people, distributed throughout the population, with the segments showing radically different patterns relating sensory magnitude to liking. The paper discusses the use of this analytic approach both for science (viz., to understand how people transform the sensory information into liking ratings), and for practical product development (viz., how to create highly acceptable products, targeted to a specific sensory segment).

Introduction

Quite often the first response to a food, perfume or to a simple chemo-sensory stimulus is 'I like it' or 'I dislike it'. Hedonics, likes and dislikes, constitute an exceptionally important aspect of taste and smell. The hedonic tone of a chemo-sensory stimulus may drive consumption if the stimulus is a food, or purchase and use if the stimulus is used for health and beauty aids.

A recurrent research problem concerns the 'drivers' of liking – or, more specifically, the variables of a chemo-sensory stimulus that change the degree of liking. One key driver is the sensory quality. We like the taste of sweet, we dislike the taste of bitter, and we may like or dislike the sour and salt tastes. Whether working in taste or in smell, researchers do not know why some sensory qualities are liked, whereas others are disliked, and thus this issue is not one that is empirically addressed, except perhaps by large scale polling questions (viz., where the issue is to measure the proportion of people who like or dislike a particular stimulus).

One way to look at the drivers of liking uses an equation relating overall liking to each of the separate liking attributes (e.g., liking of appearance, liking of aroma, liking of taste, liking of texture, etc.). The equation describes a straight line, since in most cases there is a positive relation between attribute liking and overall liking. A simple correlation coefficient will not do, however, because correlation analysis often shows that all attribute liking ratings correlate with overall liking. A more productive analysis fits the linear equation to the results (Overall Liking = $k_0 + k_1$(Attribute Liking)), and then looks at the slope (k_1). The higher the value of k_1, the more important is the particular sensory input to the overall liking rating (1).

A more tractable aspect of drivers of liking deals with the change in degree of liking as the physical stimulus magnitude changes. A typical sensory-liking curve appears in Figure 1. As the physical magnitude of the stimulus changes (producing a perceived change in the sensory magnitude), the degree of liking also changes. The particular pattern is a function of the stimulus being tested, whether the stimulus is a model system or a food/perfume, and whether there is an accompanying cognitive input, such as a description of what a stimulus is supposed to be. [For instance, we may hate the bitter taste of a liquid, until we find out that the liquid is a bitter aperitif, meant to be consumed before a meal]. The key metric here will be the area under the sensory-liking curve.

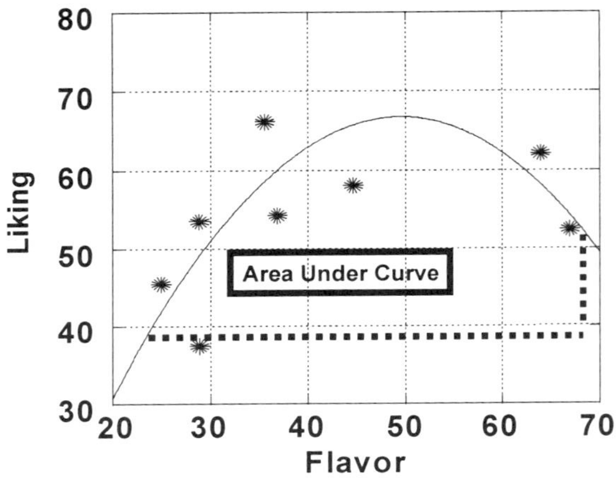

Figure 1: The relation between sensory level and liking. The area under the curve is a measure of relative importance.

What Determines The Sensory Liking Curve?

Figure 1 shows a quadratic function fitted to the empirical point. The specific nature of the curve will vary as a function of the products evaluated, the sensory range considered (larger sensory ranges give more information about the real relation), the sensory attribute used as the independent variable (taste / smell / flavor attributes generate steeper curves than do texture / appearance attributes), and the range of products evaluated.

The approach to develop the sensory-liking curves follows these steps:

1. Lay out the data in the form of a scatter-gram, with sensory attribute as the independent variable, and liking (or other evaluative attribute) as the dependent variable. Use average data only, rather than individual data.
2. Fit a quadratic function to the data, using this expression: Liking = $k_0 + k_1$(Sensory Level) + k_2(Sensory Level)^2
3. Estimate the goodness of fit (e.g., the multiple R^2)
4. Show the estimated quadratic function
5. It is worth noting that the quadratic function is the most parsimonious expression of the relation between sensory attribute and liking. The quadratic function allows for a non-linear relation (but does not force that non-linear relation).

Individual Differences In The Drivers Of Liking

Individual differences have long been recognized as pervasive in the chemical senses (2,3,4,5). For the most part these differences have been relegated to the realm of unexplained or unexplainable variability. Occasionally the differences might be traced to body state and ingestion (6,7), but for the most part the variability is so pronounced and the organizing principle underlying the variability so absent that the inter-individual differences have been ignored. Typically, they are averaged out. Consumer researchers recognize this variability, leading them to tabulate the data from many people by different key subgroups that can be easily determined in the population (e.g., age, income, gender, product usage, etc.). The tabulations do not shed much light on the relation between the variability and membership in a key subgroup.

In the mid 1980's the author suggested an alternative way to look at inter-individual variability (8). The key was the sensory-liking function. A given individual might show a peak on the function at some specific sensory level. Rather than looking at variability due to ratings of liking, the suggestion was made to look at the variability of the optimum level of an individual. Each individual generated a specific optimum level somewhere on the curve. The height of the curve (corresponding to the highest degree of liking) was not relevant, since this could be due to scaling biases. [Viz., some individuals might naturally assign higher numbers, whereas other individuals might assign lower numbers. Yet, the individuals might agree on the sensory level at which the liking ratings would reach its optimum].

Using The Sensory-Liking Optimum To Create Operationally Defined Segments

The sensory-liking curve generates a single point for each individual, for each attribute. For a complex product that excites several sensory systems, and has multiple sensory attributes (even within the same sensory system) one can develop a number of different sensory-liking curves. Each sensory attribute will generate one specific sensory-liking curve for a particular panelist. With M sensory attributes (distributed across appearance, aroma, taste, and texture) and with P people, there will be MxP different curves, and consequently MxP different sensory optima. Many of these sensory optima will correlate with each other, simply because many of the sensory attributes correlate with each other. Consequently, one must factor analyze the matrix of the M different optima, and reduce this matrix to a smaller size, comprising uncorrelated variables. [The specific methods have been previously presented(8)].

The analysis generates a set of variables, and the factor scores of the panelists. One then clusters the panelists together, based upon their factor scores. Panelists falling into the same cluster typically have similar profiles of factor scores, and thus show similar profiles of sensory optima. Although this approach is empirical, it has worked for 20 years (9) and appears to generate segments of panelists with radically different sensory-liking curves. Furthermore, the sensory-liking curves for panelists in different sensory segments diverge from each other, whereas the sensory-liking curves for panelists divided by more traditional means (e.g., age, brand used most often) converge, and are almost identical to each other.

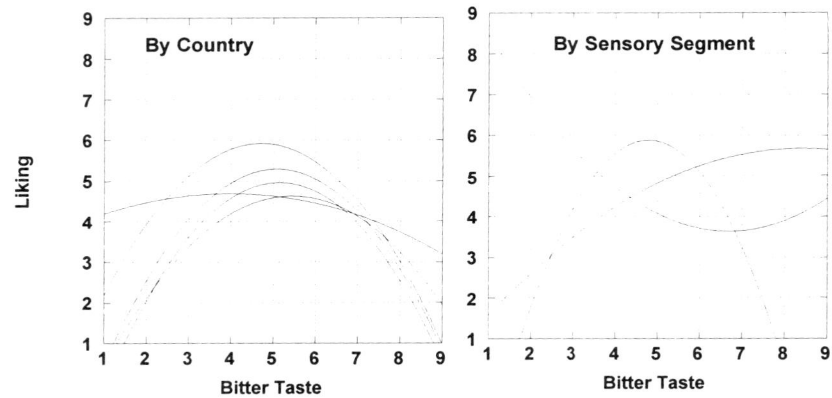

Figure 2: Relation between the bitter taste of coffee from the expert panel (x-axis), and overall liking by the consumer panel (y-axis). The graph on the left is from five countries on a country-by-country basis, and the graph on the right represents segment analysis results.

Figure 2 shows the results of a large-scale study by the European Sensory Network, as reanalyzed by the author. The independent variable is the bitter taste, as profiled by expert panels. The dependent variable is liking, assigned by the consumer panelist. The left panel shows the results for the five European countries. The right panel shows the results after the data were subjected to sensory preference modeling and segmentation. The five countries show similar results, whereas the sensory preference results show three segments, distributed in different proportions (see Table I).

Table I: Distribution of three sensory segments for coffee as a function of the country. Note the differences in proportions for each country.

	Segment 1	*Segment 2*	*Segment 3*
Denmark	10%	51%	39%
France	8%	40%	53%
Germany	12%	45%	42%
Poland	38%	29%	34%
United K.	29%	46%	25%
Total	20%	41%	38%
Male	25%	41%	35%
Female	19%	42%	40%
Age 16-24	17%	46%	37%
Age 25-34	19%	40%	41%
Age 35-44	27%	36%	37%
Age 45-54	14%	45%	41%
Age 55-64	22%	50%	28%
Age 65+	27%	27%	47%

Results – Looking At A Complex Product That Excites Different Senses

One way to understand these drivers of liking and sensory segments is through an analysis of a single product. The optimal conditions for that product are that the product represent either many different in-market products with varying sensory characteristics, or that the product be systematically varied on a set of ingredients, which variation will generate a wide variation in sensory attribute levels. Since sensory preference segmentation is a post-hoc analysis, it is critical that the researcher starts off with a wide range of sensory levels.

The specific product is a pasta sauce. During the past decade and a half the pasta sauce market has expanded with many new products exciting different appearance attributes, aromas, tastes and textures. As a consequence, any evaluation of the product category will generate a plethora of different sensory impressions. Thus pasta sauce makes a perfect stimulus with which to investigate the drivers of liking and the sensory segments.

Research Protocol

Tests with multiple products are often run in a central location, well supervised, in an extended test session (9). The extended session allows the panelists to evaluate a number of different products (here 18 of 50), over several days (here two sessions, each lasting four hours, on two consecutive days). The protocol has been used in other studies for a variety of different products (10).

Each panelist from a group of 225 individuals (75 in each of three geographically dispersed markets) evaluated a randomized 9 products per day from the total set of 50. The randomization was not complete, since it was impossible to prepare 50 pasta sauces and keep these sauces fresh (no older than 10 minutes). Consequently, the set of samples was broken into blocks, the blocks rotated, and the panelists evaluated on product per block.

Panelists rated each of the sauces on a set of sensory and liking scales, anchored at both ends. Liking ratings were anchored at 0 by the phrase 0=hate, and the phrase 100=love. This is a simple scale to use. Panelists could use any number within the range. Sensory ratings were anchored by the appropriate terms. For instance, spiciness was anchored by the phrase 0=not spicy, and by the phrase 100=extremely spicy.

Results – What Sensory Inputs Are Most Important

The analysis of attribute liking versus overall liking (by a linear equation) suggests that liking of aroma and texture are extremely important, with slopes around 1.4-1.5, that liking of appearance and liking of flavor are important, but less so (see Table II). However, these results do not give any idea of what specific attributes drive liking. They show that some sensory inputs are more important than others.

Table II: Slope of the relation between attribute liking and overall liking. The higher the slope the more important the sensory input.

Attribute	*Slope*	*Goodness Of Fit - R*
Aroma	1.47	0.78
Texture	1.43	0.89
Appearance	1.00	0.77
Flavor	0.96	0.98

Results – Sensory Liking Curves For The Full Panel And Usage Subgroups

Figure 3 (left) shows how a visual attribute (size of vegetable pieces), and how a flavor attribute (strength of onion flavor) drives overall liking for two groups of users – those who use Product A and those who use Product B. These groups have some overlap in the population, but not much. The curves are fitted. Note the similarity of the sensory-liking relation for the two user groups.

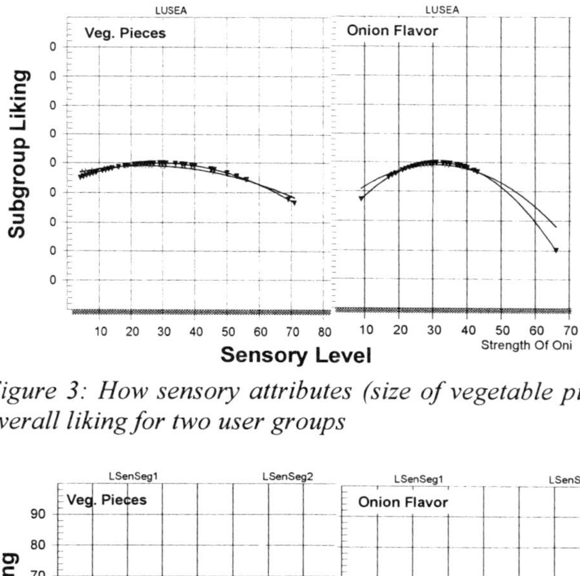

Figure 3: How sensory attributes (size of vegetable pieces, onion flavor) drive overall liking for two user groups

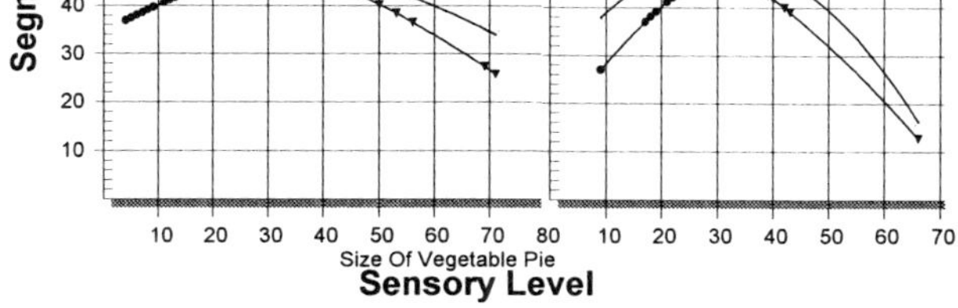

Figure 4: How sensory attributes (size of vegetable pieces, onion flavor) drive overall liking for three sensory segments

A quite different story emerges when we look at the sensory-liking curves for the three segments that emerged from this data set. The three segments were high flavor impact, low flavor impact, and high texture seekers, respectively. These three groups differed in the products that they liked, and in the nature of the sensory-liking curve. Figure 4 shows the sensory-liking curves for size of vegetable pieces (left panel), and the sensory-liking curves for onion flavor (right panel). Whereas the user groups show similar patterns, the sensory segments show different patterns. Furthermore, these patterns evidence themselves on a variety of sensory dimensions, not just chemo-sensory ones.

Quantifying The Magnitude Of The "Driver"

Thus far the analysis has focused on the segments. What is missing, however, is a measure of how important is a specific attribute as a driver of liking. That is, we see the different sensory-liking curves, and we can recognize their difference. An important additional analysis is a single index of relative importance, much as we used the slope of the linear function as a measure of importance. One plausible index is the area under the sensory liking curve. This is defined as the definite integral of the quadratic function minus the actual rectangular area not involved in the function. Figure 1 shows this area, as that subtended by the sensory-liking curve.

The definite integral takes into account two things – the actual sensory-liking curve (a quadratic function), and the sensory range tested. Table 3 presents these areas under the curve for the different sensory attributes and the different key subgroups. We see from Table 3 that the area under the curve varies by attribute, with appearance (e.g., brown color of sauce, amount of tomato pieces) being very important, and other attributes (e.g., those involving mushrooms) being least important. Furthermore, we can compute the highest area under the curve for the user groups (use product A, use product B), and for the sensory preference segments (1,2,3 respectively). The ratio (maximum area for the segments divided by maximum area for user groups) is always greater than 1.0, meaning that for each attribute one of the sensory preference segments always shows the greatest area.

It is also worth noting that the chemical senses are not always the sensory inputs that are the key drivers of liking. Appearance is as important as taste/flavor in driving liking. Texture, however, is not particularly important as a driver of liking. It is also worth nothing that the importance of an attribute is a function of the specific attribute, the product being tested, the range of sensory levels encountered in that attribute, and the specific sensory segment under consideration. Depending upon the particular sensory segment an attribute can have a lot of area, or only a little area.

Table III. Relative Importance Of Attributes As Drivers Of Liking, Based Upon Area Under The Sensory-Liking Curve (Partial List)

Attribute	*Total*	*Use A*	*Use B*		*Seg1*	*Seg2*	*Seg3*	*Ratio*
Larger Area								
Appearance – Brown	2452	2599	2368		1988	2844	2089	1.09
Appearance - Tomato Pieces	1856	1971	2047		2195	2106	1310	1.07
Appearance – Chunky	1824	2161	1855		2571	2244	1312	1.19
Flavor – Herb	1758	2012	1709		2118	1902	1290	1.05
Flavor – Pepper	1689	1892	1742		1736	1428	1906	1.01
Flavor – Tomato	1585	1680	1600		1331	1848	1325	1.10
Appearance – Flecks	1574	1737	1547		927	1854	1985	1.14
Flavor – Strength	1513	1684	1509		1721	1537	1303	1.02
Smaller Areas								
Flavor – Garlic	882	1055	777		989	1123	531	1.06
Appearance – Oily	809	746	777		713	990	725	1.27
Flavor – Sour	806	866	757		936	942	614	1.09
Appearance – Size of Vegetable Pieces	506	670	540		1292	761	772	1.93
Flavor – Mushroom	405	488	400		401	444	526	1.08
Appearance – Size of Mushroom Pieces	374	463	365		592	567	526	1.28
Flavor – Salty	267	314	268		45	304	390	1.24
Flavor – Vegetable	248	298	194		702	412	800	2.68
Appearance – Amount Mushrooms	212	287	204		404	350	353	1.41

[Note: Ratio = the ratio of the largest area for an attribute among the three sensory segments to the largest area for the same area amount the two user subgroups.]

Discussion

The Link Between Model Systems And Real Products

The non-linear relation between sensory magnitude and liking was noticed more than a century ago by Wilhelm Wundt, the founder of experimental psychology (11). Wundt speculated that as a sensory impression increased in strength, the corresponding liking of that impression increased, peaked, and then dropped beyond an optimal level. It fell to later researchers to show that this type of sensory-liking relation pervaded the chemical senses (2, 12, 13). Most of that early work to establish the sensory-liking relation came from studies that dealt with model systems, such as sugar or salt dissolved in water, at different concentrations. Although it might seem that a consumer would have a problem rating 'liking' of a sugar solution, nonetheless from study to study it appeared that the sensory-liking relation existed, and that the optimum sensory level for a caloric sweetener lay at the sweetness corresponding to 9% sucrose. It is no surprise that many carbonated sweetened beverages possess this level of sweetness.

Importance Of The Analysis For An Understanding Of How We React To Chemo-Sensory Stimuli

At a basic science level the understanding of 'drivers of liking' tell the researcher a lot about the behaving organism. Although a great deal of work has been done on the perception of sensory magnitude (e.g., thresholds, suprathreshold scaling), to a great degree it is the hedonics of the stimulus that motivate behavior, such as ingestion. Indeed, it is the pleasantness of a stimulus, rather than its sensory quality or magnitude that is often the determining factor for the usefulness of that stimulus. [This is especially true in taste and smell]. One of the interesting results of this study is that appearance attributes, along with texture attributes, can be drivers of liking. For model systems it is rare that appearance or texture (kinaesthesis) drive liking, except in the most artificial of circumstances, when the panelist is required to scale perceived liking (13, 14, 15). For model systems, which have no context of food, liking ratings can be assigned, but the task may require a stretch of the panelist's imagination. In contrast, when the stimulus is food panelists have no trouble assigning ratings of liking. Furthermore, other attributes besides those resulting from taste and smell can emerge and become important.

On Sensory Segmentation As An Organizing Principle For Future Research

Researchers in the chemical senses and in food science often seek organizing principles by which to understand human perception. Sensory segmentation provides a useful organizing principle. It divides people by the sensory attribute levels that they find most acceptable. Furthermore, the sensory segmentation approach presented here does not just cluster people on the basis of the magnitude of liking (which could generate an artifact, since it could be influenced both by the actual sensory preferences and by the numbers used by a specific panelist). Rather, the sensory segmentation approach uses a well-established relation between liking and sensory magnitude (the quadratic function).

Sensory segmentation can be used to explore a number of different aspects of sensory perception.

1) How do sensory segments, as shown here, distribute themselves in different countries? The European Sensory Network project on coffee (16) showed the existence of the same sensory segments in five different countries. The

juice study (17) showed the existence of the same sensory segments in three European countries.

2) Is membership in a sensory segment constant across a person's life, or does a person change? This requires longitudinal studies.
3) Is membership in a sensory segment familial (inherited) or can it be influenced by advertising and experience?
4) Are there more overarching sensory segments, applying to a variety of stimuli, or is membership in a sensory segment for one product category (e.g., high impact for sauces) independent of member in a sensory segment for another product (e.g., coffee)?
5) Do people in different sensory segments differ also in their responsiveness to communications (words, phrases), colors (e.g., package designs), etc? Up to now the sensory segmentation procedure has operated only within the realm of products, for sensory research, and only in the realm of concepts and communications for concept research (18). Is there a connection between membership in a sensory segment for a product category (e.g., coffee) and membership in a segment for communication (e.g., concepts about coffee)? In other words, does the 'mind know what the tongue likes'?

Importance Of The Analysis For The Development Of Consumer Products

A great deal of research in industry deals with the creation of products for consumers. Traditionally, sensory researchers as well as market researchers, divided the population in ways that reflected easy to measure variables, such as age, gender, product used more often, etc. The data shown here, as well as extensive data in other product categories, shows that the traditional method for dividing consumers generates products that are similar. That is, a product created for users of Product A will be very similar to a product created for users of Product B. [Similar outcomes occur when the development is focused on products created for different geographical markets, or even countries]. Sensory segmentation provides a way to divide the consumer population into truly different groups, and generate products for each group that are maximally acceptable. In a sense, the product development is guided by the preference patterns of the sensory segment, so that the product developer truly creates meaningfully different products to segments who likes and dislikes differ and are established.

References

1. Moskowitz, H.R., & Krieger, B. Food Quality and Preference, (1995), 6, 83.
2. Ekman, G., & Akesson, Report 177, (c.a. 1964), Psychological Laboratories, University Of Stockholm, Sweden.
3. Moskowitz, H.R. Journal Of The Society Of Cosmetic Chemistry, (1986), 37, 233.

4. Pangborn, R.M. Psychonomic Science, (1970), 21, 125.
5. Pangborn, R. M. Criteria Of Food Acceptance: How Man Chooses_What He Eats, (ed. J. Solms & R.L. Hall), (1981), Forster Verlag, Zurich, 177.
6.Moskowitz, H.R., Kumaraiah, V., Sharma, K.N., Jacobs, H.I., & Sharma, S.D. Physiology & Behavior, (1976) 16, 471.
7. Rodin, J., Moskowitz, H.R., & Bray, G.A. Physiology & Behavior, (1976), 17, 591.
8. Moskowitz, H.R., Jacobs, B.E., & Lazar, N. Journal Of Food Quality, (1985) 8, 168.
9. Moskowitz, H.R. New Directions In Product Testing And Sensory Evaluation Of Foods, (1985), Food And Nutrition Press Inc. Trumbull,CT.
10. Moskowitz, H.R.Food Quality and Preference, (2000), In Press.
11. Boring, E.G. A History Of Experimental Psychology, (1929), Appleton Century, New York.
12. Engel, R. Pfluegers Archiv fur die Gesamte Physiologie, (1928), 64, 1.
13. Moskowitz, H.R. Journal Of Food Quality, (1981), 109-138.
14. Beebe-Center,J.G. The Psychology Of Pleasantness And_Unpleasantness, (1932), Van Nostrand Reinhold, New York.
15. Moskowitz, H.R. American Journal Of Psychology, (1971), 84, 387.
16. European Sensory Network, A European Sensory and Consumer Study: A Case Study on Coffee. Published by the European Sensory Network, (1996), Available from Campden & Chorleywood Food Research Association, Chipping Campden, Gloucestershire, GL55 6LD, UK
17. Moskowitz, H.R. & Krieger, B. Food Quality and Preference, (1998), 9, 443.
18. Moskowitz, H.R. Proceedings Of The 49th ESOMAR Congress, Istanbul, (1996), 35.

Chapter 18

Internal and External Preference Mapping: Understanding Market Segmentation and Identifying Drivers of Liking

Jean-Xavier Guinard

Department of Food Science and Technology, University of California, Davis, CA 95616

Traditional methods relating consumer and sensory data (e.g., response surface methodology, the regression method) regress averaged hedonic ratings onto mean analytical ratings. By contrast, preference mapping techniques examine the preferences of each consumer. Internal preference mapping analyzes hedonic ratings by consumers for a product set by principal component analysis (PCA) of the covariance matrix, and provides a summary of the main preference directions. Using a number of regression models (from linear to quadratic ones), external preference mapping regresses the preferences of each consumer onto the first two principal components of a PCA of the products' sensory characteristics (derived from descriptive analysis or instrumental measurements). A case study with vanilla ice cream shows that consumer preferences can vary broadly for a given product type and that different regression models are required to relate individual consumer preferences to product sensory characteristics.

Introduction

Successful product optimization requires good knowledge of the market and a solid understanding of which variables drive consumer liking. Among those variables that drive consumer liking are sensory attributes of the product (so-called 'drivers of liking'). To identify those, product developers have traditionally related consumer and sensory descriptive or instrumental data for a set of products using methods such as the Regression Method (*1*), stepwise and/or multiple regression, and Response Surface methodology (RSM) (*2*). These methods regress averaged hedonic ratings onto mean analytical ratings for a product set. By taking averaged hedonic ratings as the dependent variable, these methods assume that all consumers exhibit the same behavior, and that a single mean value is representative of all the consumers. Consumers actually are heterogeneous in their likes and dislikes of most products. This is called 'market segmentation'. It results that conclusions drawn from averaged hedonic ratings across the consumer population may not be accurate.

Furthermore, traditional methods relating consumer and sensory data such as the Regression Method (*1*) assume a linear relationship between degree of liking and perceived intensity of a sensory attribute. While it may occur with some attributes, in most instances, the relation between liking and attribute intensity is curvilinear with an inverted-U shape, and a polynomial regression with a quadratic term is required to describe it.

Finally, consumers have a limited vocabulary to describe their sensory perceptions of products. They typically produce comments about products which are hedonically based rather than related to specific attributes, so that a product developer is at a loss when attempting to have the consumer define what it is about the product that leads him to like or dislike it. Indirect methods (that regress consumer data onto sensory data) are therefore required to get at why a product is liked or disliked by a consumer.

To remedy the limitations of traditional methods and to get at market segmentation and sensory drivers of liking, a variety of techniques have recently been developed that examine the preferences of each consumer, and in some instances relate them to product characteristics, by regressing hedonic ratings onto a set of analytical (sensory or instrumental) variables. These techniques include internal and external preference mapping (*3, 4, 5, 6, 7, 8, 9, 10*), multiple factor analysis (*11, 12*) and partial least squares regression (*13*). They require that hedonic ratings by consumers and descriptive ratings by experts and/or instrumental measurements be collected for a set of products representative of the segment under study.

We will use a case study with vanilla ice cream to illustrate the principles of internal and external preference mapping. Vanilla ice cream was manufactured according to a 3^2 factorial design with 8, 13 or 18% sucrose and 10, 14 or 18% butterfat for a total of 9 samples that met ice cream standards of identity and were representative of commercially-available products, as shown in Table I.

Table I – Composition of the ice cream samples.

Products	*Sugar (%w/w)*	*Fat (%w/w)*	*Total Solids (%w/w)*
P1	8.94	8.73	32.49
P2	10.85	14.28	39.90
P3	11.61	17.68	45.29
P4	13.65	9.94	39.32
P5	13.54	14.99	43.95
P6	13.29	18.75	47.31
P7	17.65	11.40	44.15
P8	18.81	15.08	49.19
P9	17.91	19.30	53.16

The sensory properties of the ice creams were measured by a trained panel of 15 judges, using a modified Quantitative Descriptive Analysis (QDA) method (*14*). Fifteen attributes of appearance (color), flavor (milky/dairy, buttery, vanilla, custard/eggy, almond, sweetness, coolness/cooling) and texture/mouthfeel (melting rate, fatty, creamy, fluffy/aerated, doughy/pasty/elastic, ice crystals, mouthcoating afterfeel) were evaluated in duplicate across the samples.

A consumer test was carried out with 146 users and likers of ice cream, 73 each men and women. Consumers tasted the 9 ice cream samples in one session and rated their overall degree of liking and degree of liking of the texture/mouthfeel and of the flavor of the samples on the 9-point hedonic scale from 1= 'dislike extremely' to 9= 'like extremely', with 5= 'neither like nor dislike' (*15*).

Internal Preference Mapping

Internal preference mapping is a principal component analysis of the matrix of consumer hedonic ratings across the products, based on the covariance matrix. In that PCA, the consumers are the variables, and the products are the objects. The outcome of the analysis is a biplot of the consumers (the internal preference map) and a corresponding biplot of the products.

Figures 1 and 2 show the distribution of the consumers and of the products in the internal preference map for the ice cream data. The combination of PCs 1 and 2 accounted for 48.8% of the variance in the data. The main conclusion to be drawn from that internal map is that consumer preferences were spread over 180 degrees on the biplot, with two clusters of consumers in the upper-left and lower-left quadrants. The first cluster liked P7, P8 and P9 – the high-sugar samples – the most. It can be characterized as a segment of the population that will favor highly-sweet vanilla ice cream, regardless of its fat content. The second cluster clearly liked P5 the most. P5 had medium sugar and medium fat contents. In this instance, the manufacturer would have to decide between marketing an ice cream like P5 with medium levels of sugar and fat (a decision he would have made had he/she consulted only the mean hedonic of the ice creams across all consumers – P5 had the highest), or manufacturing and marketing two types of vanilla ice cream, each targeting one of the two market segments. If electing to produce an ice cream to please the upper-left cluster, the manufacturer would need to make an ice cream high in sugar, yet different from P7, P8 and P9 in that it would need to be in the upper-left corner of the biplot. The product developer must then resort to external preference mapping to figure out what the characteristics of that ideal product should be (besides high sweetness).

The first two principal components in an internal preference map may only account for a limited amount of the variance in the data (in this example, 49%). Indeed, the analysis reduces the variation in the data from as many original dimensions as there are consumers (146 in this case) to only two dimensions... That means that some information is lost, but the main preference directions usually are identified.

Internal Preference Clustering

The best way to uncover and characterize market segmentation in the hedonic ratings of a consumer population is to carry out a preference clustering

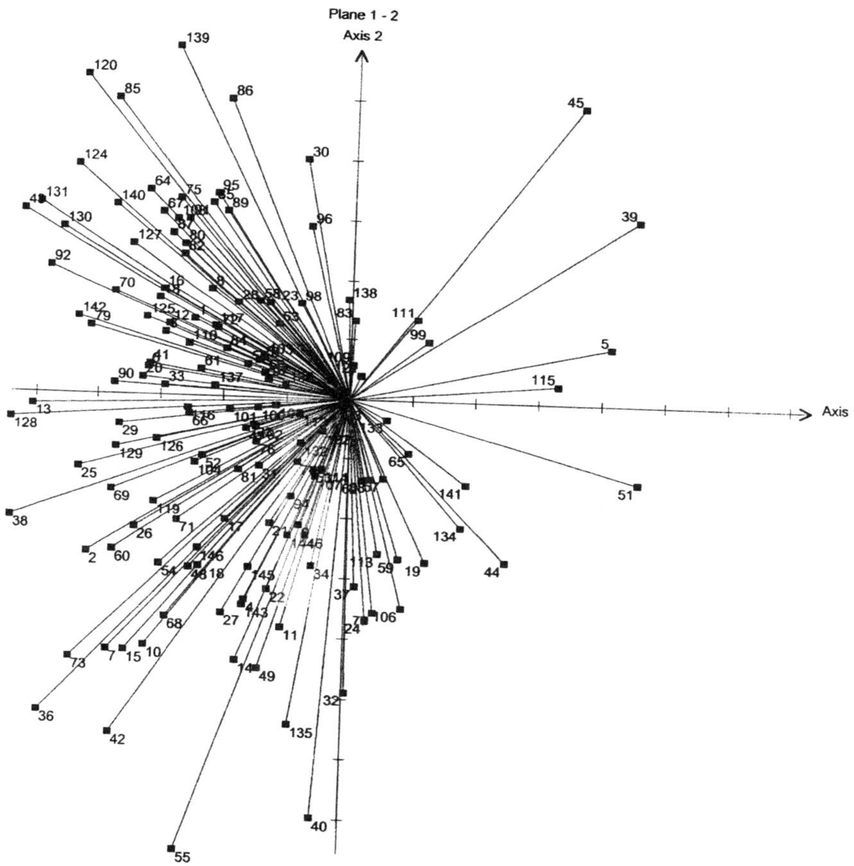

Figure 1. Internal Preference map matrix of hedonic ratings for the 9 ice cream samples showing the consumers (n=146).

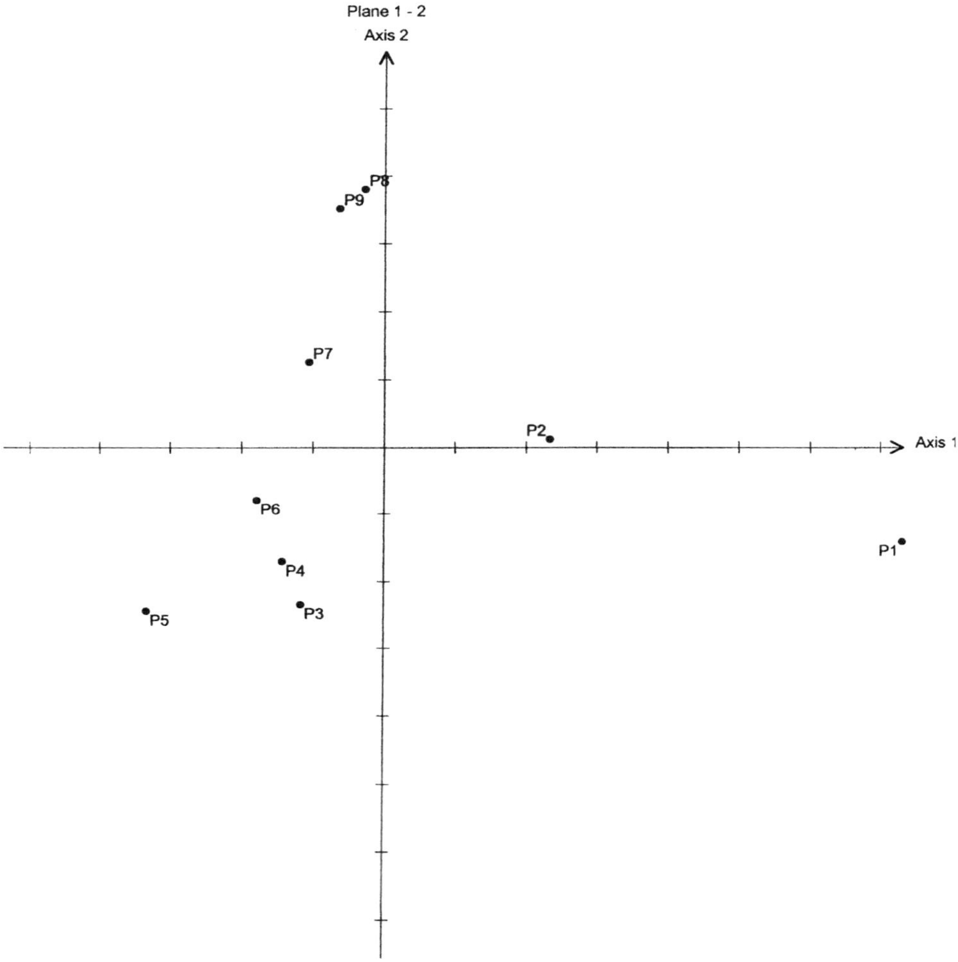

Figure 2. Internal Preference map matrix of hedonic ratings for the 9 ice cream samples showing the products (n=9).

analysis. On the same matrix of consumer ratings across a set of products, one can run a cluster analysis which readily identifies segments or clusters in the population on the basis of similarity of their likes and dislikes for the products. A cluster analysis of the matrix of hedonic ratings for the ice cream samples confirmed the existence of two main clusters of 42 and 53 consumers, respectively, corresponding to the upper- and lower-left quadrants in Figure 1. Once market segments (clusters) have been identified, one can obtain mean hedonic ratings of the samples for each cluster, and confirm which samples are most- and least-liked by that cluster of consumers.

External Preference Mapping

With external preference mapping, the hedonic ratings for each consumer are regressed onto the product coordinates obtained from the multivariate analysis (e.g., principal component analysis) of the sensory descriptive or instrumental data. The models used to regress the hedonic ratings onto these coordinates may be linear or involve squared and/or interactive terms. They are the vectorial, circular, elliptical (with maximum or saddle point) and quadratic models (*10*). The equation relating DOL (Y) for a consumer to PC1 (X_1) and PC2 (X_2) of a PCA of the descriptive or instrumental data therefore ranges from a simple, linear one, e.g.,

$$Y = a + bX_1 + cX_2 \text{ (vectorial)}$$

To a complex, second-order one with quadratic and cross-product effects, e.g.,

$$Y = a + bX_1 + cX_2 + dX_1^2 + e\,X_2^2 + f\,X_1X_2 \text{ (quadratic).}$$

The first step is to carry out a principal component analysis of the descriptive analysis data, again based on the covariance matrix. The use of the correlation matrix would only be warranted if the variables in the matrix had different units of measurements. That would be the case if a matrix of instrumental measurements were used.

The PCA of the descriptive analysis data for the ice cream samples is shown in Figure 3. The first two PCs accounted for 92.3% of the variance in the data. The ice cream samples differed along a dimension (PC1) which contrasted the attributes doughy, fatty, sweet and creamy, with the attributes ice crystals and cooling. P1 and P2 are found at the right end of that dimension, whereas P8

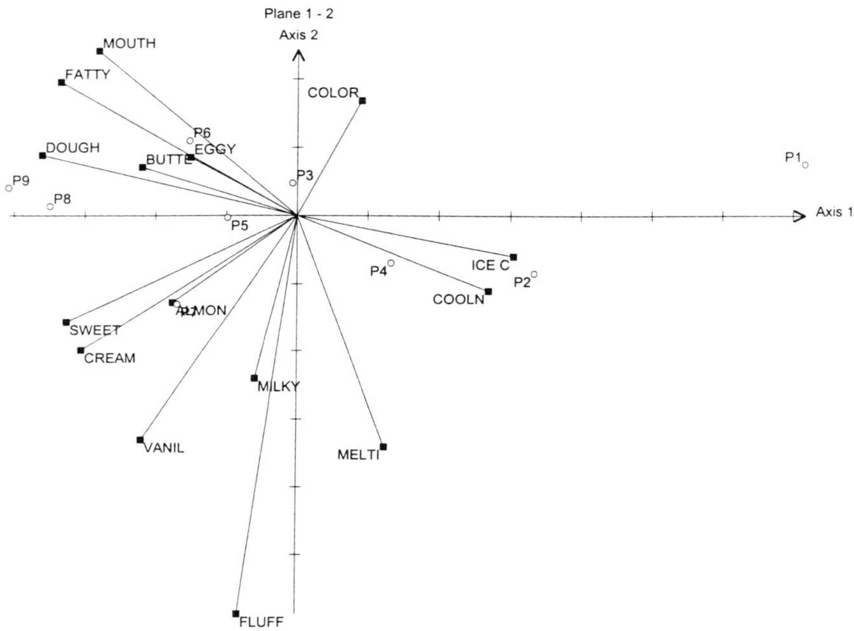

Figure 3. Principal component analysis (PCA) of the matrix of mean descriptive ratings acrosee the 9 icecream samples

and P9 are found at the left end. The second dimension (PC2) contrasted a yellowish color with a fluffy mouthfeel and high melting properties, but it did not quite separate the products because of its limited contribution to the overall variance (4.5%).

The external preference map shows how the preferences of each consumer relate to the two dimensions (PC1 and PC2) of the sensory map described above. Examples of consumers fitted by vectorial, elliptical and quadratic models are shown in Figures 4, 5 and 6, respectively.

There are two ways to display an external preference map. One is to show all the consumers fitted by one of the models – in this example, 104 out of 146 consumers. In Figure 7, the consumers whose preferences were fitted by one of the regression models are displayed with their best model. Straight lines represent vectorial models; triangles represent circular models (with positive ideal point for those triangles pointing up, and negative ideal points for those pointing down); elliptical models are shown as X's; and quadratic models are shown as squares. It can be seen that most of the fitted consumers are fitted best by the vectorial model. And a majority of those consumers' preferences point to the center, bottom of the map, an area not covered by any of the 9 samples in the design, which would be that of an ice cream high in the fluffy/aerated and melting properties and off-white in color (as per Figure 3).

Another way to show the results of the analysis is to display a response surface of number of 'satisfied consumers' on the sensory map of PC1 and PC2 as shown in Figure 8. For a given point on the sensory map (a hypothetical product), one can view how many consumers would have given this product a hedonic rating higher than the mean hedonic rating for the set of 9 samples plus 50% of the range of ratings. But the criterion for inclusion in the response surface may be anything set by the analyst (number of consumers who would give a hedonic rating above 7 to the product on the 9-point hedonic scale, for example). From Figure 8, it appears that the area of the sensory map that would satisfy the majority of consumers is the lower-left corner of the map, with coordinates of –8.5 on PC1 and –3.0 on PC2.

External preference mapping requires a high number of products, to achieve sufficient power for quadratic and cross-product regression. A minimum of 6 products is required, but we recommend at least 10 products to run a meaningful analysis. The selected products should also provide adequate representation of the sensory characteristics of the product segment. If preference mapping is being carried out for reformulation purposes, we recommend designing a sample set that includes both existing (commercial) products and prototypes. This is an excellent way to find out whether the prototypes meet the needs/preferences of consumers, or at least of the main consumer segment, as shown by preference mapping or clustering.

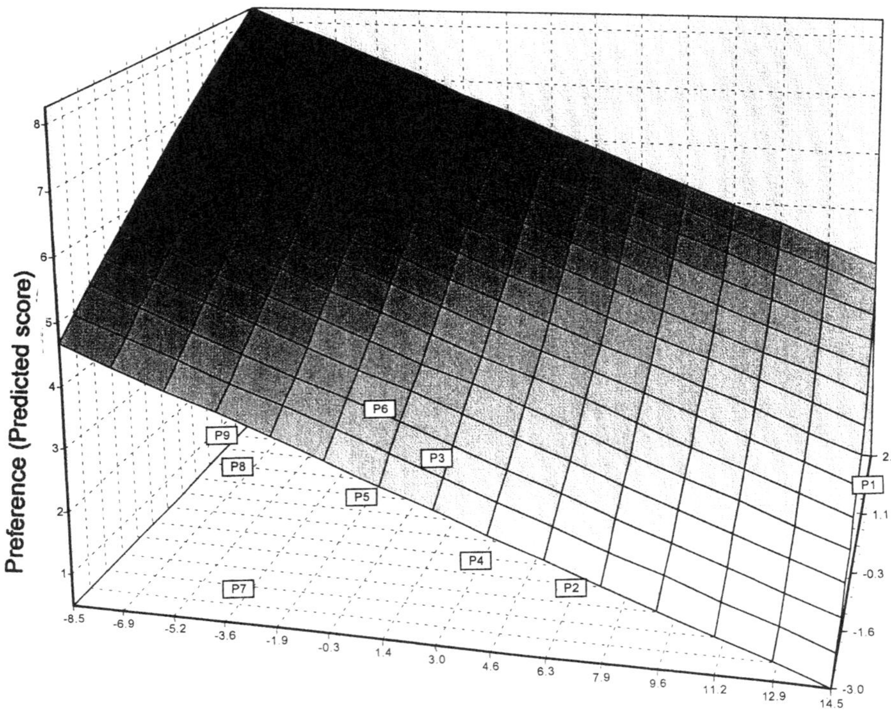

Figure 4. Example of a consumer fitted by a vectorial model

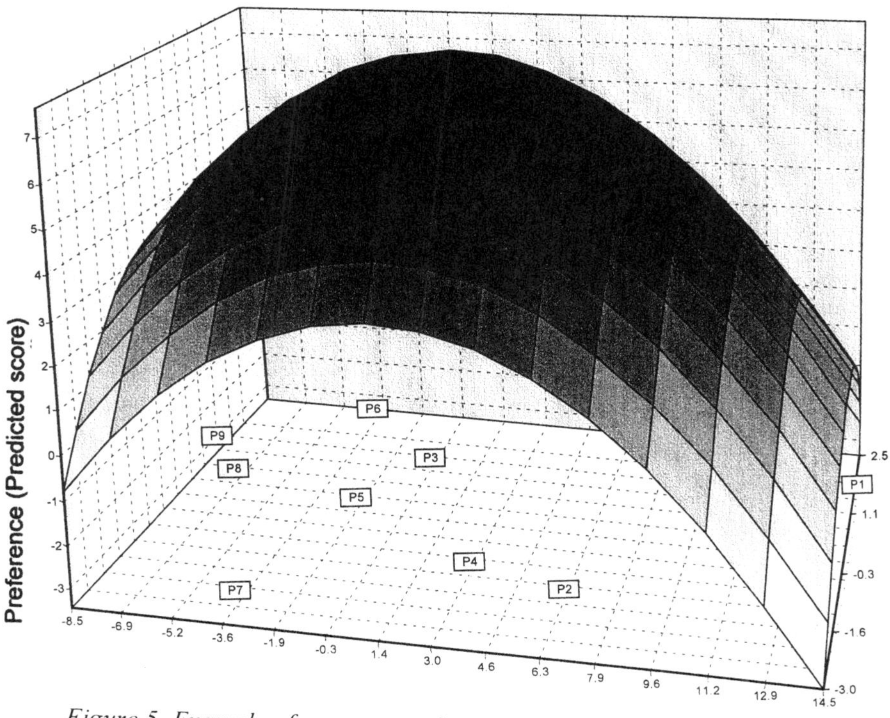

Figure 5. Example of a consumer fitted by an elliptical model

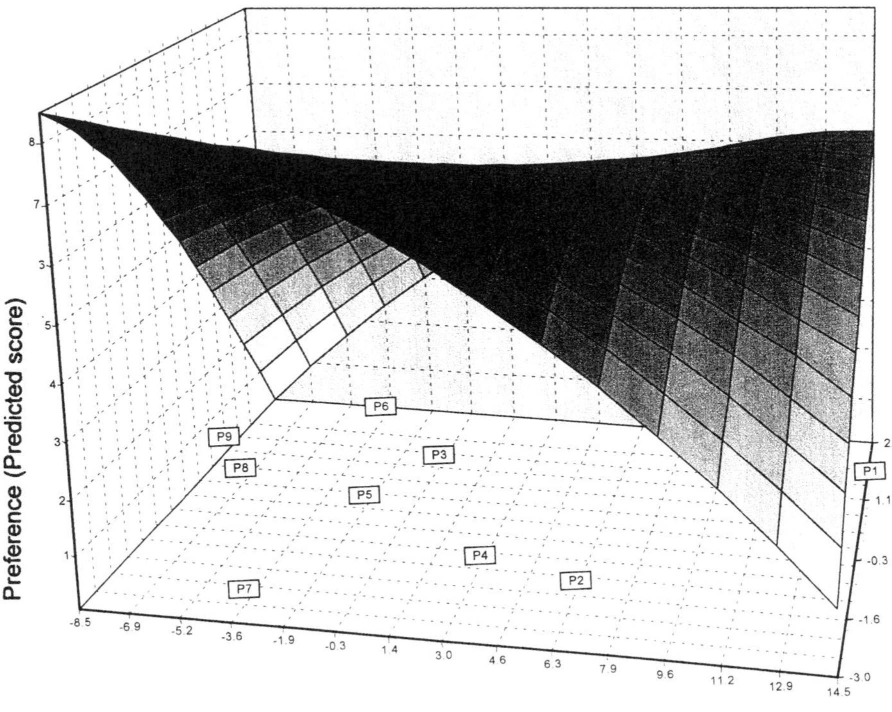

Figure 6. Example of a consumer fitted by a quadratic model

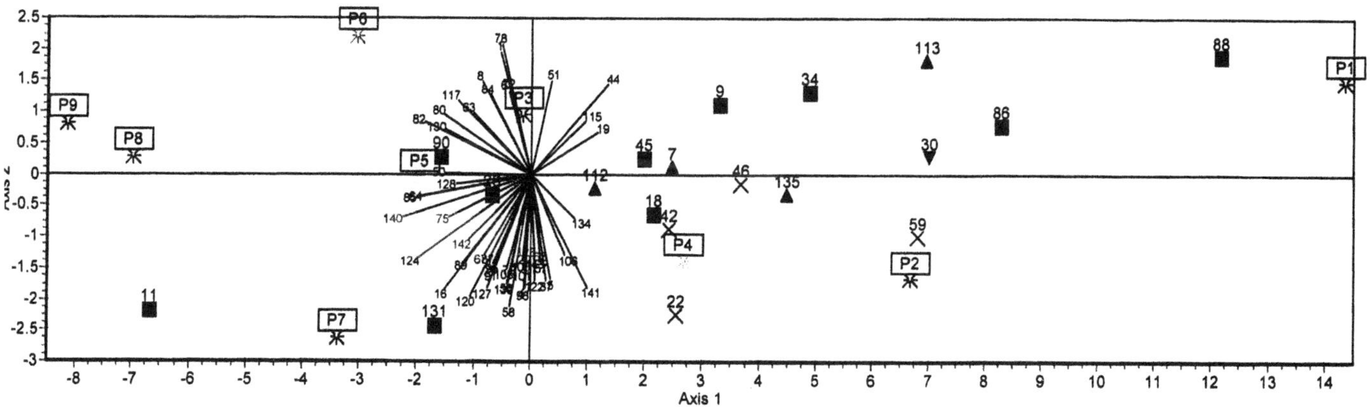

Figure 7. Example of an external preference map

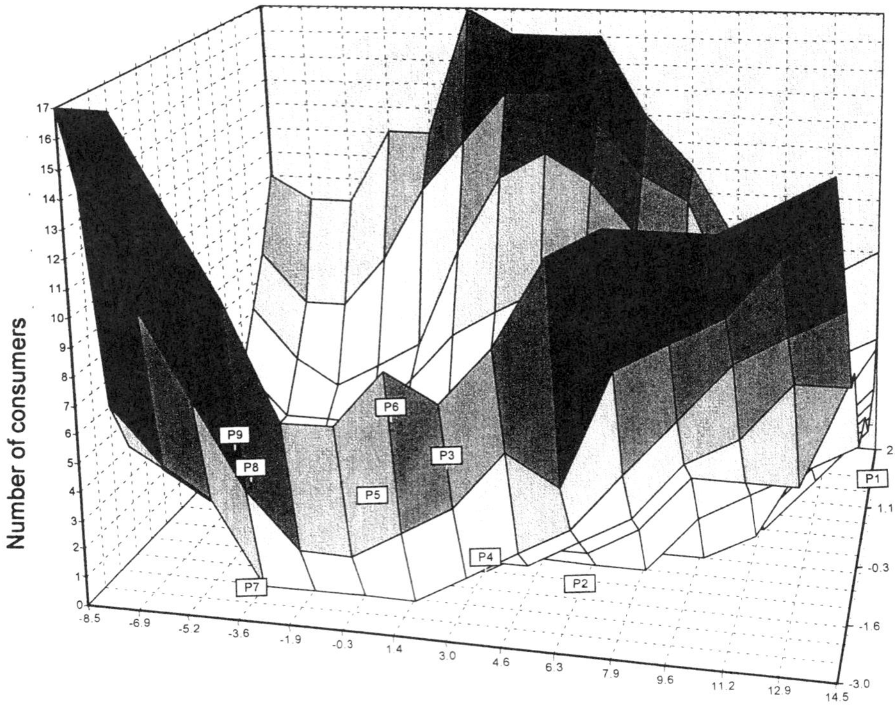

Figure 8. Example of an external preference map showing response of "satisfied" consumers

One limitation of external preference mapping is that some consumers may not be fitted by any of the models. This occurs for a variety of reasons. A low number of samples means less degrees of freedom for the regression models. By increasing the number of samples in the design and the power of the regressions, more consumers are usually fitted by a model. Mathematical models may not always fit actual behavior. There clearly are some consumers whose degree of liking for a set of products does not follow any preset predictice model. Even though some information is lost with 'unfitted consumers', the main market segments and drivers of liking are sorted out with external preference mapping.

The question of whether principal components from a PCA of descriptive or instrumental data are 'actionable' is often raised. Is the manufacturer, in the process of optimizing the sensory quality of a product, able to adjust the product formulation on the basis of principal components from a PCA of sensory or instrumental data? The answer typically is 'yes' because these principal components usually relate to ingredient or process variations (which can be acted upon). There may even be cases when principal components are more actionable than a set of individual sensory attributes which have been identified as drivers of liking.

Conclusions

Internal and external preference mapping, and internal preference clustering, as applied to hedonic ratings by consumers and descriptive ratings by trained judges for a set of 9 ice cream samples varying in fat and sugar, proved useful methods for uncovering market segmentation and identifying ways to optimize product sensory quality to satisfy a majority of consumers.

Additional information about the consumers, such as that gathered from uses and attitudes (U&A) measures, may be factored into these analyses to characterize the market segments in terms of consumer characteristics. Furthermore, individual sensory attributes may also be confirmed as drivers of liking by regressing mean liking for a market segment onto each sensory attribute using a polynomial regression, and examining the curvature of the ensuing function. Knowing what characterizes a market segment and where its preferences lie in terms of product sensory characteristics, makes for a very powerful product optimization tool.

References

1. Schutz, H. *Food Technol.* **1983**, *37*(11), 46.
2. Giovanni, M. *Food Technol.* **1983**, *37*(11), 41.
3. Arditti, S. *Food Qual. Pref.* **1997**, *8*, 323.
4. Dalliant-Spinnler, B.; macFie, H. J. H.; Beyts, P. K.; Hedderley, D. *Food Qual. Pref.* **1996**, *7*, 113.
5. Greenhoff, K.; MacFie, H. J. H. Preference Mapping in Practice. In *Measurement of Food Preferences*; MacFie, H. J. H.; Thomson, D. M. H., Eds.; Blackie Academic & Professional, London, 1994; p.137.
6. Jaeger, S. R.; Andani, Z.; Wakeling, I. N.; MacFie, H. J. H. *Food Qual. Pref.* **1998**, *9*, 355.
7. McEwan, J. A. Preference Mapping for Product Optimization. *In Multivariate Analysis of Data in Sensory Science*; Naes, T.; Risvik, E., Eds; Elsevier, Amsterdam, 1996; p. 71.
8. Monteleone, E.; Frewer, L.; Wakeling, I.; Mela, D. J. *Food Qual. Pref.* **1998**, *9*, 211.
9. Hough, G.; Sanchez, R. *Food Qual. Pref.* **1998**, *9*, 197.
10. Schlich, P. Preference Mapping: Relating Consumer Preferences to Sensory or Instrumental Measurements. In Bioflavour' 95. Analysis/Precursor Studies/Biotechnology, Etievant, P.; Schreier, P. INRA Editions, Versailles, 1995.
11. Belin-Batard, E.; Huon de Kermadec, F.; Barthelemy, J. *Psychol. Francaise* **1996**, *41*, 301.
12. Escofier, B.; Pages, J. *Analyses Factorielles Simples et Multiples, Objectifs, Methodes et Interpretation.* Bordas: Paris, 1990.
13. Huon de Kermadec, F.; Durand, J. F.; Sabatier, R. *Food Qual. Pref.* **1997**, *8*, 395.
14. Guinard, J.-X.; Zoumas-Morse, C.; Mori, L.; Uotani, B.; Panyam, D.; Kilara, A. *J. Food Sci.* **1997**, *62*, 1087.
15. Peryam, D. R.; Pilgrim, F. J. *Food Technol.* **1957**, *11*(9), 9.

Chapter 19

Measurement of Emotion in Olfactory Research

Stephen Warrenburg

International Flavors and Fragrances, 1515 State Highway, Union Beach, NJ 07735

Odors have a special link to the emotions that can be assessed by careful measurement. This chapter reviews research carried out to measure the emotional effects of odors both physiologically and using self-report methodology. Physiological monitoring of blood pressure, brain waves, various measures of the autonomic nervous system, and the startle reflex have demonstrated that pleasant versus unpleasant odors generally evoke a positive versus a negative emotional response. However, these methods have not successfully demonstrated a finer discrimination among different positive emotions, such as *stimulation* from *relaxation*. Various approaches to measuring moods, or self-reported emotions, are described. Issues relating to the dimensions and categories of mood are considered, as well as whether mood effects of fragrances are best measured in a before-after, a retrospective, or a forced-choice methodology. A simple, forced-choice technique is highlighted, called Mood Mapping®. It measures a respondent's mood association to a flavor or fragrance chosen from a set of 8 mood categories. Comparisons are made among different fragrances or flavors based on the distribution of mood votes across respondents. It has been used to screen hundreds of perfumery materials, as well as many more Living® flowers, fruits, and spices, commercial fragrances, and fragrances in development. A large database has thus been cataloged and incorporated into a perfumer's tool called the Consumer Fragrance Thesaurus. This tool gives guidance to IFF's creative staff in developing fragrances for aromatherapy based products, where moods are a principal concern, as well as other fragrances, where other attributes are of key interest (e.g., clean, fresh, etc.).

Introduction

Smelling an odor engages a perceptual and cognitive process that is especially linked to human emotions, as compared to other sensory modalities. Unlike other senses the olfactory neuroanatomy has extensive structural overlap with the limbic system, which plays a major role in emotion in both humans and other mammals[1]. A series of research studies by Rachel Herz and her colleagues has shown that odors elicit memories that are more emotional, compared to visual, auditory (linguistic), and tactile stimuli (2-8). This special ability of odors to evoke emotional memories has been anecdotally noted as the "Proust effect," from that author's reference to the joyful childhood memories evoked as an adult by the smell of a Madeleine biscuit dipped in tea. From ancient times aromatic plant extracts, or essential oils, have been used in what is called "aromatherapy" for ameliorating mental and physical ailments. In the west this tradition became established in France with Gattefosse in 1928 (9). More recently, Tisserand (10), Valnet (11), and Davis (12) have popularized aromatherapy in Europe and the US. Now there are hundreds of books and hundreds of web sites on the topic. Although aromatherapy has a long tradition, and aromatherapy is in the process of standardization in many countries (13), there is relatively little empirical research literature on the effects of the essential oils used in its practice. Stephan Jellinek has cogently raised the point that aromatherapy effects may be due to one or more of the following mechanisms: quasi-pharmacological, semantic (context- or learning-based), hedonic valence, or placebo (14) (cf. also (15)). There is virtually no existing research that has attempted to tease apart which of these mechanisms are active in purported aromatherapeutic effects. Aside from studies of essential oils, however, there is a growing research literature on the human psychological effects of odors and fragrances in general.

The term *Aroma-Chology* was coined in 1992 by the New York-based Olfactory Research Fund to refer to the scientific study of the effects of odors (fragrances) on human emotions (16). There are a number of reviews of the psychological effects of fragrance, including those focusing on emotion, or more generally the effect of odors on human behavior (17-20). A general characterization of this research is that pleasant odors tend to improve mood and induce approach behaviors, and unpleasant odors tend to disturb mood and induce avoidance behaviors, although there are a number of studies that fail to

---

[1] Lorig has argued that although the limbic system is known to be anatomically co-extensive with neural sites of olfactory processing, it is erroneous to conclude that there is a functional interplay olfaction and emotion based solely on anatomical proximity [1]. He does not deny the link between emotion and olfaction, however, but rather the justification of this link based upon neural proximity.

find significant effects of odors. Some of the positive findings include a reduction of anxiety in medical patients undergoing an MRI diagnostic scan who were exposed to the aroma chemical, heliotropine (21), enhanced altruistic behavior in shoppers who were exposed to pleasant odors of baking cookies or roasting coffee (22), and improved mood on a standard mood profile (POMS) in middle aged men and women in a daily cologne use study (23).

Physiological Effects

We at IFF became interested in the mid 1980s in whether certain aromatherapy oils could be proved to be stress-reducing. We studied the physiological effects of these odorants in the laboratory in subjects undergoing several stress analog tasks (17). Nutmeg oil was shown to reduce the blood pressure (BP) response during stress, regardless of whether it was administered by itself or as part of another "carrier" fragrance (24). When we divided the subject pool by their degree of trait (i.e., general level of) anxiety, we found that nutmeg was effective in reducing the diastolic BP response to stress in both high and low anxious subjects. The apple carrier fragrance without nutmeg was also effective in reducing the diastolic BP stress response in low anxious subjects, but was ineffective in high anxious subjects (25). While this finding is consistent with an interpretation that the nutmeg effect may be quasi-pharmacological, it is not necessarily so. It is still possible that these fragrances are stress-reducing due to their ability to distract subjects from the stress. Perhaps nutmeg is more distracting than the apple fragrance for high anxious subjects, who are not so easily distracted as their low anxious peers. More work in this area is clearly needed before a pharmacological vs. non-pharmacological explanation can be determined.

The significant effects of nutmeg on blood pressure in stress conditions led us to investigate the physiological effects of nutmeg and other aromatherapy oils in subjects at rest. We found, however, that other fragrances exert minimal effects on a variety of peripheral autonomic nervous system measures when non-stressed, resting subjects were tested. In a study of 28 college students, we examined physiological changes in facial muscle tension, galvanic skin responses, heart rate, and skin temperature as subjects smelled apple plus nutmeg, neroli, or galbanum, as well as a non-odor control (26). The three odorants were chosen to generate a range of hedonic preference. Apple/nutmeg was evaluated as pleasant, galbanum as unpleasant, and neroli gave a mixed response with half of the subjects saying it was pleasant, and the other half saying it was unpleasant. However, the three odor conditions could not be distinguished on the basis of their physiological patterns. The lack of differentiation among widely different odorants led us to conclude that peripheral physiological measures probably are not sensitive enough to discriminate the effects of odors that are of commercial interest. At the time we speculated that effects of all but the most unpleasant odors are probably too subtle to register appreciable consequences in the peripheral nervous system. A similar case was obtained in a study of spontaneous EEG brain wave responses to seven odors varying widely in character, hedonics, and perceived intensity (27). Although odor-related EEG findings were obtained, they were not interpretable in terms of these three major characteristics of odor. Indeed, individual differences in subjects' EEG responses to the odors were remarkably large.

More recently, investigators have succeeded in using physiological measurements to distinguish pleasant odorants from unpleasant ones, or from a neutral odorant or control condition in resting subjects. Alaoui-Ismaili, Vernet-Maury and their colleagues have developed a rather sophisticated pattern analysis method of six autonomic measures to distinguish among different emotional states. In an earlier study this method was able to distinguish among the visually evoked emotions: happiness, surprise, anger, fear, sadness, and disgust (28). The most relevant study from this group was of five odorants: lavender (LAV), ethyl acetoacetate (EAA), camphor (CAM), acetic acid (AA), and butyric acid (BA) (29). Two of these odors were perceived as pleasant (LAV and EAA), one as relatively neutral (CAM), and two as very unpleasant (AA and BA). Using their autonomic pattern analysis measure, the authors were able to distinguish the very unpleasant odors from the others, and the neutral odor from the two pleasant odors. They were not able to distinguish between the two pleasant odors, however. This exact same pattern was also obtained in an another study by this same group, in which two pleasant odors, vanillin and menthol, were distinguishable from a neutral hedonic odor, eugenol, and also from two unpleasant odors, propionic acid and methyl methacrylate (30). As in the above study, there was no distinguishable response between the two pleasant odors. Other research studies measuring physiological responses to odorants have produced similar findings. Ehrlichman et al. distinguished between pleasant odor, unpleasant odor, and no-odor control conditions in a study of the modulated startle reflex (31), but they were not able to distinguish among different pleasant odorants using this paradigm (32).

One initially promising finding purported to show that the brain measure, contingent negative variation, or CNV, could distinguish between *stimulating* and *relaxing* fragrances (33). The investigators asserted that an the enhanced CNV obtained to the essential oil, jasmin, was an indication of its stimulating quality, and the diminished CNV to lavender indicated its relaxing effects. However, another study, attempting to replicate this finding, found that the CNV brainwave patterns were strongly influenced by subjects' experimentally manipulated expectations about the fragrances (34). Furthermore, the *relaxed* CNV pattern obtained to lavender was apparently due to its *distracting* effects, and *not* to due to its *relaxing* effects. It is well known that CNV can be reduced by such distracting conditions as fatigue or stress, as well as by relaxation. Another promising approach would seem to monitoring of facial electromyogram (EMG) to detect subtle facial expressions of emotional reactions to odors. Unfortunately, the face does not react to odors when a person is not aware of being observed, unless the odor is very unpleasant (35,36).

Our review of the research literature and our own work all indicate to us that the use of psychophysiological methods cannot be used to discriminate among different pleasant odors in a way that could be used to characterize the emotional reaction to the odor. Thus, as a commercial enterprise, IFF has concluded that these methods are not fruitful for the task of determining the "best" or most appropriate fragrance to employ for a given product, whatever the category. Physiological measures can distinguish between grossly unpleasant and pleasant odors, it is true. But the interesting question to us was whether these methods distinguish among rather pleasant odors differing in some qualitative way, associated with odor character or with some ingredient having special aromatherapeutic properties. In essence, it appears that the effects of odors, while varying widely in perception (mentally), are just too subtle be detectable using physiological sensors currently available.

Theoretical Approaches to Emotion

The most basic processes are difficult to define in psychology, which is nowhere more true than for "emotion." As paraphrased from Wolman's Dictionary of Behavioral Science (37), emotion is a complex reaction involving physiological changes from a state of equilibrium which: a) is subjectively experienced as feeling, and b) causes bodily changes that prepare the individual to act overtly. "Mood" is defined here as the subjective aspect of emotion, which is a transient phenomenon lasting from several seconds to several hours. The term "affect" is frequently used in the psychological literature as a synonym for "emotion," but in certain schools of psychology and psychiatry it's meaning is more specific. For our purposes, we are interested in emotional states of short duration, on the order of several minutes, and in particular in the subjective, or mood, component of those states. It is these fleeting subjective responses that we believed would be sensitive to odor stimuli, particularly fragrances.

Most current theorists conceptualize emotion as having two dimensions, but the description of these two dimensions has fallen into two distinct theoretical camps. One camp, led by James Russell, asserts that emotion is most logically viewed as having a valence dimension (pleasure-displeasure) and an activation dimension (38,39). A second camp, led by Watson and others, views the two dimensions as positive affectivity and negative affectivity (40,41). Both approaches are closely related to each other, in that they both recognize that emotions have two orthogonal dimensions, and that this space contains a positive-negative factor and an activation or energy factor. They differ in how they draw these two factors into the two dimensional space. In my view Russell's valence and activation model is simpler and easier to conceptualize. Watson's approach is to define the pole of the high activation plus positive valence as "high positive affect," (or positive energy) and the high activation plus negative valence pole as "high negative affect" (negative energy). There is much to be said on either side of the debate, which is a lively one, even after 20 years running. Clearly any approach to measuring mood in relation to fragrance must consider these valence and activation dimensions.

Psychological Self-Report Methods

The major approach to the study of mood has been through the use of mood adjectives. Lists of such terms are either checked off ("yes"/"no") or are rated on simple scales with 3 to 9 numbered categories. Terms are chosen to represent the domain of mood states that are of theoretical or practical interest to the researcher. Three methods of data collection have been used. By far the most frequent technique has been to administer the mood list to relatively large numbers of subjects who complete the form once only, and to compute relationships among terms across all subjects (called the R-technique). The second technique is to have a small number of subjects complete the form on many occasions, and to compute relationships across occasions for each subject individually (P-technique). The third method is to administer the mood list twice to each subject, generally before and after an experimental manipulation, and to calculate the change scores for each term, which then become the raw data in computing relationships across subjects (dR-technique). We began our research on fragrance by using this latter technique, and have adapted it so that each subject provides before/after evaluations of multiple fragrances.

Once the relationships among the mood terms have been calculated, usually in the form of an inter-correlation matrix, a statistical method of simplification is applied, in order to reveal the number and identity of groupings of similar terms. The most frequent procedure applied to this problem has been factor analysis (FA). With one exception, all of the studies reviewed below have used this technique. Unfortunately, as described below, FA has a number of shortcomings that hinder its usefulness. An alternative, more recent, and more effective approach is multidimensional scaling. First, we will review several studies that have employed FA.

Nowlis is one of the early pioneers in this area of research, and his mood adjective checklist (MACL) became the prototype for all who followed him (42). His earliest work in the 1950s was designed to describe mood changes following administration of psychoactive drugs. He reasoned that there were four bipolar dimensions to mood: activation/deactivation, in control/out of control, socially oriented/withdrawn, and pleasantness/unpleasantness. Accordingly he formed the MACL using 130 terms, and administered it to over 400 students in a variety of situations (R-technique). His rating scale had four points, which he assumed (incorrectly, as shown in later research) to form a linear scale: "definitely describes," "slightly applies," "cannot decide," and "does not apply." FA was applied to these data, and 12 monopolar factors were yielded: aggression (anger), anxiety, surgency (pleasantly excited), elation, concentration, fatigue, social affection, sadness, skepticism, egotism, vigor, and nonchalance. "Monopolar" denotes that all terms grouped in the factor contribute to higher, or more positive scores on that factor. In contrast, "bipolar" factors have some terms that contribute to lower, or more negative scores on that factor. For example, a bipolar "happy/sad" factor would have both "happy" terms adding positively and "sad" terms adding negatively to the factor score.

Other researchers using similar approaches obtained a different set of factors. McNair, Lorr and their colleagues developed the Profile of Mood States (POMS) based initially on studies of drug effects in psychotherapy patients (43). A similar methodology was applied and six monopolar factors were obtained from FA: tension/anxiety, anger/hostility, depression/- dejection, vigor/activity, fatigue/inertia, and confusion/bewilderment. Izard investigated emotion as revealed in facial expression, and thereby defined ten "fundamental emotions" (44,45) He then developed a 67 item mood adjective list with a 5 point scale, called the Differential Emotions Scale. Using the R-technique and FA, he obtained results yielding 11 monopolar factors. Subsequent revisions led to a 33 item list that yielded ten monopolar factors corresponding to his ten fundamental emotions: joy, interest/excitement, surprise, sadness, anger, disgust, contempt fear, shame/shyness, and guilt. Cattell used P-factoring and dr-factoring on mood terms and derived 8 - 13 monopolar factors at a "first-order" level and 4 bipolar factors at a higher, "second-order" level (46). That is, factor scores were calculated from the original data by averaging terms on each factor, and then a second FA was done on these factor scores. The results showing a higher order factor structure is meaningful in that it indicates that whereas a number of lower order mood states are distinguishable, these states are related to each other in a structured, higher order manner. We will return to this concept of higher order structure later in this chapter.

The problems with this FA research on moods is the diversity in number and content of the factors obtained. These differences are a function of several variables. First, the list of mood adjectives varies widely from study to study, thus affecting the number and types of moods available for representation in the FA. In particular, Izard "handpicked" near synonyms within each category and excluded "in-between" terms (45). On the other hand, McNair et al. picked primarily negative mood terms, appropriate to their psychopathological orientation (43). Others, like Nowlis, picked from a broad range of mood terms (42). Other problems are response bias tendencies of the subjects, and difficulties with the rating scale itself. These problems have operated to yield monopolar, as opposed to bipolar factors, and many small factors as opposed to few large factors.

"Common sense" of the layman suggests that mood states are related along bipolar dimensions, such as "happy"/"sad" rather than monopolar dimensions in which emotional states are largely independent. The failure of the majority of the above studies to obtain bipolar structure would seem to indicate a weakness of this statistical approach. An exception to this research finding was the higher order FA of Cattell, suggesting that ordinary FA may be missing something (46). According to Meddis (47), Russell (38) and others, the tendency for subjects to vary in the degree to which they endorse terms in general (i.e., independently of term content) is a key problem in analyzing mood lists. This tendency is termed "response acquiescence," and it can be calculated, and statistically removed from a subject's response profile. When this is done, previously monopolar factors tend to coalesce into fewer bipolar factors. Meddis and Russell identified a second major variable influencing factor number and polarity: the rating scale used in marking a response to a term. Nowlis' rating scale, described above, contained a "cannot decide" category in the middle of the scale (42). However, examination of the distribution of responses to this scale indicates fewer people check "cannot decide" than the scale positions on either side. This non-symmetric response characteristic also tends to break apart large bipolar factors into smaller monopolar ones. Indeed, any non- symmetry in the response distribution, like having the most frequent response being at the low end of the scale (e.g., "not at all"), impedes the statistical basis for FA.

Fortunately, an alternative to FA has been developed, multidimensional scaling (MDS), which is not influenced greatly by the above variables that interfere with FA. The primary advantage of MDS over FA is that the former does not require the scale being measured (e.g., "happiness/sadness") to be a linear scale, as does FA. MDS only requires the scale to vary monotonically, i.e., to be continuous and not fold back on itself. Curves in the scale are permitted. This easier requirement in MDS tend to keep terms at the extreme end of a scale (e.g., "ecstatic" on a happiness scale) together with intermediate terms, whereas FA tends to break these terms into two factors. MDS also requires fewer decisions to arrive at a final solution. In FA, decisions must be made at many more points, concerning method of FA, number of factors to be extracted, and the orientation, or rotation, of the dimensions. More decisions leads to more arbitrariness. The net result is that MDS usually allows data to be plotted in fewer dimensions than FA. Typically, good solutions are found in two dimensions, thus allowing readily visualizable results. Russell applied MDS to subjects ratings of the semantic similarity of 28 mood terms. This task asks subjects not how they feel themselves, but rather how they think the mood terms actually relate to each other in terms of their meaning. Two bipolar dimensions were revealed: pleasantness/unpleasantness, and arousal/drowsiness. A plot of these 28 terms in the two dimensional map is presented in Figure 1. Notice from this figure that the terms fall in a circle, or "circumplex."

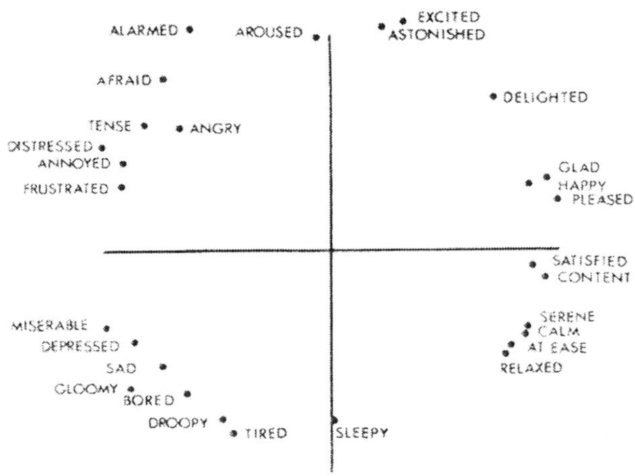

Figure 1. MDS solution for 28 affect words (From Russell (38)).

Mood Measurement of Fragrance Effects

At IFF we developed a mood checklist based on 44 adjectives chosen from the mood questionnaires described above, plus additional items that tapped whether subjects felt more sensuous, romantic, etc. in response to fragrances that they smelled. The checklist asks panelists to rate how they felt "right now" using a 5 point scale. The questionnaire was administered before and after smelling each fragrance, with controls for order effects, and sufficient time between samples to allow for return to baseline. Fragrances consisted of a range of types from perfumery ingredients to fine fragrances, administered in separate panels, conducted in a central location facility. Multiple fragrance evaluations from 18 panels were pooled together in the form of change scores from the pre-fragrance baseline to the post-fragrance condition. An inter-correlation matrix of all 44 items was formed across all responses ($n > 2000$). MDS was applied to this matrix, and a two dimensional solution was obtained. The MDS map looked quite similar to Figure 1, in that the mood adjectives fell into a circumplex shape with the positive mood terms (happy, relaxed, etc.) on the right side and the negative terms (anxious, irritated, etc. on the left. The terms at the top were higher in arousal (stimulated, irritated, stressed, etc.) and those at the bottom were lower in arousal (apathetic, relaxed, etc.). Semantically similar terms fell near each other, forming natural clusters, from which we identified 8 categories of mood adjectives. Schematically, this is shown in Figure 2. These clusters of terms were examined using various measures of reliability (internal, split-half, test-retest) and found to be psychometrically robust.

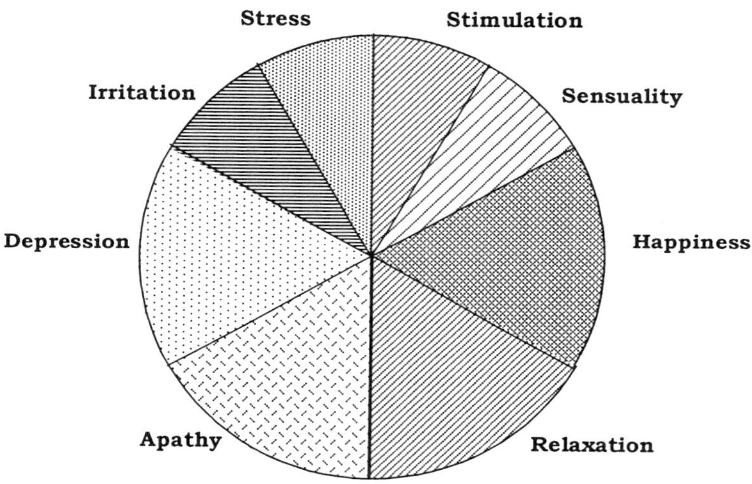

Figure 2. Mood categories formed by MDS of 44 mood adjectives.

The clusters were then used to form mood category scales and the panelists' responses to various fragrances were examined and compared. The mood category profiles that were formed by this method were also examined for reliability. Unfortunately, these analyses revealed that the average before-after mood effects of fragrances are very small in general, namely less than 0.75 points on the 5 point scale used. Furthermore, the test-retest reliability of the profile for a given fragrance was often quite inadequate, and the statistical power of this method to reveal small mood differences between fragrances was quite low for sample sizes of less than 100 panelists.

Mood Profiling, Version 2

Given the low "signal-to-noise ratio" of the before-after method, we can tell that the direct mood effects of fragrance are rather small on average, when panelists are evaluated in a central location facility. For this reason, we developed an alternate technique of mood profiling. In this procedure panelists smell, then evaluate a fragrance retrospectively, indicating how it made them feel by marking on a line how much a given mood category has changed. Each of the mood categories is rated separately on a scale that ranged from -50 to +50. We presented mood profiles of five Living Flowers® in our 1993 article, "Mood Benefits of Fragrance" (17). Pleasant smelling fragrances have profiles in which the positive moods increase up to an average of 40% of the maximum, and the negative moods decrease up to 40% of maximum. The test-retest reliability of this method was much better than that of the previous one, as was its statistical power in discriminating between fragrances in panels of n = 50 - 75. In particular, it was quite easy to distinguish among fragrances that were especially relaxing versus stimulating, or sensuous. The more pleasant fragrances (higher hedonic ratings) tended to have mood profiles that took the form of a stereotyped "halo" response, however. This limited the discriminability of the technique at the higher end of the hedonic spectrum.

Mood Mapping®

We sought yet another technique, therefore, that would yield a highly reliable, yet discriminating method for mood profiling fragrances. This technique uses a forced choice procedure in which a panelist picks the single "mood category that best fits [or matches] the fragrance" being evaluated. We discovered that this simple "voting" technique allowed panelists to focus their attention better than either of the previous methods, resulting in improved discriminability among fragrances as well as higher reliability of the profile. The "votes" for each fragrance are tallied, and can be analyzed for significance using a chi-square test. By compiling a database for hundreds of fragrances, we can map these using MDS. Figure 3 on the next page displays how two similar materials, Living® versus picked mint, result in slightly different mood distributions and consequent locations in the MDS "Mood Map®." Living® fragrances are reconstituted formulae based upon IFF's technique of chemical analysis of the headspace above a living plant (48). This technique recreates an aroma that is closer to the living entity than traditional methods of perfume extraction.

For both fragrances the modal response is "stimulated," as would be expected for mint. Unlike living mint, picked mint has a slight "off-note" due to a decay metabolite, which is apparently detected by a minority of panelists who choose "irritated" or "stressed" rather than "stimulated." Each mood category, in effect, "pulls" in a given direction from the origin to determine the map location of a fragrance based on its mood profile. The mood "stimulated" pulls towards the one o'clock position, "relaxed" towards the four o'clock position, "apathetic" towards seven o'clock, "irritated" towards nine o'clock, etc. The resulting map location is the vector sum of these individual mood effects, thus small, consistent mood profile differences can result in a very noticeable effect in the Mood Map®.

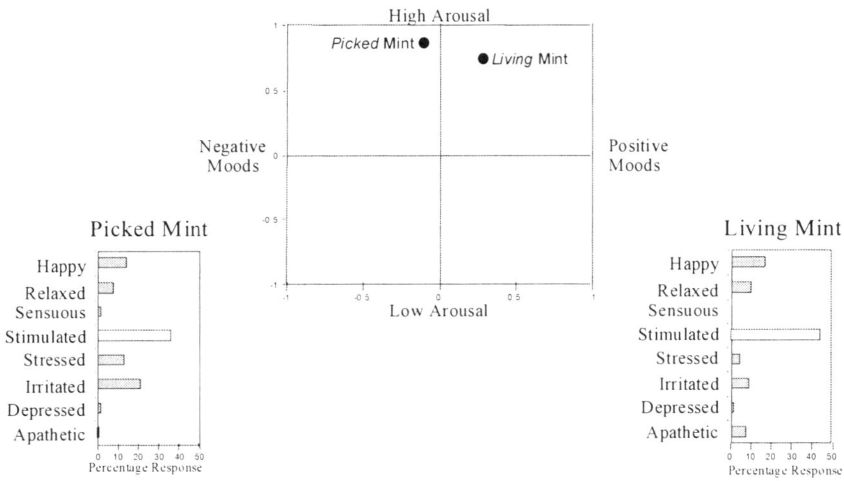

Figure 3. Mood Map® of Living® versus picked mint

Figure 4 on the next page presents IFF's Mood/Ingredient Map. The map was obtained by using MDS to map hundreds of perfumery ingredients on the basis of the mood distributions provided by our panelists. The X axis is highly correlated with pleasantness ratings of the fragrance materials (valence dimension), while the Y axis can be thought of as the arousal or activation dimension. The drawn mood areas bound all materials whose modal mood response was as indicated. Finished fragrances can be mapped into this same space using similar methods. Perfumery materials falling into the negative region tend to be "nuance" fragrance notes that occur at low concentration in a finished fragrance to give it specific fragrance "characters." We have established mood mapping databases in 10 countries around the globe.

IFF's perfumers use information from these Mood Mapping® databases to create fragrances having specific mood properties. Our mood research has assisted perfumers in the creation of fragrances for many different aromatherapy products, as well as for products ranging from women's fine fragrances to personal care (e.g., shampoo, shower gel), and home care, such as laundry detergents, candles and room sprays. Mood Mapping® allows confirmation that the fragrance fits the "mood" of the product concept, including support of "soft" product claims for aromatherapy products. A "soft" claim alludes to mood benefits without actually stating them outright (e.g., "revitalizing, energizing fragrance..."). Over the past 10 years there has been a blossoming of public awareness of and interest in aromatherapy (49,50).

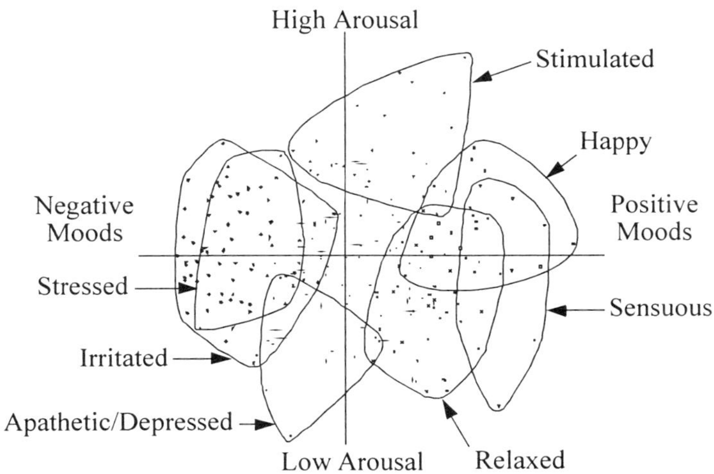

Figure 4. Mood Map® of fragrance materials

Mood Mapping® of flavors

The Mood Mapping® protocol can also be applied to both to the evaluation of flavor aromas and tastes. We have used this approach with a variety of beverage products, gums, candies, and other food products. Other than an occasional puzzled look initially, panelists usually have no trouble

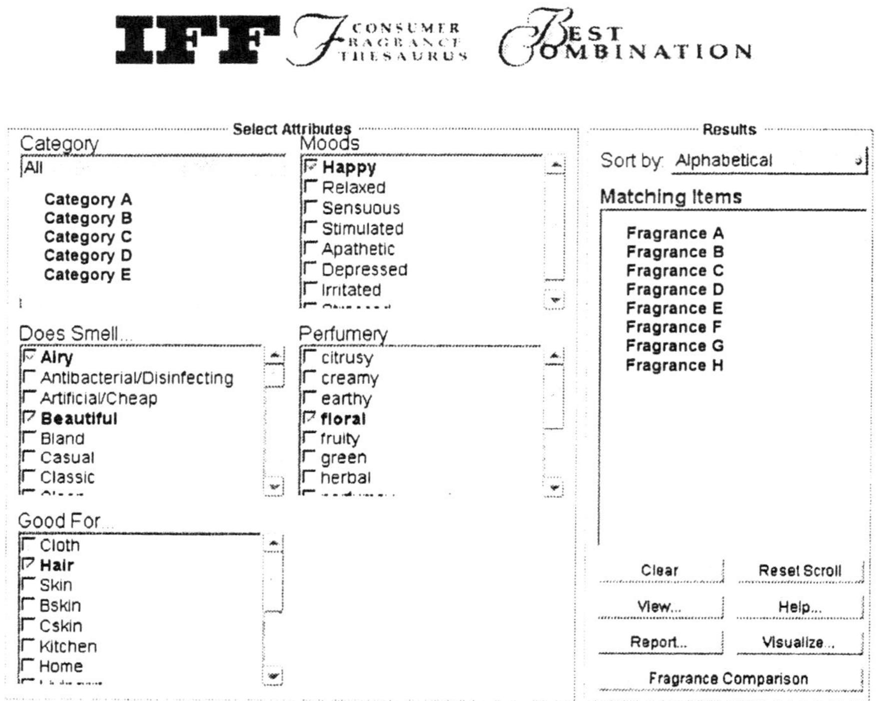

Figure 5. Consumer Fragrance Thesaurus - Best Combination page

responding to the "best fitting mood" question. Data analysis and mapping can be handled in the same way as with fragrances. Beverages, gums, and candies lend themselves well to this approach.

Beyond Mood Mapping®: The Consumer Fragrance Thesaurus

Moods are just one type of response to a fragrance, however. What about the imagery and other mental associations evoked by fragrance? Our creative staff wanted much more information about the various ways that fragrance affected the consumer. Furthermore, they wanted the information to be easier to access and use. Recently we have described our ongoing program of research that incorporates Mood Mapping® into a more extensive evaluation of the mental imagery evoked by fragrances (51). Using this approach, we developed an intranet-based database of consumer evaluation results for hundreds of fragrances and fragrance components. We call this application the Consumer Fragrance Thesaurus (CFT), whose main interface is shown in Figure 5.

The user clicks on the combination of attributes, moods, applications (Good For), and/or perfumery descriptions that capture his/her desired features of the fragrance. The Matching Items box lists the fragrances offering the best combination of these characteristics. Each fragrance's complete profile can be viewed, as well as the expert description of the fragrance character, and other technical information. The resulting fragrance profiles in the database have also been mapped using our Mood/Attribute Mapping™ method, that allows visualization of where fragrances lie in the multidimensional space formed by the attributes and moods that comprise the personality of a fragrance. The valence and activation dimensions described above figure prominently in the Mood/Attribute Mapping™ of the CFT database.

Directions for Future Research

Much remains to be learned about how odors and fragrances affect human emotions. It is very possible that future research will discover physiological measures that are sensitive to the subtle differences in emotional impact of fragrances that now are only distinguishable by self-report. Several self-report methods have been described here, one of which appears best suited to measuring the faint mood associations that are evoked or *primed* by olfactory stimuli. We believe that odors prepare, or *prime*, the mind to make learned cognitive associations and images that are linked to these moods. Many of these linked odor/mood/cognitive associations originate in early experience, and some may even be innate. Further research on the developmental processes involved promise to provide a fascinating avenue into how humans derive meaning from their olfactory world.

References

1. Lorig, T., *On the similarity of odor and language perception.* Neuroscience and Biobehavioral Reviews, 1999. **23**: p. 391-398.
2. Herz, R.C. and G.C. Cupchik, *An experimental characterization of odor-evoked memories in humans.* Chemical Senses, 1992. **17**(5): p. 519-528.
3. Herz, R. and G. Cupchik, *The emotional distinctiveness of odor-evoked memories.* Chemical Senses, 1995. **20**: p. 517-528.
4. Herz, R. and T. Engen, *Odor memory: Review and analysis.* Psychonomic Bulletin & Review, 1996. **3**(3): p. 300-313.
5. Herz, R., *The effects of cue distinctiveness on odor-based context-dependent memory.* Memory & Cognition, 1997. **25**(3): p. 375-380.
6. Herz, R., *An examination of objective and subjective measures of experience associated to odors, music, and paintings.* Empirical Studies of the Arts, 1998. **16**(2): p. 137-152.
7. Herz, R., *Are odors the best cues to memory? A cross-modal comparison of associative memory stimuli.* Annals of the New York Academy of Science, 1998. **855**: p. 670-74.
8. Herz, R., *A naturalistic study of autobiographical memories evoked to olfactory versus visual cues: Testing the Proustian hypothesis*, . 1999, Monell Chemical Senses Center.
9. Gattefosse, R.M., *Aromatherapie.* 1928, Paris: Giradot editeur.
10. Tisserand, R.B., *The art of aromatherapy.* 1977, Rochester, Vermont: Healing Arts Press.
11. Valnet, J., *The practice of aromatherapy.* 1980, Essex, England: C.W. Daniel Co., Ltd.
12. Davis, P., *Aromatherapy: An A-Z.* 1988, Saffron Walden: D. W. Daniel.
13. Price, S. and L. Price, *Aromatherapy for Health Professionals.* 2nd ed. 1999, New York: Churchill Livingstone. 391pp.
14. Jellinek, J.S., *Psychodynamic odor effects and their mechanisms.* Perfumer & Flavorist, 1997. **22**: p. 29-41.
15. Lawless, H., *Effects of odors on mood and behavior: Aromatherapy and related effects*, in *The human sense of smell*, R.L.D. D.G. Laing, and W. Breipohl, Editor. 1992, Springer-Verlag: New York. p. 361-386.
16. Fund, O.R., *URL http://www.olfactory.org*, .
17. Warren, C. and S. Warrenburg, *Mood benefits of fragrance.* Perfumer & Flavorist, 1993. **18**: p. 9-16.
18. Jellinek, J.S., *Aroma-Chology: A status review.* Perfumer & Flavorist, 1994. **19**: p. 25-49.

19. Ehrlichman, H. and L. Bastone, *Olfaction and emotion*, in *Science of Olfaction*, M.J. Serby and K.L. Chobor, Editors. 92, Springer-Verlag: New York. p. 410-438.
20. Ehrlichman, H. and L. Bastone, *The use of odour in the study of emotion*, in *Fragrance: the psychology and biology of perfume*, S.V. Toller and G.H. Dodd, Editors. 1992, Elsevier: Barking, Essex, England. p. 143-159.
21. Redd, W.H., *et al.*, *Fragrance administration to reduce anxiety during MR imaging.* Journal of Magnetic Resonance Imaging, 1994. **4**: p. 623-626.
22. Baron, R., *The sweet smell of...helping effects of pleasant ambient fragrance on prosocial behavior in shopping malls.* PSPB, 97. **23**(5): p. 498-503.
23. Schiffman, S., *Pleasant odors improve mood of women and men at 3mid-life*, in *Compendium of Olfactory Research*, A.N. Gilbert, Editor. 1995, Kendall/Hunt: Dubuque, Iowa. p. 97-103.
24. Warren, C.B., *et al.*, *Method of Causing the Reduction of Physiological and/or Subjective Reactivity to Stress in Humans Being Subjected to Stress Conditions*, in *US Patent*. 1987: United States.
25. Warrenburg, S. and S. Warren, *unpublished data*, . 2000.
26. Warrenburg, S. and G.E. Schwartz, *A psychophysiological study of three odorants.* Aroma-Chology Review, 1995. **4**(1): p. 5-7.
27. Klemm, W.R., *et al.*, *Topographical EEG maps of human responses to odors.* 92. **17**(3): p. 347-361.
28. Collet, C., *et al.*, *Autonomic nervous system response patterns specificity to basic emotions.* Journal of the Autonomic Nervous System, 1997. **62**: p. 45-57.
29. Alaoui-Ismaili, O., *et al.*, *Odor hedonics: Connection with emotional response estimated by autonomic parameters.* Chemical Senses, 1997. **22**: p. 237-248.
30. Alaoui-Ismaili, O., *et al.*, *Basic emotions evoked by odorants: Comparison between autonomic responses and self-evaluation.* Physiology and Behavior, 1997. **62**(4): p. 713-720.
31. Ehrlichman, H., *et al.*, *Startle reflex modulation by pleasant and unpleasant odors in a between-subjects design.* Psychophysiology, 1997. **34**: p. 726-729.
32. Ehrlichman, H., *et al.*, *Startle reflex modulation during exposure to pleasant and unpleasant odors.* Psychophysiology, 1995. **32**: p. 150-154.
33. Torii, S., *et al.*, *Contingent negative variation (CNV) and the psychological effects of odour.* 88: p. 107-120.
34. Lorig, T.S. and M. Roberts, *Odor and cognitive alteration of the contingent negative variation.* 90. **15**(5): p. 537-545.

35. Gilbert, A.N., A.J. Fridulund, and J. Sabini, *Hedonic and social determinants of facial displays to odors.* Chemical Senses, 1987. **12**(2): p. 355-363.
36. Jancke, L. and N. Kaufmann, *Facial EMG responses to odors in solitude and with an audience.* Chemical Senses, 1994. **19**(2): p. 99-111.
37. Wolman, B.B., *Dictionary of Behavioral Science.* Second ed. 1989, New York: Academic Press.
38. Russell, J., *A circumplex model of affect.* Journal of Personality and Social Psychology, 1980. **39**(6): p. 1161-1178.
39. Barrett, L.F. and J.A. Russell, *The structure of current affect: Controversies and emerging consensus.* Current Directions in Psychological Science, 1999. **8**(1): p. 10-14.
40. Watson, D. and A. Tellegen, *Toward a consensual structure of mood.* Psychological Bulletin, 1985. **98**(2): p. 219-235.
41. Cacioppo, J.T. and G.G. Berntson, *The affect system: Architecture and operating characteristics.* Current Directions in Psychological Science, 1999. **8**(5): p. 133-136.
42. Nowlis, V., *Research with the mood adjective check list*, in *Affect, Cognition, & Personality*, S.S.T.C.E. Izard, Editor. 1965, Springer: New York. p. 352-389.
43. McNair, D.M., M. Lorr, and L.F. Droppleman, *Manual: Profile of mood states.* 1971, San Diego, CA: Educational and Industrial Testing Service.
44. Izard, C.E., *Patterns of emotion.* 1972, New York: Academic Press.
45. Izard, C.E., *The face of emotion.* 1971, New York: Appleton-Century-Crofts.
46. Cattell, R.G., *Personality and mood by questionnaire.* 1973, Champaign, IL: IPAT.
47. Meddis, R., *Bipolar factors in mood adjective checklists.* British Journal of Social and Clinical Psychology, 1972. **11**: p. 178-184.
48. Mookherjee, B.D., R.W. Trenkle, and R.A. Wilson, *The chemistry of flowers, fruits, and spices: live vs. dead - a new dimension in fragrance research.* Pure & Applied Chemistry, 1990. **62**(7): p. 1357-1364.
49. Klepacki, L., *Altering the mood market - aromatherapy to the mass market.* Women's Wear Daily, 1998. **175**(58 (March 27)): p. 4.
50. Butcher, D., *Aromatherapy - its past and future.* Drug & Cosmetic Industry, 1998. **162**(3): p. 22-24.
51. Warrenburg, S., *The Consumer FragranceThesaurus: Putting consumer insights into the perfumer's hands.* The Aroma-Chology Review, 1999. **8**(2): p. 4-7.

Analytical Approaches

Chapter 20

In Vitro Taste Sensors: Technology and Applications

Glenn Roy[1] and John T. McDevitt[2]

[1]Ingrediant and Flavor Technology, Pepsi-Cola Company, 100 Stevens Avenue, Valhalla, NY 10595

[2]Department of Chemistry and Biochemistry, University of Texas, Austin, TX 78712

Specific attributes of the taste, aroma and texture of a food product or beverages drives consumer preference and purchase intent. Generally, in product development stages, products are tested in sensory panels to determine which variant of a chosen set is the ideal candidate for further consumer testing leading to commercialization. However, sensory tasting panels are fatiguing and require commitment on the part of panelists to attend and be consistent in their evaluation of randomly presented samples. In the absence of a willing, precise and trained human taste panel, as well as the need for a high through-put screening tool capable of real time assessment of product quality, the need for *in vitro* taste sensors correlated to human sensory perception is growing. In the industrial and academic limelight, one such technology is affectionately called an electronic tongue.

Quality Control measurement in the food and beverage industry requires a correlation of some specific analytical finding to some specific sensorially derived standard. For the senses of taste and smell, many kinds of chemical substances are received simultaneously by taste and olfactory cells. Taste and odor sensors could make it possible to automate quality control procedures for foods, drugs and cosmetics and facilitate taste modification efforts towards optimal sensory acceptance. The subject technique of this chapter is *in vitro* Taste Sensors or the Electronic Tongue and was recently overviewed *(1)*. The general utility of taste sensor technology is applied and tabulated/referenced for applications in Food and Beverage Quality Control. Unfortunately, the taste sensor devices' use in pharmaceuticals is poorly referenced but most likely the highest potential for use. Would you sit on a QC panel for liquid codeine formulations day after day?

A commercial QC device (SA 402) is available from Anritsu Corp. (Japan) although no US installations are reported. The Anritsu Corp. has described their innovation of multichannel taste sensors, multichannel electrodes, quantification of food taste, and quality control of foods *(2)*. Alpha M.O.S. (France) provides US installations of their liquid and taste analyzer (α-Astree). Recent work in the area of bead arrays as taste sensors developed at The University of Texas at Austin appears to be promising due to the large range of analytes that can be detected *(3)*. It should be mentioned, this chapter is not addressing the in-mouth pH and conductivity measurement of Univ. of Nottingham *(4)* or the development of The Electronic Nose *(5)* (gas sensors) despite their similar configuration to a liquid phase sensor. We will also not focus on hybrid modular sensor systems for odors and flavors *(6)*. "Instrumentation and Sensors for the Food Industry" recently became a Food Technology Reference book edited by Erika Kress-Rogers of ATI Sensor Applications Ltd., in the United Kingdom. These techniques for food measurement, analysis and evaluation are not taste sensors per se, but specific devices to detect temperature, moisture by near infrared resonance absorption (NIR), color, microbial growth, particle size, dielectric constant (ionic strength), flow rate for rheology, pH, enzymatic activity and individual ingredient concentrations.

The Technology of Sensing Devices

This section details the numerous materials of sensor composition and how liquids effect a detectable change in electricity or light. Several artificial membranes include lipids, polyvinyl chloride (PVC) and dioctyl phenyl phosphonate (DOPP), as well as other phospholipid films. The primary technique of membrane preparation is known as Langmuir-Blodgett (LB) filming. Fibers are constructed by transferring mono-layers floating on a water surface to a solid substrate, and are widely applied to molecular electronic and bioelectronic devices because of characteristics such as the thickness and molecular arrangement control at the molecular level *(7)*. Glass and metal electrodes, optical fibers and Light Addressing Potentiometers (LAP) offer visual cues to the presence of key susbtances. Surface Photovoltage, Surface Plasmon Resonance (SPR), and QSFET will be referenced techniques. The simplest device is comprised of Water/Oil/Water Liquid Membranes or Agar-Gel Salt Bridges. Most recently an innovative approach is to utilize natural sensors with selective enzyme LB films and even adzuki bean roots.

The electrical or optical identifications of numerous membrane types and other surface active substrates are identifying certain taste modalities by treating them with multivariate analysis and pattern recognition to correlate to sensory differences and, in some instances, preferences. In each compositional test the measures and interpretations of the effects on the basic tastes are the changes in electric potential due to partitioning of ions or hydrophobicity of tastant near the sensor giving a result as a principal component analysis commonly seen in sensory circles. The principles and applications of taste sensors based an lipid/polymer membranes and multichannel electrodes are only recently published in English. After many years of Japanese publications, various aspects of the taste recognition and signal processing are finally discussed, along with practical examples *(2,8,10)*. This chapter makes an effort to provide English references. The mathematical treatment of the membrane potential and membrane electrical resistance for lipid membranes are derived according to the theoretical framework of the charged membrane-aqueous electrolyte system. Experimental results on the interaction between taste substances and a lipid membrane demonstrate theoretical fits to the partition coefficient of a taste substance between the membrane and aqueous phases, the association constant between a taste substance and lipid molecules, and the mobility of a taste substance. The agreement was satisfactory except for high concentration ranges of sweet and bitter substances *(9)*. A transducer plays a role of transforming taste information generated by chemical substances into electrical potential changes. Some evidence suggests lipid membranes are stable more than half a year at room temperature. The lipid/polymer membranes for transforming information of taste substances into electrical signals are interpreted by computer. The resultant Principal Component Analysis with eight membranes (channels) can group taste modalities (see Fig. 1 below). The sensor responds quantitatively to different tastants and interprets data when synergistic and suppression effects are present. The sensitivity, reproducibility, and durability are said to be superior to those of human taste *(10)*.

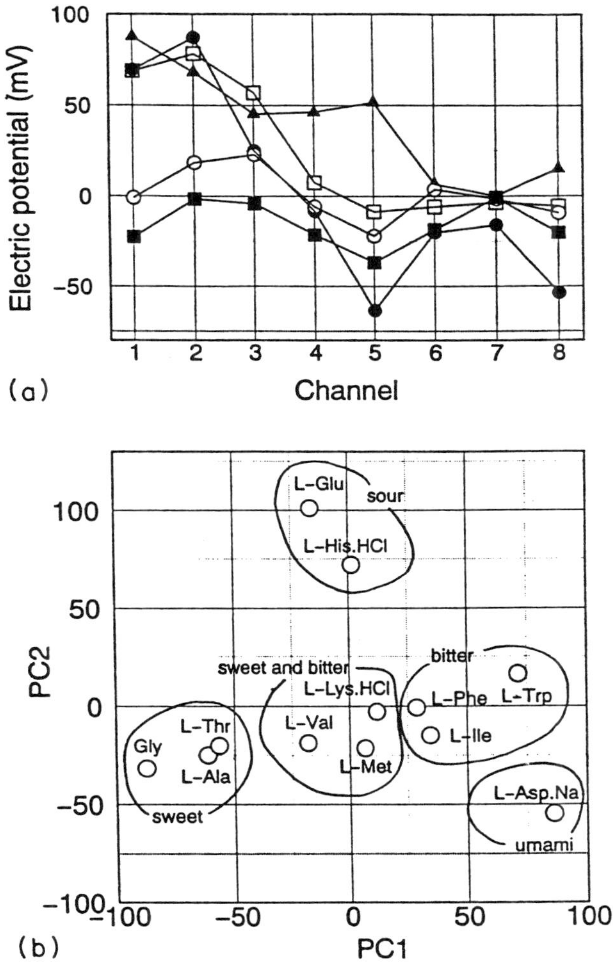

Figure 1. Taste modality differentiation by principle component analysis of electronic responses from a lipid/polymer membrane. (a) Response patterns for amino acids and (b) principal component analysis. Symbols ▲,□,●,○, and ■ denote L-glu (10mM), L-trp (10mM), L-asp. Na (100mM), L-val (100mM) and L-ala (100mM), respectively. Reprinted from Figure 9 (a) and (b), reference 40 with permission from Elsevier Science.

Adsorption of quinine ions onto the membrane surfaces and then into the membrane helps quantify bitterness and also fabrication of receptor membranes for the taste sensor *(11)*. Electrical characteristics of monoolein PVC membranes alone have a slow time response, while mixing plasticizers into the membrane increases the time response and also lowers the detection threshold *(12)*. Static and dynamic elec. responses to various kinds of taste substances in lipid/polymer membranes are different *(13)*. Responses to quinine showing bitterness decreased systematically with increasing the quantity of negatively charged lipids with PVC and plasticizer contained in the membrane, whereas the response did not depend on differences in the lipid hydrocarbon chain length *(14)*. Charged membranes do make a difference in that lipid/PVC/DOPP membranes change from the weakly charged state to the fully charged state by dissociation of H^+ from the lipid with increasing NaCl concentration *(15)*. Reports have examined dynamic responses of chaotic self-contained oscillations of Millipore membranes infiltrated with dioleyl phosphate (DOPH) and dioleoyl phosphatidylethanolamine, and DOPH and cholesterol to taste substances. Apparently, even more information for taste sensing can be obtained from the dynamic responses of those membranes in addition to the membrane infiltrated with only DOPH *(16)*. Characterization of the dynamic behavior of the membrane potential using chaotic attractors was shown to be an effective method of measuring taste *(17)*. Taste sensing based on impedance measurement uses different information from potentiometric measurements, and hence will be effective for the detection of nonelectrolyte taste substances such as sucrose *(18)*. Similarly, the bead-based chip array methodologies which combine receptors designed for small molecules, electrolytes, acid-bases, protein, and DNA with a highly parallel method of optical readout has been shown recently to extend the possibilities for multianalyte detection of complex fluids *(19)*. The four basic taste substances: sour, salty, bitter and sweet, can all elicit distinct sensor stimulations. Umami substances have been investigated as the fifth independent taste *(20)*. An improved method of determining the taste of a sample solution by using taste sensors involves alternating the immersion between the sample and a reference standard at least twice and then measuring the difference *(21)*. Washing membranes with ethanol (30%) containing HCl or NaOH and NaCl is recommended for compositions of dioctylphosphate, cholesterol, oleic acid, decyl alcohol, trioctylmethyl ammonium chloride, oleylamine, lecithin, or phosphatidylserine (especially when samples are coffee, tea, wine, beer, or milk) *(22)*. The effect of sour (HCl), sweet (sucrose) and salty (NaCl) substances on the electrical characteristics of a lecithin-cellulose membrane construct is reported *(23)*. A popular plasticizer material is dioctyl phenylphosphonate (DOPP). It is known that the membrane made of PVC and DOPP shows selectivity to cations. Yet, his membrane should have no negative electrical charge to cause cation selectivity. Its mechanism was clarified by the identification of the charged impurity, phenylphosphonic acid monooctyl ester, from commercially available DOPP *(24)*. Astringent substances and pungent substances were studied using a multichannel taste sensor with lipid membranes. Pungent substances, such as capsaicin, piperine, and allyl isothiocyanate, had no effect on the membrane potentials of the lipid membranes. On the other hand, astringent substances such as tannic acid, catechin, gallic acid, and chlorogenic acid changed the potentials remarkably. A principal component analysis of the patterns (in electrical potential changes) caused by the taste substances revealed that astringency is located between bitterness and sourness *(25,26)*. A gradual dilution system enhances the high potential response of a pungent sensing electrode *(27)*. Various voltammetric techniques, such as large and small amplitude pulse voltammetry, can generate information when combined with multivariate analysis. A prototype of an electronic tongue was designed with a double working electrode of gold and platinum, and employing principal component analysis. This electronic tongue is able to classify various samples such as fruit juices, still drinks and milk and follow aging processes of milk and orange juice when stored at room temp *(28)*. Certainly a visual cue to specific taste substances would complement the interpretation of sensor analyses. The Sekisui Chemical Co., Ltd. has a test element for detecting bitter flavor of substances such as quinine consists of a base plate and a membrane structure containing lipids and fluorescent substances (e.g., 3,3'-dioctadecyl-2,2'-oxacarbocyanin perchlorate or 3,3'-dioctadecyl-2,2'-thiacyanine). A taste substance is adsorbed on the membrane and irradiated with an excitation light at 495nm. The fluorescence from the excited substance is measured for bitter flavor detection *(29)*. As electrical conductance of the membrane increased for the basic tastes the electrical selectivity and order of the detection-threshold of the membrane was quinine hydrochloride $<$ HCl $<$ mono-sodium L-glutamate $<$ NaCl $<$ sucrose. This is the same detection threshold order as a human taste panel would assess for bitterness and is reproducible. As a result, the response patterns showed the different shape among the taste solutions such as NaCl and sucrose. The authors then developed taste response patterns as six-axis spider plots for discrimination of basic tastes. Thus, the taste was identified by the elecetrical and optical characteristics of the membranes *(30)*. An alternative optical method uses an arachidic acid LB film doped with voltage-sensitive rhodamine B (RB_{C18}) dye with a long hydrocarbon chain in a mixture ratio of 75:1 and transferred onto a substrate. One measures both the fluorescence intensity around λ= 600 nm and the membrane potential of the (RB_{C18})-LB film in various taste substance solutions. As a result, the fluorescence intensity increased as the NaCl concentration increased but on the other hand the membrane potential decreased. Thus, optical fiber sensors may be useful *(31)*. An optical fiber chemical sensor of an immobilized liposome membrane with protein and/or biotin-avidin or biotin-streptavidin pair mimics taste or smell-sensing and may have use in food and cosmetics. In the example, phosphatidylserine and/or phosphatidylcholine liposome encapsulated with bis-(1,3-dibutylbarbiturate)-pentamethine oxonol and alamethicin, biotin, and streptavidin was prepared and immobilized on a quartz optical fiber *(32)*. A surface photovoltage (SPV) technique has been applied to construct a taste sensor by combining modified Langmuir-Blodgett (LB) immobilized taste-sensitive membranes. Several kinds of artificial lipid membranes may be monolithically integrated on a semiconductor surface, and the surface potential change caused by the reactions with taste substances is detected by scanning a light beam along the semiconductor surface. The uniformly oriented lipid membranes exhibited different responses to the five basic taste substances with high sensitivity and fast response rate for commercial drinks *(33)*. A new differential measurement method for a light addressable potentiometric (LAP) sensor has been developed and applied to fabricate an integrated taste sensor with artificial lipid membranes as the ion-sensitive material. Sensitivity was further enhanced by at least two orders of magnitude compared to a conventional LAP system. The sensor shows highly sensitive responses to various taste substances through pattern-recognition analyses. Miniaturization of the LAP system is possible by using a small metal pseudo-reference electrode instead of a glass electrode *(34)*. Water/octanol/water liquid membranes work effectively to evaluate papaverine hydrochloride, maltose, malic acid, magnesium chloride and disodium guanosinate. There is a particular oscillation mode for bitterness, sweetness, sourness, saltiness and umami. Erythritol exhibited an oscillation mode similar to maltose and other sugars. The oscillation mode for sodium saccharin was similar to that for bitter substances characterized by a decrease in amplitude. Bitter and salty, sweet and salty, sour and salty, and umami and umami substance combinations were examined for their capacities to modify taste quality *(35)*. Transient and steady membrane potentials in response to a bitter substance have been measured in a membrane filter impregnated with phospholipid and 1-octanol. Quinine was used as the bitter substance. The internal and external solutions both contained 0.1 mol/dm^{-3} KCl. A cation exchange liquid membrane was formed in a Teflon filter by impregnating it with a 1-octanol solution of dihexadecyl hydrogenphosphate (DHP). By fitting the theoretical curves to the experimental results, large values of the partition coefficient and the equilibrium constant of the complex formation with DHP were obtained for monoprotonated quinine ion relative to potassium ion *(36)*. Oscillations of electric potential across a liquid membrane consisting of picric acid in nitrobenzene between two aqueous layers were found to be characteristic of the structural class of a tastant present in the liquid membrane *(37)*. Fourier analysis establishes a "finger-print" of the substance that can be correlated with its taste index. The electrical oscillations consisted of a number of weak damped oscillators. The Fourier spectra of these signals showed a correlation between the frequency of the first peak with the taste index for bitter substances, whereas salty substances correlated with the amplitude of the first two peaks of the Fourier spectrum *(38)*.

Applications for Food and Beverage QC

An enormous literature bank describes applications of various taste sensor devices in food, beverages and pharmaceuticals. The most facile way to reference the utilities are in tabular form illustrated below:

Taste Sensor Application	**References**
Amino acids	*40,45-47,52,59*
Analgesics (anionic)	*48*
Beer; ripening of alcoholic beverage	*39,40,50-52,55,59*
Bitter suppressants	*45*
Carbonated soft drinks, soft drinks	*39,55*
Coffee	*40,49,52,59*
Drugs: bitter suppression	*58-60*
Fruit purees and juice	*55*
Milk pasteurization	*56,59*
Mineral water	*40,59*
Miso (soybean mash)	*40,59*
Rice (boiled)	*40,59*
Sake: acidity, EtOH conc, aging	*40,41,44,52,59*
Soy sauce	*43*
Soybean seeds (bitterness)	*57*
Sweeteners (artificial and sucrose)	*45,48*
Tea	*55*
Vegetables	*40,52,59*
Wine (apple, red), cider vinegars	*53,54,59*
Yogurt	*42*

Future perspective

The sensor technologies described herein make possible improved finished product Quality Control, process control monitoring, or monitoring product aging events.

The utility and advances of taste sensors has the potential ability to screen for bitter (and other noxious) tasting substances without recruiting a willing and human taste panel. Discrimination and quantification of standard taste stimuli permits distinguishing when one parameter is suppressed, e.g., bitterness suppressants. The ideal sensor must correlate to human sensing. Simple membrane systems of chemoreceptors in biological membranes have been examined. Changes of membrane potential and intracellular Ca^{+2} levels were measured simultaneously by optical imaging analysis in mouse taste buds. The results suggest that this method can be useful for the analysis of various taste responses *(61)*. The changes in electrical potential of biological membranes and liposomes in response to the addition of various bitter substances, amino acids, and umami taste substances are reviewed *(62)*. Similar results to artificial lipid membranes and other biological systems have been observed in electrical responses from the roots of adzuki bean. Solutions of five (sour, salty, bitter, sweet, umami) basic taste substances were injected into the aqueous solution around the root of adzuki bean. The root responded to the substances in different ways by the depolarization of the membrane potential at the root surface. Little change of electric potential was noted by nonelectrolytic taste substances *(63)* Since remote, non-human sensing efforts can be directed towards quantifying bitter perception, the knowledge gained may ultimately provide assay avenues to universal bitter reduction and inhibition. With the genetics of the human taste receptor nearly in hand, one could imagine a cloned human taste receptor as the sensor. The purported universal bitter inhibitor, phosphatidic acid and β-lactoglobulin complex (PA-LG), will eventually find use in suppressing bitter taste response in humans *(64)*. A more improved taste sensor could rapidly aid the development and food applications of new bitterness-masking agents.

References

1. *Modifying Bitterness: Mechanism, Ingredients, Applications.* Ed. Roy, G.; Technomic Publishing Co.: Lancaster PA. 1997.
2. Toko, K. *Electronic sensing of tastes.* Sens. Update 1998, 3, 131-160, Wiley-VCH.
3. Lavigne, J.J.; Savoy, S.; Clevenger, M.B.; McDaniel, B.; Yoo, S-J.; Anslyn, E.V.; McDevitt, J.T.; Shear, J.; Niekirk, D. *Solution-based analysis of multiple analytes by a sensor array: Toward the development of an electronic tongue.* J. Am. Chem. Soc., 1998, 120, 6429-6430 and SPIE Conf. On Chemical Microsensors and Applications, 1999, Vol. 3539, 17-26.

4. Davidson, J. M.; Linforth, R. S. T.; Taylor, A. J. *In-mouth measurement of pH and conductivity during eating.* J. Agric. Food Chem. 1998, 46(12), 5210-5214.
5. Mielle, P. *Electronic noses - towards the objective instrumental characterization of food aroma.* Trends Food Sci. Technol. 1996, 7(12), 432-438 (Eng), Elsevier.
6. Ulmer, H.; Mitrovics, J.; Noetzel, G.; Weimar, U.; Gopel, W. *Odors and flavors identified with hybrid modular sensor systems.* Sens. Actuators, B 1970, 43(1-3), 24-33 Elsevier Science S.A.
7. Petty, M. C. *Langmuir-Blodgett Films, An Introduction.* February 1996; 0-521-41396-6.
8. Toko, K. *Electronic sensing of tastes.* Electroanalysis 1998, 10(10), 657-669. Wiley-VCH Verlag GmbH
9. Nomura, K., Toko, K. *A theoretical consideration of interactions between lipid membranes and taste substances.* Sens. Mater., 1992. 4(2), 89-99.
10. Toko, K. *Taste sensor.* Charact. Food: Emerging Methods, 1995, 377-401. Edited by Gaonkar, A.G., Elsevier.
11. Chishaki, K.; Nakagawa, Y.; Toko, K. *Effect of bitter substances on transducer membranes of taste sensor.* Res. Rep. Inf. Sci. Electr. Eng. Kyushu Univ. 1997, 2(2), 297-303.
12. Misawa, K.; Ohashi, H.; Arisawa, J. *Electrical characteristics of plasticizer-mixed synthetic lipid membrane for taste solutions.* Hokkaido Kogyo Daigaku Kenkyu Kiyo 1996, 24, 295-302 (English).
13. Toko, K.; Nakagawa, Y.; Obata, M.; Yahiro, T. *Discrimination of taste qualities using static and dynamic responses of multichannel taste sensor.* Sens. Mater. 1997, 9(5), 297-306.
14. Iiyama, S.; Azuma, Y.; Nagaishi, M.; Toko, K.. *Change in electric characteristics of membranes in response to taste stimuli with increasing amount of lipids in membrane matrix of PVC and plasticizer.* Biophys. Chem. 1996, 61(1), 23-27 (English).
15. Oohira, K.; Toko, K. *Electrical characteristics of lipid/PVC/DOPP membrane and PVC/DOPP membrane used as transducers in chemical sensors.* Sens. Mater. 1997, 9(1), 57-68 (Eng), Scientific Publishing Division of MYU K.K.
16. Saito, M. *Dynamic responses at artificial membranes composed of a mixture of lipids to taste substances.* Sens. Mater., 1992, 4(2), 73-9.
17. Saida, Y., Matsuno, T., Toko, K., Yamafuji, K. 1992. *Taste detection using chaos in excitable lipid membrane.* Sens. Mater., 4(3), 135-44.
18. Toko, K.; Akiyama, H.; Chishaki, K.; Ezaki, S.; lyota, T.; Yamafuji, K. *Detection of taste substances using impedance change in lipid/polymer membranes.* Sens. Mater. 1997, 9(5), 321-329 English) Scientific Publishing Division of MYU K.K.
19. Lavigne, J.J.; Metzger, A.; Niikura, K.; Cabell, L.A.; Savoy, S.; Yoo, S-J.; McDevitt, J.T.; Niekirk, D.; Shear, J.; Anslyn, E.V. *Single analyte to*

multianalyte fluorescence sensors. Proc. SPIE - Int. Soc. Opt. Eng., 1999, Vol. 3602, 220-231.

20. Iiyama, S.; Iida, Y.; Toko, K. *Measurement of umami substances using multichannel taste sensor with lipid membranes.* Sens. Mater. 1998, 10(8), 475-485 (English) Scientific Publishing Division of MYU K.K.
21. Ikezaki, H. *Method of determining food taste with gustatory sensor made from lipid membrane.* PCT Int. Appl. WO 9630753; 1996.
22. Ikezaki, H. *Taste biosensor and washing of sensing membrane.* JP 08,271,473; 1996.
23. Jin, L.; Sun, W.; Bai, Z.; Fang, Y. *Studies on electrochemical taste sensors. The response of an excitable lipid membrane to taste substances.* Huadong Shifan Daxue Xuebao, Ziran Kexueban 1996, (2), 66-71 (English)
24. Watanabe M.; Toko, K.; Sato, K.; Kina, K., Takahashi, Y.; Iiyama, S. *Charged impurities of plasticizer used for ion-selective electrode and taste sensor.* Sens. Mater. 1998, 10(2) 103-112.
25. Iiyama, S., Toko, K., Matsuno, T., Yamafuji, K. *Responses of lipid membranes of taste sensor to astringent and pungent substances.* Chemical Senses, 1994, 19(1), 87-96.
26. Iiyama, S., Ezaki, S., Toko, K., Matsuno, T., Yamafuji, K. *Study of astringency and pungency with multichannel taste sensor made of lipid membranes.* Sens. Actuators, B 1995, 24-25(1-3), 75-9.
27. Kawaguchi, A.; Magaino, S.; Sobue, K.; Sakurai, M. *Evaluation of pungent sensing electrode using gradual dilution system in flow analysis.* Chem. Sens. 1997, 13 (Suppl. A, Proceedings of the 24th Chemical Sensor Symposium, 1997), 181-184.
28. Winquist, F.; Wide, P.; Lundstrom, I. *An electronic tongue based on voltammetry.* Analysis Chim. Acta 1997, 357(1-2), 21-31 (Eng), Elsevier Science B.V..
29. Minami, K., Takazawa, Y. 1992. *Test elements for detection of substance with bitter flavor.* JP 04,215,042 and JP 04,340,444 to Sekisui Chemical Co., Ltd., 1992.
30. Misawa, K.; Sudo, N.; Arisawa, J. *Electrical and optical response patterns of artificial lipid membrane for taste solutions.* Hokkaido Kogyo Daigaku Kenkyu Kiyo 1997, 25, 301-308 (English).
31. Morisawa, M.; Yonezaki, Y.; Maekawa, K.; Vishnoi, G.; Muto, S. *Optical sensing of taste substances using rhodamine dye-doped LB film.* Proc. SPIE-Int. Soc. Opt. Eng. 1997, 3105 (Chemical, Biochemical, and Environmental Fiber Sensors IX), 353-360 (English)
32. Umibe, K., Myamoto, H., Saito, M., Kato, M. 1995. *Optical fiber immobilized liposome encapsulated with dye sensitive to membrane electric potential for biosensor.* JP 07 35,691 to Oki Electric Ind., Co. Ltd.

33. Kanai, Y., Shimizu, M., Uchida, H., Nakahara, H., Zhou, C.G., Maekawa, H., Katsube, T. 1994. *Integrated taste sensor using surface photovoltage technique.* Sens. Actuators, B, 20(2-3), 175-9.
34. Sasaki, Y., Kanai, Y., Uchida, H., Katsube, T. 1995. *Highly sensitive taste sensor with a new differential LAPS method.* Sens. Actuators, B, 25(1-3), 819-22.
35. Arai, K.; Kusu, F.; Takamura, K. *Electrical potential oscillation across a water/octanol/water liquid membrane in the presence of two taste substances.* Analysis Chim. Acta 1998, 365(1-3), 279-284 (English) Elsevier Science B.V.
36. Yata, T.; Nomura, K. *Transient Membrane Potential in Response to a Bitter Substance in a Membrane Filter Impregnated with Phospholipid and 1-Octanol.* Bull. Chem. Soc. Jpn. 1998, 71(11), 2589-2595 (English)
37. Shaw, P., Coddington, J.M. 1995. *Possible prediction of taste quality using a liquid membrane.* Biophys. Chem. 55(3), 209-13.
38. Cucu, D.; Mihailescu, D.; Mihailescu, G.; Nikolelis, D. P.; Flonta, M.-L.; Frangopol, P. T. *Fourier analysis of potential oscillations of a liquid membrane for the discrimination of taste substances.* Biophys. Chem. 1996, 63(1), 47- 54 (English).
39. Santo, K., Toko, K., Hayashi, K., Ikezaki, H., Higashikubo, R., Sato, K. *Organic membranes for taste sensor.* JP 07 05,147 (1995) and *Taste sensor.* JP 05 34,311 (1993) to Anritsu Corp..
40. Toko, K. *Taste sensor with global selectivity.* Mater. Sci. Eng., C 1996, C4(2),69-82 (English), Elsevier.
41. Imamura, T.; Toko, K.; Yanagisawa, S.; Kume, T. *Monitoring of fermentation process of miso (soybean paste) using multichannel taste sensor.* Sens. Actuators, B 1996, B37(3), 179-185 (English), Elsevier.
42. Takahashi, N.; Hatamoto, F.; Kawai, N. *Flavor estimation of plain yogurt using artificial neural network.* Nippon Shokuhin Kagaku Kogaku Kaishi 1998, 45(7), 408-414.
43. Iiyama, S.; Ikeda, T.; Toko, K.; Yahiro, M. *Measurements of shoyu (soy sauce) with multichannel taste sensor.* Nippon Shokuhin Kagaku Kogaku Kaishi 1997, 44(9), 615- 622 (in Japanese)
44. Arikawa, Y.; Toko, K.; Ikezaki, H.; Shinha, Y.; Ito, T.; Oguri, I.; Baba, S. *Analysis of sake mash using multichannel taste sensor.* J. Ferment. Bioeng. 1996, 82(4), 371-376 (English).
45. Kurihara, K. *Taste sensors.* Shirizu: Nyu Baiofijikkusu, 1997 Volume 6, 31-43. Editor(s): Tsuda, Motoyuki. Kyoritsu Shuppan: (in Japanese).
46. Toko, K.; Fukusaka, T. *Measurement of hydrophobicity of amino acids using a multichannel taste sensor.* Sens. Mater. 1997, 9(3), 171- 176 (English)

47. Toko, K.; Mori, S.; Nakabayashi, Y.; Kanda, M.; Fukunaga, T. *The taste sensor.* Colloq. Sci. Int. Cafe, [C. R.] 1995, 16th(Vol. 1, Seizieme Colloque Scientifique International sur le Cafe, 1995, Vol. 1), 292-9 (English).
48. Schnierle, P.; Kappes, T.; Hauser, P.C. *Capillary Electrophoretic Determination of Different Classes of Organic Ions by Potentiometric Detection with Coated-Wire Ion-Selective Electrodes.* Analysis Chem. ACS ASAP (English). American Chemical Society.
49. Komai, H.; Naito, Y.; Sato, K.; Ikezaki, H.; Taniguchi, A.; Toko, K. *Measurement of coffee taste using lipid membrane taste sensors.* Colloq. Sci. Int. Cafe, [C. R.] 1995, 16th- (Vol. 1, Seizieme Colloque Scientifique International sur le Cafe, 1995, Vol. 1), 300-8 (English).
50. Toko, K. *The taste sensor.* Olfaction Taste XI, Proc. Int. Symp., 11th 1993 (Pub. 1994), 686-9 (Eng). Edited by Kurihara, Kenzo; Suzuki, Noriyo; Ogawa, Hisashi. Springer: Tokyo, Japan.
51. Pearce, T.C.; Gardner, J.W. *Prediction of organoleptic flavor notes using multi-sensor arrays.* Semin. Food Analysis 1997, 2(4), 275-281 (English) Chapman & Hall
52. Toko, K. *The Taste Sensor.* In Olfaction and Taste XI. Ed. by Kurihara, K., Susuki, N., Ogawa, H. 1993. Springer-Verlag NY.
53. Gerbi, V.; Zeppa, G.; Beltramo, R.; Carnacini, A.; Antonelli, A. *Characterization of white vinegars of different sources with artificial neural networks.* J. Sci. Food Agric., 78(3), 417-422 (English) 1998. Publisher: John Wiley & Sons Ltd.
54. Baldacci, S.; Matsuno, T.; Toko, K.; Stella, R.; De Rossi, D. *Discrimination of wine using taste and smell sensors.* Sens. Mater. 1998, 10(3) 185-200 (English)
55. Legin, A.; Rudnitskaya, A.; Vlasov, Y.; Di Natale, C.; Davide, F.; D'Amico, A. *Tasting of beverages using an electronic tongue.* Sens. Actuators, B 1997, B44(1-3), 291-296 (English) Elsevier Science S.A.
56. Yamada, H.; Mizota, Y.; Toko, K.; Doi, T. *Highly sensitive discrimination of taste of milk with homogenization treatment using a taste sensor.* Mater. Sci. Eng., C 1997, C5(1), 41-45 (English) Elsevier
57. Okubo, K., Iijima, M., Kobayashi, Y. 1992. *Components responsible for the undesirable taste of soybean seeds.* Biosci., Biotechnol., Biochem. 56(1), 99-103.
58. Takagi, S.; Toko, K.; Wada, K.; Yamada, H.; Toyoshima, K. *Detection of suppression of bitterness by sweet substance using a multichannel taste sensor.* J. Pharm. Sci. 1998, 87(5), 552-555 (Eng), American Chemical Society.
59. Toko, K. *A taste sensor.* Meas. Sci, Technol. 1998, 9(12), 1919-1936 (English) Institute of Physics Publishing.

60. Arai, K.; Fukuyama, S.; Kusu, F.; Takamura, K. *Electrical responses of a water/octanol/water liquid membrane to taste substances.* Denki Kagaku oyobi Kogyo Butsuri Kagaku 1995, 63(12), 1183-8 (English).
61. Hayashi, Y. *Optical imaging analysis of taste response in mouse taste buds.* Nippon Aji to Nioi Gakkaishi 1996, 3(2), 141-143 (in Japanese).
62. Kumazawa, T. *Initial reception of taste. Receptor and change in membrane potential.* Denki Kagaku oyobi Kogyo Butsuri Kagaku, 1994, 62(3), 196-201 in Japanese.
63. Ezaki, S., Iiyama, S., Toko, K. *Electrical response of a root of the higher plant to taste substances.* Kinki Daigaku Kyushu Kogakubu Kenkyu Hokoku Rikogaku-hen 1993, 22, 31-6 in Japanese.
64. Katsuragi, Y. *Masking agents of bitterness. Selective control of bitterness by phosphatidic acid.* Shokuhin Kogyo 1998, 41(2), 75-78 (in Japanese).

Chapter 21

Mimicking the Mammalian Sense of Taste Through Single-Component and Multicomponent Analyte Sensors

Sheryl L. Wiskur, Axel Metzger, John J. Lavigne, Stephen E. Schneider, Eric V. Anslyn*, John T. McDevitt, Dean Neikirk, and Jason B. Shear

Department of Chemistry and Biochemistry, University of Texas, Austin, TX 78712

An array for the simultaneous detection of multiple analytes in a solution has been developed. Single analyte receptors on resin beads in conjunction with fluorescent and colorimetric signaling strategies are the basis of the sensing ensemble. The pattern created is specific for each particular set of stimuli, allowing for analyte detection, analogous to the mammalian tongue.

Of the five mammalian senses, vision, hearing, touch, smell, and taste, the first three are physical senses based on forces such as sound waves, light waves, etc. The other two, taste and smell, are often referred to as the lower senses, and are based on our response to chemical stimuli (*1*).

Over the years the number of primary tastes has been disputed and at one point even included "tastes" such as fatty, astringent, sharp, and nauseous. Wund eventually would narrow it down to only sweet, sour, salty, and bitter (*1*). Even with only four primary tastes, mixing these components in different combinations create unique tastes, much like the mixing of the primary colors.

Taste or gustation is a result of chemical interactions or molecular recognition events on the surface of the tongue through "taste buds," shaped like and named for an unopened flower (*2*). The concept of taste receptors has been around since the time of Aristotle, but taste buds were first identified in the mid nineteenth century by Leydig. Schultze (1863) proposed the idea that they were chemosensory structures (*2*). Taste buds exhibit selectivity for a particular taste and are clustered in different regions of the tongue in depressions called taste pores according to their selectivity. The tip of the tongue is where salt is sensed, the sides nearest the tip is sweet, further back on the sides is sour, and across the back of the tongue seems to have the greatest sensitivity for bitterness. The center of the tongue appears to be insensitive to all taste. When chemical stimuli enter the taste pores, the epithelial cells within the taste buds respond to the stimuli and activate nerves (*2*). Sweetness results from sugar, sugar derivatives, sweet amino acids like most D-amino acids, and some simple salts (beryllium and lead). Bitterness is thought to be a safety mechanism against poisons by being directly related to the gag reflux and is attributed to a diverse amount of stimuli including, hydrophobic amino acids, basic heterogeneous compounds, divalent salts, and some peptides (*3*). Sour is dependent on the acidity of the food, while saltiness stems from halide salts such as sodium chloride (*4*). Due to the diverse types of stimuli for one particular taste, multiple receptors are required. The response of each of these receptors is combined to create a pattern that is specific for a certain taste. This pattern can be stored in the brain so that the same taste can be identified upon future exposure. This pattern may also be used to classify similar tastes as pleasant or unpleasing.

Single Analyte Molecular Recognition

To mimic the mammalian sense of taste, the detection of a specific analyte in the presence of competing analytes is the first obstacle to overcome. The simplest form of this detection is molecular recognition of a single analyte. Molecular recognition begins with the design and synthesis of a binding site that is preorganized and complimentary to the analyte one wishes to bind. With this in mind, a chemosensor for the tris-carboxylate compound citrate (**1**) has been developed that binds even in a competitive aqueous media. The base of the molecular scaffold consists of a 1,3,5-trisubstituted-2,4,6-triethylbenzene unit. The six substituents point alternately up and down around the ring (*5*). Therefore, the 1,3,5-binding sites are preorganized on one face of the benzene ring, sterically enforced by the 2,4,6-ethyl groups (*6*). This preorganization has been shown to exhibit positive effects on binding (*6*, *7*, *8*). This principle has been demonstrated with the citrate host molecule (**2**) and a similar molecule lacking the ethyl groups. The binding affinity dropped in half for the host lacking preorganization.

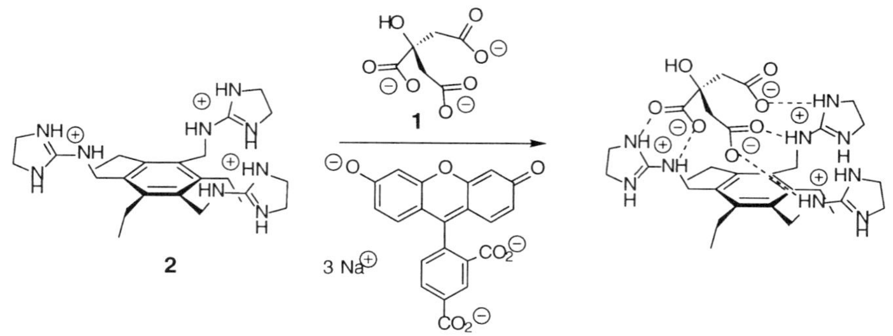

3

Citrate is a common ingredient in citrus beverages, and is trianionic at neutral pH. Therefore, this necessitated the development of a receptor that would recognize and bind anionic guests, in particular carboxylates. Anion binding sites such as ammonium ions have a high localization of charge and function primarily by charge pairing, but have a poor hydrogen bonding geometry for binding carboxylates. Guanidiniums were chosen as the recognition units, since they have been used for anionic binding through hydrogen bonding and charge pairing interactions both in natural and abiotic systems, particularly for carboxylate and phosphate groups (*9*). The guanidinium groups remain protonated over a larger pH range as compared with the ammonium species, and their geometry is more conducive for the binding of carboxylates (*10*, *11*, *12*).

A sensor was created from this system by employing a competition assay using the fluorophore 5-carboxyfluorescein (**3**) to signal binding. The competition assay functions by binding the indicator molecule to the receptor, which is then displaced by the analyte causing a signal modulation (*13*). It was expected that the fluorophore would bind to the host through interactions between the carboxylate and phenolate groups and the guanidiniums. Since the indicator's absorbance and fluorescence are sensitive to changes in pH, it was expected that when the probe was bound to the guanidinium, the phenol's pK_a would be lowered by the positively charged microenvironment of the host, resulting in a higher protonation state when the probe was free in solution. Upon addition of **2** to a constant concentration of **3** (25% v/v water in methanol, 5 mM HEPES buffer, pH 7.4), the absorbance and fluorescence increases as more of the probe is bound to the host (Figure 1A). As citrate is added to the solution of **2** and **3**, the absorbance and fluorescence decreases, as more of the probe is free in solution (Figure 1B).

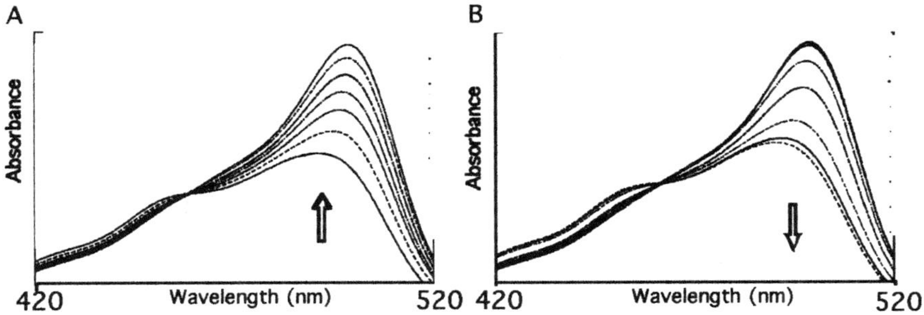

Figure 1. UV/Vis spectra, λ_{max} = 498 nm. (**A**) The increase in absorbance as **2** was added to a constant concentration of **3**. (B) The decrease in absorbance as citrate was added to a constant concentration of **2** and **3**. (25% v/v water in methanol, 5 mM HEPES buffer, pH 7.4)

Binding studies were conducted using ^{1}H NMR and UV/Vis spectroscopy under Benesi-Hildebrand conditions (*14*). The host was found to have a binding constant for citrate of 2.9×10^5 M^{-1} (by UV/Vis) in a 25% (v/v) water in methanol solvent system (5 mM HEPES buffer, pH 7.4). Calibration curves for citrate (Figure 2A) and other similar analytes were generated by addition of the analyte to a solution containing constant concentrations of **2** and **3**. There was a small decrease in absorbance (fluorescence) of the sensing ensemble upon the addition of succinate, but there was no change upon the addition of acetate, salt, or sugars (Figure 2B). These curves were used to analyze a variety of beverages, most of which contained citrate (Table I). The beverage to be analyzed was

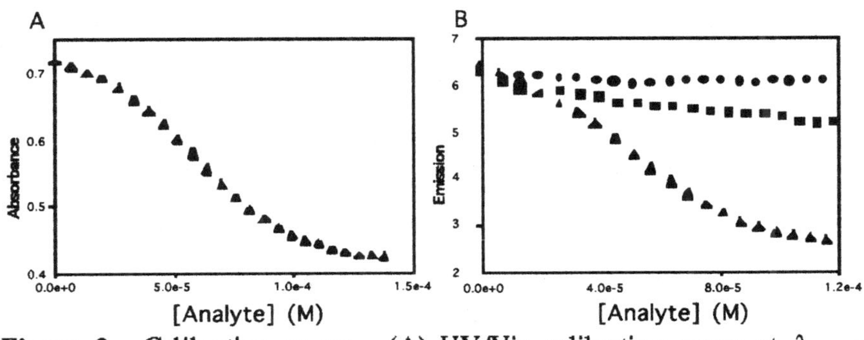

Figure 2. Calibration curves. (**A**) UV/Vis calibration curve at λ_{max} = 498nm. (B) Fluorescence calibration curve at λ_{max} = 525nm. (▲ = citrate, ■ = succinate, ● = acetate) (25% (v/v) water in methanol, 5mM HEPES buffer, pH=7.4)

added to the sensing ensemble, the fluorescence or absorbance was measured, and the concentration was determined from the calibration curve. Their concentrations were also determined by NMR methods, showing that competition method generally was in agreement within 10% and worked well in the presence of sugars and phosphates.

The study reveals that the host is selective for citrate over di- and mono-carboxylates, phosphates, and sugars, and shows that a selective receptor can be designed for hydrophilic guests in aqueous media. The crystal structure of the host-guest complex suggests five hydrogen bonds and three ion pairs demonstrating the necessity of all three guanidinium groups.

Table I. Analysis of citrate concentration in beverages determined by NMR and competition assay.

	by NMR	**2** plus **3**, absorbance	**2** plus **3**, emission
citrate model solution		30.3	29.9
orange juice	43.1	44.1	44.7
Gatorade	16.0	15.1	15.1
Powerade	12.4	11.1	11.3
All Sport	7.4	7.1	8.1
Mountain Dew	8.0	5.5	5.4
tonic water	21.0	21.2	20.8
Coca Cola	0	0	<0.5
Diet Coke	<0.2	<0.4	<0.7

Tartaric acid (**4**) is a second analyte that was studied. This bis-carboxy diol compound is commonly found in wines and influences not only their taste, but color, aroma, and astringency, as well. To design a receptor that is selective for tartaric acid, a binding site that has affinity for 1,2-diols needed to be incorporated. Boronic acids are known to bind 1,2- and 1,3-diols through a reversible covalent interaction. The diols form cyclic boronate esters with tetrahedral borons under basic conditions (*15*, *16*). A nitrogen adjacent to the boronic acid can also coordinate with the boron atom thereby changing the hybridization and allowing for the formation of the boronate esters at or near neutral pH.

A receptor similar to **2** was designed to be selective for tartaric acid by replacing one of the guanidinium groups with a phenyl boronic acid adjacent to an amine, host **5** (*17*). The two charged carboxylate groups of the guest were predicted to interact with the guanidiniums as in receptor **2**, and the diol was expected to bind with the boronic acid. A competition assay was also employed for the creation of a molecular sensor and used alizarin complexone (**6**) as the probe because it possessed the same functionalities as tartaric acid. In this assay, an observable color change was expected as the "protonation state" of the phenols changed between free

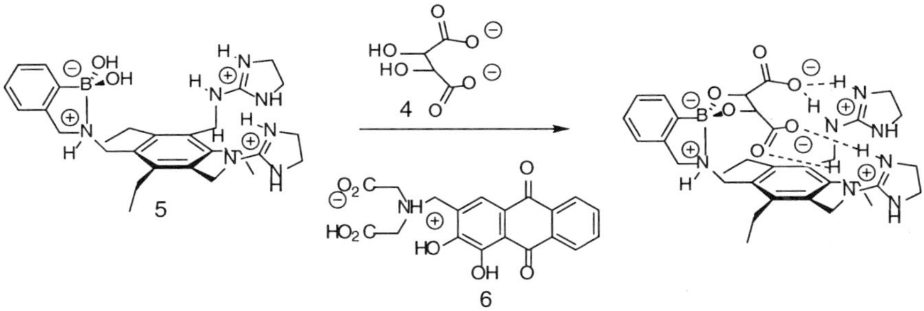

indicator and that bound to the host. The formation of the boronate ester with the phenols did indeed cause a color change from burgundy to yellow/orange (Figure 3A). When **4** was added to the solution, the probe was released from the binding site resulting in a color shift back to burgundy (Figure 3B). A binding constant of 5.5×10^4 M^{-1} was measured for tartrate and host **5** in a 25% (v/v) water in methanol solvent system (pH 7.3, HEPES buffer (10 mM)), using UV/Vis spectroscopy. The sensor was also tested with other similar analytes including ascorbate, L-malate, succinate, lactate, and sugars. Out of all the analytes, the sensor was selective for tartartic acid with the exception of malate, which had a binding constant of 4.8×10^4 M^{-1}. This suggests that interaction at all three binding sites is required for recognition.

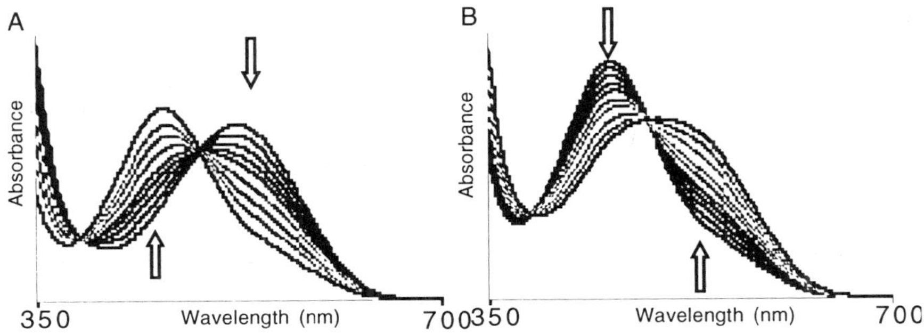

Figure 3. UV/Vis spectra for **6**. (A) The change in absorbance of a constant concentration of **6** with an increase at 450 nm and a decrease at 525 nm as **5** was added to the solution. (B) The change in absorbance as tartrate was added to a constant concentration of **5** and **6**, with a decrease at 450 nm and an increase at 525 nm. (25% (v/v) water in methanol, 10 mM HEPES buffer, pH 7.3).

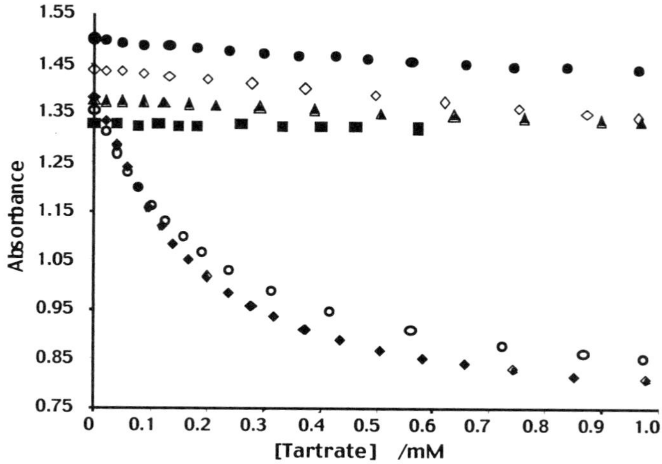

Figure 4. Calibration curves. UV/Vis calibration curves at 450 nm for the chemosensor **5** and **6** upon addition of analyte. (◆ = tartrate, ❍ = malate, ◇ = ascorbate, ▲ = succinate, ■ = glucose, ● = lactate) (25% (v/v) water in methanol, 10mM HEPES buffer, pH=7.3)

Calibration curves (Figure 4) for tartrate, malate, and other similar analytes were also generated for the tartrate chemosensor. These were used to evaluate several grape-derived beverages to determine their tartrate/malate concentrations (Table II). These values were in good agreement with values obtained from ^{1}H NMR studies, showing that this colorimetric assay can be employed in the determination of the concentration of tartrate/malate in a highly competitive media.

Table II. Tartrate and malate concentrations in grape derived beverages determined by both NMR and colorimetric assay.

	NMR	**5 + 6** *UV-Vis*
Tartaric acid model solution	51.2	50.2
Ernest & Julio Gallo Sauvignon Blanc	35.6	32.9
Ste. Genevieve Chardonnay	34.1	36.3
Henri Marchant Spumante	26.5	24.9
Talus Merlot	19.5	20.3
Santa Cruz Organic White Grape Juice	43.7	42.3
Welch's Grape Juice	69.4	71.3

Support Bound Single Analyte Sensing

Additionally, a support bound single analyte chemosensor was designed to be selective for ATP (*18*). This sensor takes advantage of a rationally designed core while utilizing combinatorial methods in the development of the recognition moieties. A preorganized cavity was formed using the 1,3,5-trisubstituted-2,4,6-triethylbenzene scaffold. One of the substituents consisted of lysine/urea linkage to Tentagel resin. The remaining substituents were identical tripeptide chains extending from guanidinium linkages to the base **7** (*19*). The fluorophore 5-carboxyfluorescein was attached to the N-terminus of the peptide chains while 7-diethylaminocoumarin-3-carboxylic acid was attached to the lysine side chain. It was expected that binding could then be signaled by some perturbation of the fluorescence resonance energy transfer (FRET) between the two fluorophores (*20*).

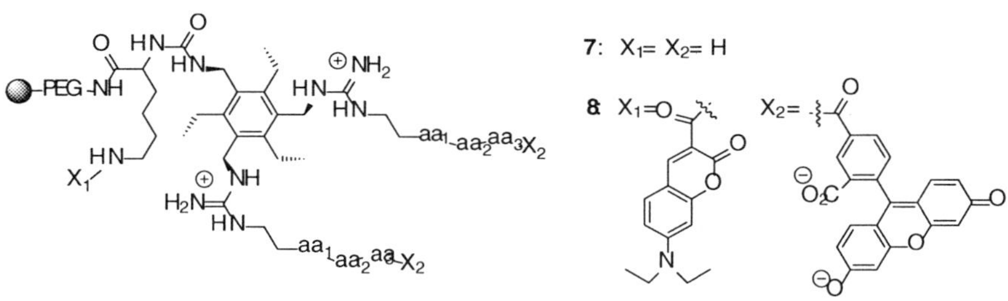

A library was generated and then screened against the desired analyte, ATP, to determine the most selective receptors. The host without the fluorophores (**7**) was used to screen the library. A fluorescently labeled N-methylanthraniloyl-ATP (MANT-ATP, 0.25 mM) was used to select the strongest binding members of the library (HEPES buffer 10 mM, pH 7.1) (*21*). Illumination of the beads at 366 nm allowed for a range of fluorescent intensities to be viewed. Several of the highly fluorescent and nonfluorescent beads were removed and sequenced to determine which peptide chains were active and inactive (Table III). Three highly fluorescent "hits" and one nonfluorescent "miss" were resynthesized with the fluorophores as in **8**.

The randomly chosen "hit" and "miss" beads were sandwiched between two layers of gold mesh on a glass slide to create a mono layer which could be studied using a standard fluorimeter to determine their ATP signaling ability. As expected, the "miss" exhibited no fluorescence modulation upon excitation of either fluorophore when exposed to ATP. Upon excitation of fluorescein all but one "hit" (Thr-Val-Asp) exhibited fluorescence modulation, however no change in the extent of FRET was observed. A large spectral change was observed with ATP and the Ser-Tyr-Ser sensor to yield a binding constant of 3.4×10^{3} M^{-1} (Figure 5). The fluorescence modulation may be due to an increase in the positive microenvironment around the fluorescein upon binding of the

Table III: Sequencing results. The peptide chains in bold were resynthesized, incorporating the fluorophores.

Active Beads	Inactive Beads
Thr-Val-Asp	**His-Phe-Gly**
Asp-Ala-Asp	Ser-Ala-Asp
Ser-Tyr-Ser	Trp-Asn-Glu
Asp-His-Asp	Thr-Phe-Ser
Met-Thr-His	
Glu-Pro-Thr	
His-Ala-Asp	

ATP. The potentially competing analytes AMP and GTP were also tested with this sensor due to their similarity in structure to ATP. Selectivity for ATP over AMP suggests the importance of the guanidinium binding. Similarly, the specificity over GTP indicates the tripeptide arms create specificity for nucleotide bases. This demonstrates that a selective resin bound chemosensor can be created through combinatorial methods.

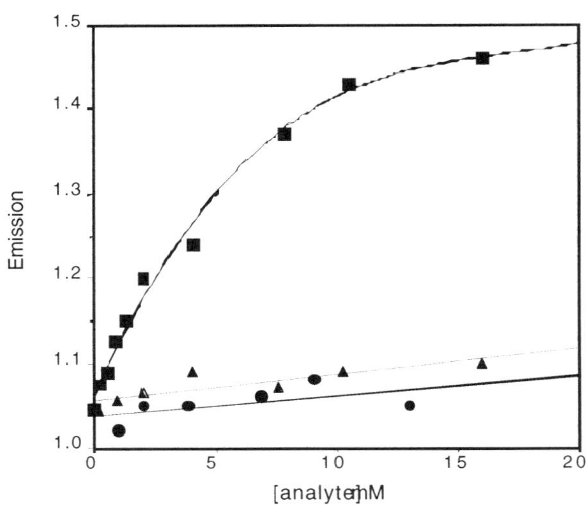

Figure 5. Binding isotherms for the Ser-Tyr-Ser **8** when combined with ■ = ATP, ▲ = AMP, and ● = GTP. (200 mM HEPES buffer, pH 7.4)

Multi-Component Analyte Sensor

The success in developing single analyte sensors creates a spring board for moving into the area of combining these to create a multi-component sensor array, much like nature uses. To mimic the sense of taste, we need to have the ability to detect single or multiple analytes through a collection of specific or general sensors utilizing a signaling scheme and pattern recognition. An array of sensors is compiled and a specific pattern emerges for each stimuli and mixture of stimuli that is tested. This pattern can then be stored in a computer to create a library of "tastes" to be utilized for future recognition. One approach uses poly-(ethylene glycol)-polystyrene (PEG-PS) resin beads with a variety of covalently attached chemical sensors to mimic taste buds (Figure 6A) (*22*). These sensors were selective for individual analytes, but not specific in there recognition ability. These derivatized beads were placed in micromachined wells (Figure 6B) in silicon wafers to immobilize the beads (*23*). A 3 x 3 array was used to detect a variety of analytes at once. A charge-coupled device (CCD) was used to determine the absorption properties of the beads, and the red, green, and blue (RGB) light intensities were studied for each individual bead upon addition of analyte. Of the four sensors chosen, one was specific and three were nonspecific. Each was compared to a control, which consisted of a resin bead where the amines were acetylated. The sensors included fluorescein as a pH indicator (*24*), *o*-cresolphthalein complexone for detection of Ca^{2+} and pH (*25*), alizarin complexone for Ce^{3+}, Ca^{2+}, and pH (*26*), and a boronic acid with a resorufin-derivatized galactose associated with it to indicate the presence of simple sugars.

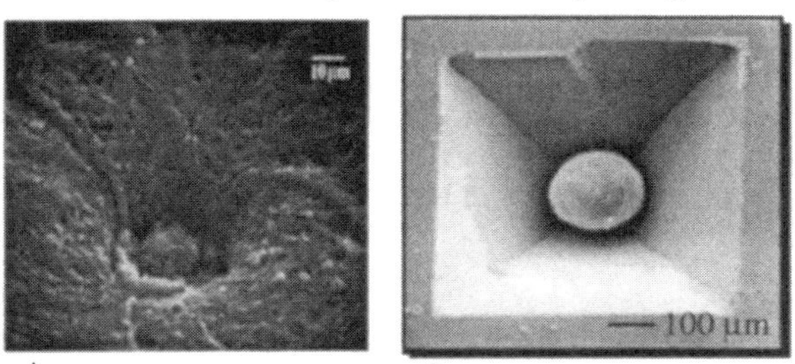

A. B.

Figure 6. (A) A rat fungiform papilla as seen by a scanning electron microscope (SEM) (*27*). (B) An SEM image of a micromachined silicon well, with a bead immobilized in it.

A proof of concept for the detection of Ca^{2+} using 0.1 M $Ca(NO_3)_2$ at various pHs is shown in Figure 7. The CCD array analyzed the change in transmitted light as the beads responded to calcium and pH. Experiments were also performed with cerium ($Ce(NO_3)_3$), a combination of cerium and calcium, and the simple sugar fructose. *o*-Cresolphthalein indicated the presence of calcium at a pH of 11.4 by a purple color. In the absence of calcium ion, the purple color was not observed until pH 12.5. Fluorescein was used as a pH sensor, changing from light yellow to orange at a pH around 6. Alizarin complexone was used for multiple roles, including a pH sensor were the beads were yellow at a pH less than 4.5, orange/red from pH 4.5-10, and a deep purple when the pH was higher than 11.5. When cerium was present, the change from yellow to orange/red occurred at a lower pH. The same effect was also observed for calcium, but the color change was not as dramatic. When fructose is added to the resorufin-β-D-galactopyranoside boronic acid complex, the bead changes from dark orange to yellow as the tagged sugar is displaced from the boronic acid and washed away due to the higher binding affinity of fructose. There was also a slight sensitivity to pH that was recorded by the CCD showing an increase in the absorbance of red light. These studies show that it is possible to detect multiple analytes at once, by analysis of the RGB patterns.

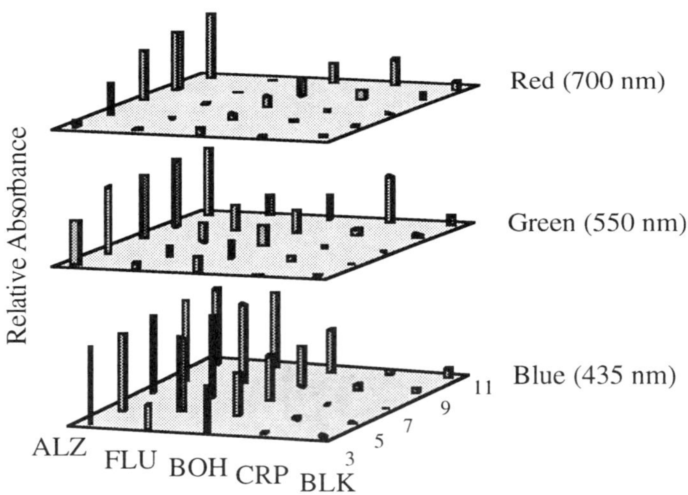

Figure 7. A series of bar graphs showing the color attenuation that was recorded by the CCD as the sensors were exposed to calcium at various pHs. (ALZ = alizarin complexone, FLU = fluorescein, BOH = boronic acid/galactose/resorufin, CRP = *o*-cresolphthalein complexone, BLK = blank)

Summary

The study of molecular recognition and the development of single analyte sensors has led to the ability to simultaneous analyze for several components in a complex solution. The creation of a pattern, or fingerprint, resulting from the response of several sensors allows for the determination and potential quantification of these analytes using pattern recognition protocols. The fact that the molecular recognition events are taking place in parallel allows for "cross-talk" or non-specific receptors to be used. A library can be created out of the patterns obtained, to be used for future analysis of unknown solutions. Finally, studies are under way to increase the complexity of the sensor array and its uses.

References

1. Shallenberger, R. S., *Taste Chemistry; Blankie Academic and Professional*: London, **1993**, Chapter 1.
2. Miller, Jr., I. J., *Handbook of Olfaction and Gustation*; Doty, R. L.; Marcel Dekker, Inc.: New York, **1995**; p. 521-547.
3. Margolskee, R. F.; *Handbook of Olfaction and Gustation*; Doty, R. L.; Marcel Dekker, Inc.: New York, **1995**; p. 575-595.
4. Mierson, S.; *Handbook of Olfaction and Gustation*; Doty, R. L.; Marcel Dekker, Inc.: New York, **1995**; p. 597-610.
5. Iverson, D. J.; Hunter, G.; Blount, J. F.; Damewood, J. R.; Mislow, K. *J. Am. Chem. Soc.* **1981**, *103*, 6073.
6. Kilway, K. V.; Siegel, J. S. *J. Am. Chem. Soc.* **1995**, *114*, 255.
7. Metzger, A.; Lynch, V. M.; Anslyn, E. V. *Angew. Chem. Int. Ed. Engl.* **1997**, *36*, 862.
8. Metzger, A.; Anslyn, E. V. *Angew. Chem. Int. Ed.* **1998**, *37*, 649.
9. Hannon, C. L.; Anslyn, E. V. *Bioorg. Frontiers* **1993**, *3*, 143.
10. Dietrich, B.; Fyles, D. L.; Fyles, T. M.; Lehn, J.-M. *Helv. Chim. Acta.* **1979**, *62*, 2763.
11 Müller, G.; Riede, J. Schmidtchen, F. P. *Angew. Chem. Int. Ed. Engl.* **1988**, *27*, 1516.
12 Echavarren, A.; Galán, A.; Lehn, J.-M.; de Mendoza, J. *J. Am. Chem. Soc.* **1989**, *111*, 4994.
13. Czarnik, A. W. *Acc. Chem. Res.* **1994**, *27*, 302.
14. Connors, K. A. *Binding Constants, The Measurement of Molecular Complex Stability*, John Wiley and Sons: New York, **1987**.
15 Wulff, G. *Pure and Appl. Chem.* **1982**, *54*, 2093.
16 James, T. D.; Sandanayake, K. R. A.; Shinkai, S. *J. Chem. Soc., Chem. Comm.* **1994**, 477.
17 Lavigne, J. J.; Anslyn, E. V. *Angew. Chem. Int. Ed.* **1999**, *38*, 3666.
18 Schneider, S. E.; O'Neil, S. N.; Anslyn, E. V. *J. Am. Chem. Soc.* **2000**, *122*, 542.
19 Furka, A.; Sebestyên, F.; Asgedom, M.; Dibó, G. Int. *J. Peptide Protein Res.* **1991**, *37*, 487.
20 Van der meer, B. W.; Cocker, G.; Chen, S.-Y. S. *Resonance Energy Transfer; Theory and Data*; VCH: New York, **1994**.
21 Hiratsuka, T. *Biochem. Biophys. Acta.* **1983**, *7442*, 496.

22 Lavigne, J. J.; Savoy, S.; Clevenger, M. B.; Ritchie, J. E.; McDoniel, B.; Yoo, S.-J.; Anslyn, E. V.; McDevitt, J. T.; Shear, J. B.; Neikirk, D. *J. Am. Chem. Soc.* **1998**, *120*, 6429.

23 Yoo, S.-J.; Lavigne, J. J.; Savoy, S.; McDoniel, J. B.; Anslyn, E. V.; McDevitt, J. T.; Neikirk, D. P.; Shear, J. B. *Micromachined storage wells for chemical sensing beads in an "artificial tongue"*; presented at SPIEs Micromachining and Microfabrication 1996 Symposium: Micromachined Devices and Components III, Proc. SPIE 322, 1997.

24 Kessler, G.; Wolfman, M. *Clin. Chem.* **1964**, *10*, 686.

25 Ray Sakar, B. C.; Chauhan, U. P. S. *Anal. Biochem.* **1967**, *20*, 155.

26 Belcher, R.; Leonard, M. A.; West, T. A. *J. Chem. Soc.* **1958**, 2390.

27 Spielman, A. I.; Brand, J. G.; Kare, M. R. *Encycl. Human Biol.* **1991**, *7*, 527.

Chapter 22

Electronic and Computational Olfaction

Alan Gelperin[1] and J. J. Hopfield[2]

[1]Monell Chemical Senses Center, 3500 Market Street, Philadelphia, PA 19104

[2]Department of Molecular Biology, Princeton University, Princeton, NJ 08544-1014

Biological gas detection and classification systems achieve a level of molecular sensitivity and robustness as yet unmatched by electronic devices. Development efforts continue to narrow the gap in functionality between electronic and biological olfaction, driven by the demonstrated utility of devices which provide objective, quantitative and reproducible odor measurements. We developed an artificial olfactory system to test whether incorporating biological design features into artificial olfactory systems would yield significant gains in functionality. Our initial system used an array of conducting polymer sensors with band pass filtering and amplification of sensor signals to optimize signal to noise in the relevant frequency range. Several pattern recognition algorithms were applied to the sensor signals. We used different cultivars of common produce items as test odor objects. Our device was able to differentiate between different cultivars of apples and oranges. We are now exploring the use of novel materials for sensor fabrication to allow implementation of new algorithms for odor identification which depend on inputs from 50-100 different classes of odor sensors having a range of binding affinities of at least three orders of magnitude.

Introduction

Biological olfactory (Bnose) systems perform impressive feats of molecular identification and mixture analysis with imprecise hardware (neurons) in circuits using a few percent of the 12 watts consumed by the entire human brain (1). The identification of minor components in complex mixtures and invariant molecular recognition over 3-4 orders of magnitude in concentration spanning detection threshold to saturated vapor represent solutions to nontrivial computational problems (2). The average value for human sensory threshold to 529 different odorants is 10 nM in air (3,4) while the silk moth *Bombyx mori* detects its mating peromone bombykol at 10^{-12} M (5). Electronic olfactory (Enose) systems which incorporate design features from Bnose systems (6) and implement algorithms from computational olfactory (Cnose) systems (7) are likely to achieve increased functionality (8). Evolution of Enose technology may yield devices able to locate some of the 10^8 buried land mines with sensitivity and reliability exceeding that of trained dogs (9,10), which currently provide the most effective olfactory detection systems for land mines.

Recent advances in understanding patterns of odorant receptor gene expression in mammals (11,12) and insects (13-15) clarify the nature of the molecular recognition events during odorant stimulation in Bnose systems. Interactions of odorants with receptor proteins in olfactory sensory neurons lead to spatial patterns of glomerular activation in the vertebrate olfactory bulb (16,17) or its invertebrate analog, the antennal lobe (18,19). Due to convergence of receptor neurons expressing the same odorant receptor gene in one or a few glomeruli (20,21), the pattern of sensory cell activation is converted to a pattern of glomerular activation. The means by which the glomerular pattern is decoded and the role of olfactory bulb/antennal lobe oscillatory dynamics in the decoding process remain to be elucidated (6,22).

Natural odor objects are complex mixtures of dozens or hundreds of different types of odor molecules. Coffee aroma has 670 components in 18 different classes. The difference in relative proportions among these components distinguishes coffee from tea aroma (23). Differentiating subtle differences in odor composition can be critical for species survival. The nine species of ermine moths in the genus *Yponomeuta* maintain species isolation by using phermone blends differing only in the proportions of a mixture of nine aldehyde and alcohol components (24). The pattern recognition task solved by the male moth is to differentiate small differences in proportions of nine similar odorants over several orders of magnitude of concentration variation while flying upwind and encountering spaced packets of pheromone in the pulsed odor plume emanating from a female moth (25,26).

Terrestrial slugs such as *Limax maximus* are totally dependent on odor cues for mate selection, food localization and homing behavior (27), and show rapid and reliable odor learning in a variety of first order and higher order conditioning paradigms (28,6). We analyze the slug's bnose with a variety of tools, including a cnose (29), which provides design principles for incorporation into an enose (8). We illustrate the synergistic interaction of bnose, cnose and enose studies in the analysis of olfaction that follows.

Electronic Olfaction

Electronic olfaction refers to odor recognition and classification by electronic systems comprised of arrays of odor sensors giving broad and partially overlapping responses analyzed by a pattern recognition system (30,31). A variety of sensor materials are in use, including conducting polymers, resonant films, ceramic materials, and films acting as chemiresistors due to swelling during odor absorption (32).

Artificial olfactory systems attempt to mimic the extraordinary chemical sensitivity and selectivity of biological olfactory systems (11,18) using an array of chemosensitive detectors and various pattern recognition algorithms applied to the detector responses (31,33,34). Electronic noses (enoses) of this generic design provide information useful in a wide variety of analytic domains (35-37), from process control to medical diagnosis (38-40). Increased understanding of biological olfactory information processing (6,35) is leading to incorporation of more biologically-inspired network design features (42-46) which will enhance both sensitivity and selectivity.

We have explored the ability of an artificial olfactory system to identify odor objects after active brief sampling by sniffing. Many biological olfactory systems actively sample the environment by initiating water or air flow over the receptor surface (47). Sniffing in mammals produces periodic airflows over the nasal epithelium (48) which vary in frequency from 2 Hz at rest to 200 Hz during prey searching in hunting dogs (10), see also (9). Our initial studies used single sniffs at widely spaced intervals as a prelude to studies of multiple sniffs analyzed by circuit elements with properties tuned to the sampling frequency, as found in biological olfactory systems (49-52).

Produce items provide convenient test objects for rapid recognition and discrimination of similar complex odor sources. A large literature exists on the volatile compounds emitted by various cultivars of commercially important species such as apples (*Malus domestica*) (53,54) and oranges (*Citrus sinensis*)

(55,56). If rapid and reliable recognition techniques based on odor pattern analysis applied to brief sensor signals can be developed, they may provide useful adjuncts or alternatives to vision-based recognition systems (57-59).

To accomplish the produce recognition task, the physical layout of the sensor head was modified. Parafilm was stretched across the sensor head aperture and punctured several times. In this way, objects smaller than the aperture could rest on the parafilm screen and still be sampled. To provide an active means of transporting air saturated with odor from the produce item into the sensor head, the unit was modified to mimic an animal's ability to "sniff." The head assembly was originally built with two ports in the side. A vacuum line controlled by an electronic valve was attached to one of these ports. The other port was closed.

The enose system was capable of recording fast responses of the conducting polymer sensors to various items of produce. Figure 1A,B, & C show examples of odor spectra. Each of the top six traces is the voltage output vs. time for one sensor element. The bottom trace is derived from a pressure sensor that indicates when a vacuum pulse has occurred. The vacuum pulse lasts for one second. Of the original twelve sensors in the sensor head, only six responded to produce odors. The segment of the wave form between the rising edge of the vacuum pulse and one second after the falling edge for the six odor-responsive sensors provided the data to which pattern recognition algorithms were applied. In figure 1D, the response of the system to room air (i.e., nothing on the sample head) can be seen.

The ability of the system to differentiate between closely related odor objects is illustrated by data obtained from Valencia and Florida oranges. Two seconds of data were taken from each of six sensors; the two second interval began at the leading edge of the vacuum pulse and terminated one second after the trailing edge. This was done to capture the biphasic transients of some sensor responses. Using two different Valencia oranges, fifty trials were performed to build the Valencia cluster in feature space. This was repeated using two Florida oranges. The test data consisted of fifty trials each of a Florida orange and a Valencia orange, neither of which had been used to construct the training set. All of the vectors were normalized to unit variance and zero mean, to minimize the effects of changing ambient conditions. With this normalization scheme, the magnitude of a sensor's response relative to that of the other sensors was preserved. We used the nearest subspace classifier because principal component analysis revealed that the clusters were highly anisotropic. The clusters were modeled as hyperplanes in feature space: vector subspaces that were defined by the principal components with the largest eigenvalues. To classify an unknown feature vector, its distance from each of the hyperplanes is calculated. The orange associated with the closest hyperplane determines the designation of the test vector. It was experimentally determined that 10 dimensional subspaces gave the best results. Using this classifier, the system achieved a 76% recognition rate.

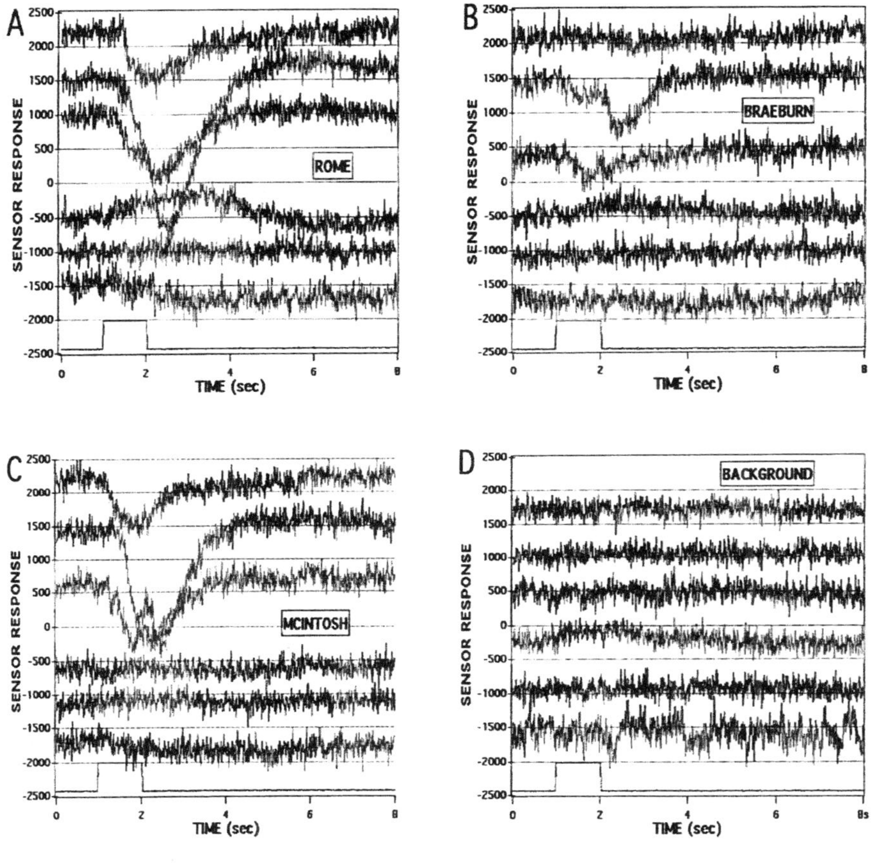

Figure 1. Patterns of sensor responses recorded during and after a 1 sec vacuum pulse applied with various apple cultivars on the perforated platform over the sensor array. A. Sensor responses to Rome apple, B. to Braeburn apple, C. to Mcintosh apple. D. Responses to sniff pulse with no apple on the perforated plate. In all cases the bottom trace indicates the onset and duration of the sniff pulse.

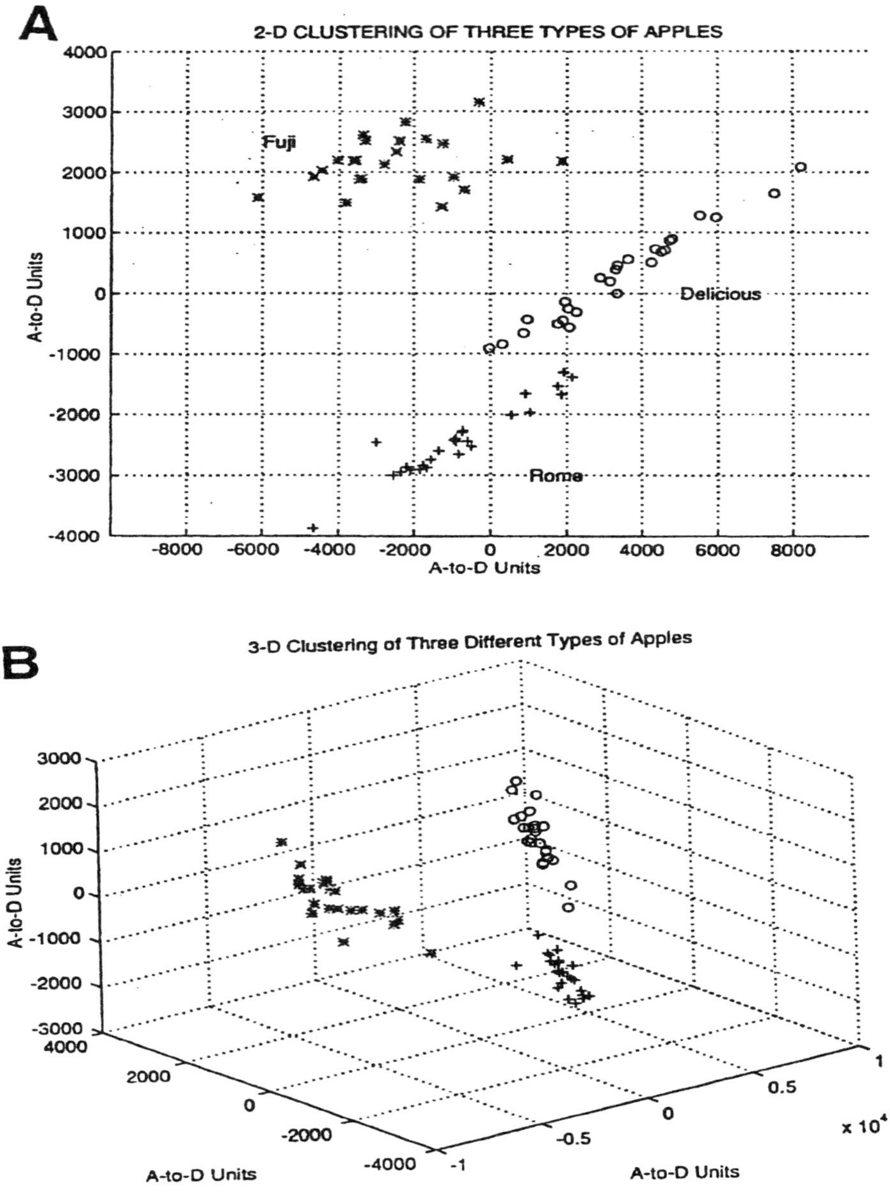

Figure 2. Examples of data clustering from three varieties of apples. Principal component analysis found the eigenvectors with the highest variance which were used to plot the data from each apple cultivar. star=Fuji, circle=Red Delicious, plus=Rome. A. Two dimensional plot using the two eigenvectors with the highest variance. B. Three dimensional plot using the three eigenvectors with the highest variance.

The application of principle component analysis to data from three apple varieties (McIntosh, Fuji, Red Delicious) is shown in figure 2A,B. These data represent multiple sniffs from four specimens of each cultivar. Clustering using 3 eigenvectors produced better separation than clustering using 2 eigenvectors. We found that several factors impacted the accuracy of subsequent identifications of test sniffs. Background odors and changes in odor features over days were found to be significant variables. Subtraction of a background odor signal from the specimen signal increased recognition accuracy but not dramatically so. This is not unexpected given the possibility for nonlinear interactions at the sensors between background and sample odors. There was also significant variation in the odor features of individual produce items across days. The production of volatile substances by various apple cultivars is known to depend on developmental state at harvest and postharvest storage conditions (60-62).

We explored five different pattern classification algorithms for their ability to assign vectors comprised of multiple odor sensor signals to the correct class. Nearest neighbor analysis, principle component analysis, nearest subspace analysis, linear discriminant analysis and a single layer of three perceptrons were applied to data obtained from 50 2-sec sniff samples of each of 2-4 representatives of different apple cultivars, such as Red Delicious, Fuji, Stayman, McIntosh, Rome and New Zealand apples. No one technique was clearly superior to the others in optimizing classification given the variability in the data caused by background odors and day-to-day variation. Variations in odor vectors due to aging and background odors may require use of a learning algorithm which can update a database of prototype odor templates when new valid exemplars of a produce category deviate significantly from the prototype template.

The results obtained to date with odor responses to 1 sec sniffs of produce items show that significant information can be obtained from the patterns of gas sensor responses to very brief odor samples obtained by sniffing. The conducting polymer sensors used in our studies normally reach asymptotic response levels in several 10s of seconds so the 1 sec limitation of odor exposure is a severe constraint. The odor sensors were not selected to be responsive to odors typically found in produce items so only a subset of the 12 sensors gave reliable responses to the 1 sec odor samples. The sensor head is not designed to optimize odor flow over the approximately 1mm^2 surface area of the conducting polymer so the efficiency of odor capture was not optimal. With these limitations it is perhaps surprising that the odor response vectors we measured for different cultivars of a given fruit species were distinctly separable under certain conditions.

The confounding factors of background odors and day to day variability in odor profile of odor objects can best be addressed by adaptive signal processing using a learning algorithm to build up odor templates for each odor object and modify the template as odor objects change over days. The issue of background odor fluctuations can be addressed by subtracting an odor vector representing background odors which is obtained periodically during a training and testing session. The adaptive signal processing required to address these issues can be implemented with a neural network. Both artificial and biologically-based neural networks have been applied to pattern recognition of gas sensor array responses to increase the accuracy of sample detection. (46) show that the temporal dynamics of odor sensor responses provides a useful source of information for pattern classification, a view supported by recent results in biological olfactory information processing (63,64).

Organic Transistors As Odor Sensors

We have recently begun to explore the performance of transistors fabricated from organic semiconductors (65) as odor sensors (66,67). This work is a collaborative effort between synthetic chemists (Z. Bao and H. E. Katz), device physicists (B. Crone, A. Dodabalapur), and ourselves at Bell Laboratories aimed at developing a new generation of enose devices with enhanced sensitivity and selectivity (68).

The sensor device is a thin-film field-effect transistor with its semiconductor fabricated from an organic polymeric material. The typical geometry of such a device arrayed for odor stimulation is shown in Fig. 3. The semiconductor material is left unprotected so that it is accessible to flowing odor molecules delivered from a multibarrel puffer controled by electrically activated solenoid valves. A constant bias voltage is applied both between the drain and source and between the gate and source. The drain current is measured in the absence of odor-containing air and in the presence of odor-containing air. By subtracton the effect on the drain current produced by the odor is determined. We typically used 5 sec pulses of saturated vapors of a variety of odors including alcohols with carbon chain lengths from C4 to C9, and representatives of several odor classes such as l-carvone, vanillin, camphor, eugenol, allyl propionate, amyl acetate, octane thiol, several nitriles, 2-heptanone, toluene and geraniol.

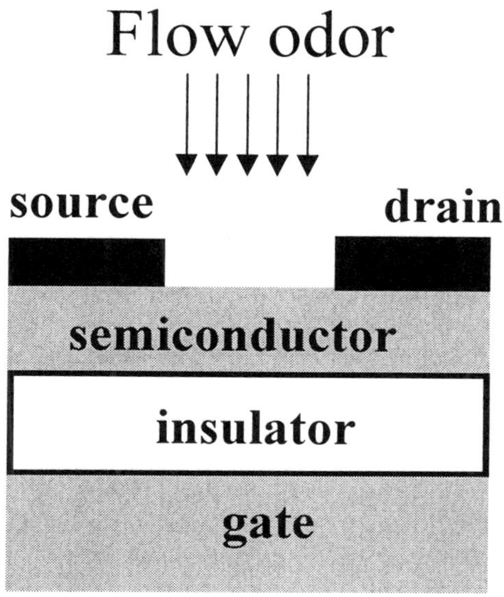

Figure 3. Diagram of a thin film field effect transistor arrayed for application of odor molecules to the semiconductor surface.

A variety of organic semiconductors have been tested for their odor sensitivity, including regioregular poly(3-hexyl thiophene) (PHT), α-dihexyl quinquethiophene (DHα5T), α-sexithiophene (α6T), didodecyl α-sexithiophene (DDα6T) and napthalene tetracarboxylicdianhydride (NTCDA). A typical response of one of these materials to a 5 sec odor pulse is shown in Fig. 4. The response to the 5 sec odor pulse outlasted the stimulus. For this reason odor responses were tested at 30 sec intervals. The response was measured at the peak of the odor-elicited change in drain current. If plastic transistors are to prove useful in fabricating a sensor array for odor identification, it is essential that different organic semiconductors have different responses to the same set of odors. Data comparing the odor responses of DHα6T and α6T to a set of alcohols from C4 to C9 plus 2-heptanone are shown in Fig. 5. Clearly these two organic semiconductors respond very differently to alcohols versus 2-heptanone, encouraging the view that a range of odor responses will be displayed by the family of organic semiconductor materials available for fabrication of discrete FET transistors (65).

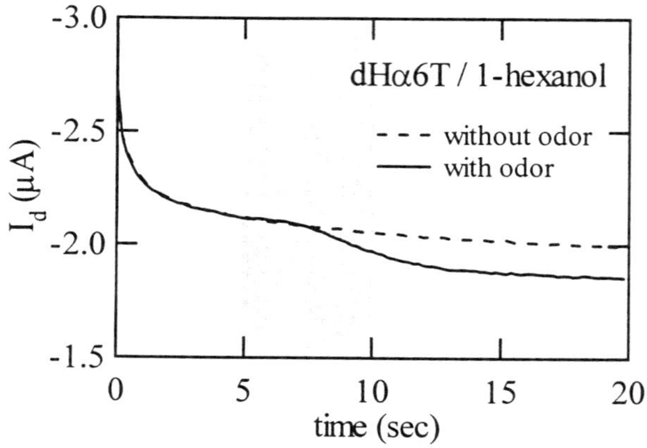

Figure 4. Typical response of an organic transistor to application of an odor. The drain current is shown as a function of time after applying a constant drain-source and gate-source voltage. The drain current is shown with a 5 sec odor applicaton and in the absence of odor application.

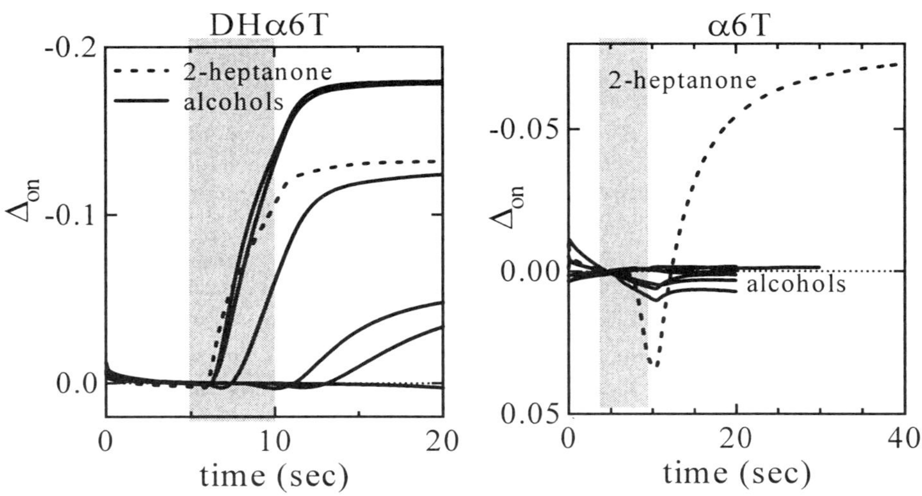

Figure 5. Responses of two different organic transistor materials to a set of straight chain alcohols with chain lengths from C4 to C9 and to 2-heptanone. In all cases a saturated vapor of the odor was applied as a 5 sec pulse and the peak response measured. The two polymeric semiconductor materials give distinctly different patterns of responses to the alcohols versus 2-heptanone.

A possible mechanism for odor-induced changes in drain current in these polymeric materials is odor intercalation into the flexible endgroups. Many of the materials have rigid thiophene cores with flexible endgroups which are aligned in the thin film of material. If odor molecules permeate the region of aligned endgroups, the mobility of charge carriers will be affected. The existence of grain boundaries within the materials also contributes to the odor sensitivity by improving odor diffusion into the materials. For the materials we have studied thus far, the threshold gate voltage at which drain current becomes nonzero is shifted for several odor classes (nitriles, alcohols, ketones), however, some shift in mobility is also found, in particular with thiols. Studies to examine these parameters using a larger set of polymeric materials and assess their contributions to odor sensitivity are in progress.

Several parameters of the FET device can affect its odor sensitivity. These include the thickness of the semiconductor layer, which can range from 5 to 200 nm. The gate voltage also provides an important control parameter which can influence the magnitude, time course and polarity of the odor-induced change in drain current. With zero gate voltage the transistor is nominally off, which allows selection from an array of FETs of the subset of sensors which will contribute to the sensor pattern for a given odor. A particularly useful aspect of organic FETs is the ability to fabricate them into circuits of multiple active elements to achieve greater selectivity and sensitivity. Large-scale complementary integrated circuits containing up to 864 transistors have recently been demonstrated (69).

Computational Olfaction

It is useful to divide computational olfaction into several levels of analysis. First, what behaviors are driven by the olfactory system? These behaviors can be either observed behaviors of animals, or desired behaviors of an engineering-based system. Second, what are the inputs to the system from which it is (apparently) able to generate these behaviors? What algorithms can be used to calculate answers which are of use in generating these behaviors? Finally, how can these algorithms be placed on the available computational 'hardware', either biological or engineering? A survey of olfactory behaviors results in a minimal list of olfactory tasks that includes:

1) Basic olfaction: learning and being able to identify an isolated odor, with appropriate generalization to previously unknown similar odors (e.g., a new variety of 'apple' on the basis of familiarity with several other varieties of apple.

2) Background elimination. When a weak known odor is thoroughly mixed with an unknown background, the known odor can be identified and its intensity measured.

3) Component separation. When a few known odors are thoroughly mixed (eliminating relative fluctuations) an animal can identify the component odors and their intensities.

4) Odor separation. When odors from different objects are mixed by air turbulence, the fluctuations of the relative contributions to the mixture can be used to separate the odors of multiple unknown objects in the environment.

Most descriptions of olfaction focus entirely on task 1. However, consideration of the complex behaviors of animals under natural conditions requires at the very least the ability to accomplish tasks 2,3,4. Simple artificial systems in highly controlled environments may be able to stop with task 1. Many chemosensing tasks will also require these further abilities. Task 4 for example is desirable in a pollution-monitoring system, to which mixtures of pollutants arrive in a time-dependent mixture of multiple sources, and it is desired to isolate the pollution footprint of a single source in order to identify the source.

The rat has about 2000 different olfactory receptor types. Such a large number of receptors is completely unnecessary to solve problems of type 1. Our rich sense of color vision and our ability to identify hundreds of hues comes from the measurement of light in 3 wavelength bands. For task 1, 10 olfactory channels would suffice in a world with millions of different odors. The reason that the generalist olfactory system in biology uses so many receptor types must have to do with the difficulty of the other tasks. The subsidiary tasks or complications of vision and of olfaction are entirely different. For example, ordinary visual objects are opaque, so that the spectrum of light arriving at a given point of the retina is that which comes from a single object. The color of the leaves which are occluded by a cardinal do not confuse the identification of the color of the bird. However, a given packet of air has a long history, and contains odorants from a variety of sources that have been mixed by turbulence and diffusion. The analogous visual situation would be that of trying to deduce the color of a semitransparent bird that is in front of other colored objects, a visual problem we cannot solve.

The available information

Each of the 2000 receptor cell types expresses a single kind of odorant binding receptor protein on its surface. The binding of a ligand to a receptor is a physical measure on a molecular piece of an object. Information about odorant-receptor binding constants comes chiefly from studies of the threshold for perceptual detection of pure compounds. In a study of 529 odorants (3), the distribution of the logarithms of threshold concentrations was approximately Gaussian (i.e., the distribution of free energies of binding is Gaussian.) The ± 2 standard deviation (± 2sigma) range within which most olfactants lay was 6.8 $\log_{10}$ units, with the extremes separated by 10 $\log_{10}$ units. The threshold for detection of a particular pure odorant will correspond to some minimum coverage cov_{min} of the odorant receptors of a given type (or more probably, a few types) by odorant ligands. The reason that the distribution of *logarithms* of binding constants is broad and smooth is a simple consequence of the exponentials involved in the physical chemistry of large binding constants. Few binding constants have been measured directly.

Odorant modeling

The pattern of excitation across the receptor cell types describes an odor. These patterns necessitate describing the binding constants of a particular ligand for receptors that do *not* bind it maximally. The best-fit ligand-receptor pairs have a ± 2s range of binding constants of more than 10^6 and thus a range of binding free energies of 10 Kcal. The same stereochemistry should produce a similar range of binding constants and binding free energies below the maximal ones. While odors differ markedly in behavior, there is a typical psychophysical range of a factor of 1000 over which the odor seems to have the same quality, and over which the relationship between the perceived intensity and the actual concentration is approximately logarithmic. An odor presented at the concentration where non-linearities begin is driving the cells having the highest affinity receptors into response saturation. At this driving strength we would expect, from the above numbers, that about 1/2 of the cell types would show responses to that odorant and about 1/2 would show no response (i.e., no response visible above noise levels). Assays are available from studying responses of many cells for an array of odorants both by conventional electrophysiology (70) and by cell biology techniques (71). For the concentrations of odorants used, the cells showed some level of response above noise levels to a particular odorant, qualitatively like the expectations described above.

For modeling odors in the absence of additional information, we presume that the receptor cell sends a signal related to its ligand coverage, and that the ligand coverages due to different ligands are additive[1]. We assume that a sensory cell has a dynamic range of 1000 in odorant concentration and that the probability distribution of binding constants of the less-than maximal cell classes for a given odor is uniformly spread (on a log scale) over a range 10^6 below the binding constant of the most sensitive cell type. Models require details, but most of the qualitative conclusions of the present model rely only on the model being consistent with the experimental fact seen by Sicard and by Malnic, namely that for unrelated odorant pairs a, b there will be hundreds of cell types which respond chiefly to a, hundreds which respond chiefly to b, and hundreds more which respond to both.

At threshold concentration c^{l}_{thresh} of ligand l, the receptors of the one cell type that responds above noise have a coverage of cov_{min}, and have a binding constant of $cov_{min}/\ c^{l}_{thresh}$. The coverages of all other receptor cell types will lie in a range of 10^6 below cov_{min}. When the concentration is raised to 10 $c_{thres,l}$, then 1/6 (2000)= 333 cell types should respond above noise. These are the cell types that have their binding constants in the range of 0.1—1 times the binding constant of the most strongly binding cell type. As the concentration is raised further more cell types respond, until at concentration 1000 (in units of the threshold concentration) about half of cell types are responding above noise.

Unrelated molecules have no particular relationship of their binding contacts and binding energies for a given receptor type. Each receptor cell type is independently assigned binding constants for each odorant by the prescription above. Real odors can be related, either because they have dominant chemicals of similar structure (e.g., ethanol and isopropranol), or because they are mixtures in different ratios of the same chemicals (orange and tangerine). While

---

[1] The real situation is somewhat more complicated. The responses may not be additive. The latency of sensory cell response can be odorant specific. The mucous layer and the proteins it contains may play some role in kinetics. The overall situation is reminiscent of the depth perception computation in vision. There are several different algorithms involved, including binocular stereopsis, shape from shading, linear perspective, occlusion, and structure from motion. Visual studies have gained much by analyzing each separately. Similarly, for olfaction it is useful to analyze how concentrations and binding can be used alone.

related odors can be described within the present context, doing so requires knowing yet unknown details of the mixtures and their binding patterns.

Olfactory tasks in a least-squared error algorithm

For simplicity, let a known odor t be a single molecular species. This molecule has binding constants K^t_i for receptors on cells of type i. When odorant t is present at concentration c^t, the fractional coverage cov^t_i of binding sites of type i will be

$$cov^t_i = c^t K^t_i \quad (1)$$

Let the fraction of coverage for some unknown odorant be given by cov^u_i. This unknown odorant should be identified as t if it is approximately true that there exists a value of c such that

$$cov^u_i \approx c K^t_i \quad (2)$$

for all values of i. (This is clearly true if u is in fact the same as t.) In what sense is this approximate equality to be viewed, in a noisy system? The simplest criterion is the mean square error, and thus to recognize the substance u as being in t if the mean square errors in the description of Eq. (2) is less than some error measure e.

This easily implemented (neurons or machine) algorithm is useful for task 1) but cannot be applied to task 2). For task 3), the algorithm works, but is very cumbersome since a comparison must be made to all possible combinations of a few known odors. However, the weakest aspect of this algorithm is its reliance on all channels with the same error measure. In both biological and artificial chemosensors, at any time it is likely that many sensor types are at the moment grossly unreliable, with bound contaminants through previous exposures to other chemicals, or adapted as a result of such exposure. A subset of channels may be telling the correct story, but we do not *a priori* know which channels. If there are 10 channels, and the possibility that up to 50% of them are in error, to compute with all the good channels we must try out ~2,000,000 possibilities. For digital machines, this is not a problem, but for neurobiology it is prohibitive. Problems with mixtures of odors are even more difficult. For neurobiology, and probably for artificial applications as well, algorithms based on a large number of channels will be much more effective.

An approach through large numbers of receptor types

To begin, we illustrate for task 2), not solvable by using the mean-square-error method. The target t is a single odorant species, and the background is taken to be a different single species b. Following the odor model above, the coverage of receptors of type i in the presence of concentrations c_t and c_b is given by

$$cov_i = cov_{min}\,(c^t/c^t_{thresh})\,(1 \text{ or } f^t_i) + cov_{min}\,(c^b/c^b_{thresh})\,(1 \text{ or } f^b_i \qquad (3)$$

In the first term parenthesis, a 1 is used if it happens that i is the cell type which maximally binds t, and otherwise an f^t_i is chosen at random in the range between 1—10^{-6} with a uniform probability distribution in the logarithmic domain. The same is true for the second parenthesis and odorant b.

If only odorant t is present, Eq.(3) reduces to Eq.(1), and thus from each cell type i which is appreciably driven the system can 'calculate' the concentration c^t using Eq.(3). When an *unknown* odor u is presented which generates a pattern of coverage cov^u_i , then for each i which would be driven by the *target* odor we can calculate the concentration c_i of t which would yield that level coverage for cells of type i, namely

$$c_i = (c_{thresh}/f^t_i\,)(cov^u_i/cov_{min}) \qquad (4)$$

c_i is the *apparent* concentration of t deduced from the level of receptor occupancy caused by the unknown odorant in a single channel i. The observation of cov^u_i is a 'vote' for the presence of t at concentration c_i.

Figure 6 a-c show smoothed histograms of all the 'votes' c_i when odor u is in fact t, presented at a concentration 10, 100, and 1000 (in units of the threshold concentration). A logarithmic scale is used because of the large range of concentrations involved. This scale is also particularly apt for systems that show an approximately logarithmic response as a function of intensity, as many sensory systems do. A gaussian noise level of ± 26% (σ = 0.1 $\log_{10}$ units) introduces width to the histograms. Because the unknown odor u is in fact t, each channel deduces (within noise) the correct concentration from its value of cov^u_i. Of the 2000 receptor cell types, about 333, 667, and 1000 are driven to observable levels by odorant t at these concentration respectively, corresponding to the total numbers of 'votes'. The concentration of t can be read from the peak positions, located as expected. As the concentration is decreased another decade, the number of responding cell types approaches one.

Figures 6d-f show histograms of the same sort, using the same 1000 channels which are potentially activated by t, but calculated for three different odors b

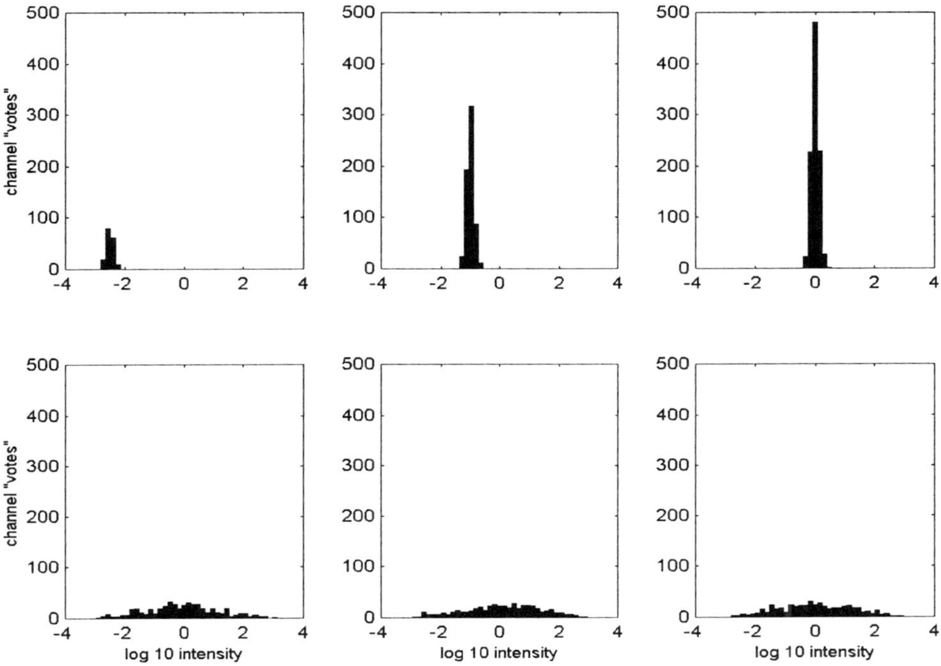

Figure 6. Histograms of the number of receptor cell types which 'vote' for various dimensionless intensities (or concentrations) of the target odor. Units of odor strengths are on a logarithmic scale; one unit is factor of 10 in intensity. Threshold intensity for an odorant corresponds to intensity 1, logI=0. a-c, the target odor itself at concentrations 10, 100, and 1000. d-f, three different non-target odors at concentrations 1000. Graphs are arranged left to right, a-c across the top, and d-f across the bottom.

unrelated to t, each at its saturation level of $1000c^{b}_{thresh}$. Only those having more than the minimal detectable coverage respond. Since there is no relationship between f_i^t and f_i^b, there is a wide spread of events within these histograms, and no sharp peak. A histogram for a weaker presentation of b would be similar, but with fewer events.

Target and non-target odors are highly distinguishable in this representation. Even when the target odor is at a strength of only $3c_{thresh,t}$ the peak in such a histogram for odor t alone has about 150 votes spread over a total concentration range of 0.4 on a logarithmic scale. The probability that an *unrelated odor u at a concentration* $1000c_{thresh,u}$ produces a total number of events greater than 100 in such a range is less than 10^{-6}. When the target odor t is at strength $10c_{thresh,t}$, it becomes astronomically unlikely to confuse t with a random strong odor. In this representation basic odor recognition can be accomplished by merely thresholding the total number of votes in a bin of appropriate width.

The problem of identifying an odor in the presence of an unknown background is examined in Fig. 7 a-c. These histograms were generated from an odor mixture in which the target odorant at strengths 10, 100, and $1000c_{thresh,t}$ is mixed with an unrelated background odorant b at strength $1000c_{thresh,b}$. The area of the peak in the histogram is decreased, but a simple threshold on the number of events in a suitable bin width will still determine when the target odor is present. The target begins to be detected when its concentration is greater than $3c_{thresh,t}$, even in the presence of the saturating concentration of an unknown odor. There is no way to solve this problem when n is small.

The statistics of large numbers and the chance occurrence of non-interference make identification in the presence of strong unknown odors

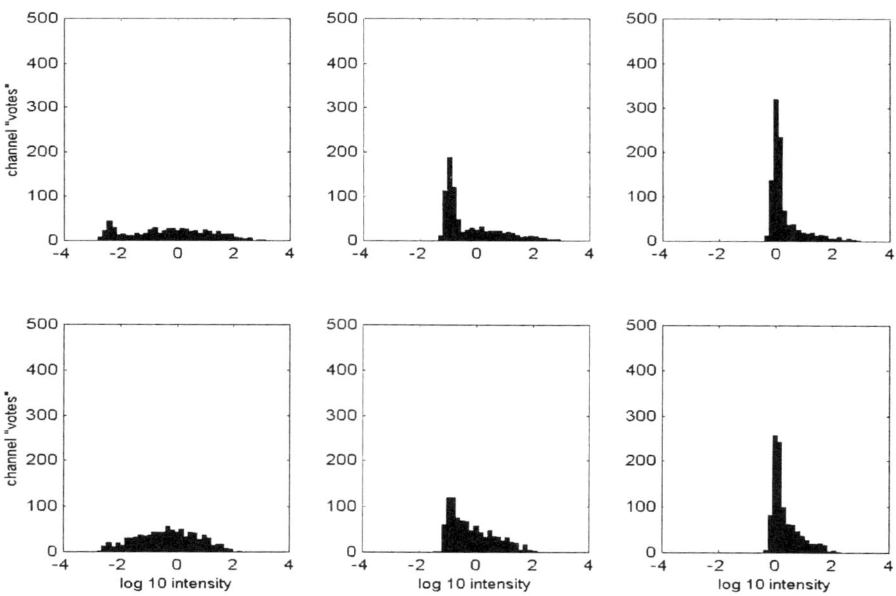

Figure 7. Histograms as in Fig. 1 a-c, but with the simultaneous presence of a background of strength $1000c_{thresh,b}$. d-f have a more complex background of total strength 400 in terms of the relevant thresholds, made up of an equal mixture of four non-target odors. Graphs are arranged left to right, a-c across the top, and d-f across the bottom

possible. No receptor cell type is specific—each has a probability of 0.5 of being at least somewhat activated by a saturating concentration of *any* odor. This translates to about 333 channels being noticably driven when odor t alone is presented at low concentration $10c_{thresh,t}$. However, each channel also has a probability of 0.5 of being negligibly driven by odor b at strength $1000c_{thresh,b}$. Thus, about 167 ± 13 channels are still expected to respond to t at concentration $10c_t$ in essentially the same way as they would have in the absence of the unknown background odor, and many more will do so at higher c_t. These channels are responsible for the lowered but appropriately located histogram peaks in Fig 7 a-c compared to Fig 6a-c. Channels that have responses which are distorted by the presence of the strong background odor produce the broad background.

Figs. 7 d-f are like figures Fig. 7 a-c except that the background odor is made up of a mixture of four unknown odorants each at a strength 100 (in terms of its threshold). The total strength of the background is now 400. It is much harder to recognize a known single-molecular species in the presence of this complex background, but quantification of a single odorant against a complex background that is several times stronger can still be done by the same method.

Separating a mixture into known components can therefore also be solved in this same fashion. The fact that we could identify a known component of strength 100 in the presence of 4 unknown chemicals each of strength 100 (shown in Fig 7e) indicates that a complex mixture of 5 known molecular components could be separated into its components. The system fails when the number is raised to 7.

The effects of erroneous data can be understood in terms of these simulations. Sensory channels which are to some extent stuck 'on' are equivalent to channels which have additional input, and have effects exactly like those of a background odor, contributing a wing to the right in Figs 7. Sensory channels that are strongly adapted contribute a wing to the left. Such effects on 50% of the channels produce little impact on performance when n is large.

Task 4: Separating unknown odors using fluctuations, covariation, and large N

The large number of independent channels can be used in conjunction with adapting neurons to solve the problem of separating independent but unknown odors. The basis for the algorithm is that all the components of a single odor will fluctuate with the same time course, while two odors that come from different positions in space will be mixed by turbulence, and the two sets of components will have different fluctuations. Using the odor model and representation already described, a 'two sniff' paradigm can discover the presence of two or more objects. Previous algorithms (7) make no use of the kind of odor model we have constructed here, and are in that sense more general. However they require a study of the fluctuating odors over a considerable time period.

Suppose two different odors x, y are in the environment, each due to its own single molecule type. A first sniff contains the combination a*x + b*y. A second sniff is due to a*x + b*y, where a,b,a,b are the intensities of x and y relative to their saturation concentrations.

Construct a target odor on the basis of the first sniff. The second sniff is then analyzed with the first sniff as the target odor, using the histogram representation as before. We anticipate the appearance of the histogram by noting that many of the channels are dominated by odor x, and will contribute just as they would have if a*x were the target and a*x the second odor, and will contribute a peak to the histogram at that a/a. Many other channels are dominated by y, and will thus generate a peak in the histogram at b/b.

Fig. 8 shows simulation results for a situation representing a weak 'object' (a=0.02) in the presence of a stronger 'background' (b = 1.0). Fig 8a shows the result when the second sniff is identical to the first. The location of the histogram peak is at 0, (representing an intensity ratio of 1 between the first and second sniffs), and its area represents the contribution of about 1200 channels. Fig 8b shows the result when the second sniff is a = .0066, b=0.33. Both odors have decreased in strength by a factor of 3. This would happen if there is really a single object (for which all components track in parallel) or if odors from two objects were thoroughly and time-independently mixed. The peak has shifted by -0.48 (or $\log_{10}(1/3)$), compared with Fig 8a, corresponding to the decrease in intensity of the second sniff compared to the first. It has less area since more of the channels are now below threshold.

Fig 8c shows the result with the same first sniff, but when the second sniff is a = 0.04, b = 0.67. The two odors have changed in intensity relative to each other by a factor of 3, and that has produced a histogram having two peaks. The smaller peak is due to channels that chiefly respond to the weaker odor, and the larger peak is due to channels that chiefly respond to the stronger odor. The position of the smaller peak at +0.27 corresponds quite well to the 0.30 = $\log_{10}(2)$ shift expected due to the increase of the intensity of target by a factor of 2, while the position of the larger peak at - 0.16 corresponds to the expected shift of $\log_{10}(0.667)$ = - 0.18 due to the increase of the second component. The channels can now be divided into two classes, corresponding to each of the two resolved peaks. When this is done, each channel responds to more sniffs by showing a single peak, whose location and height varies in the expected way with the strength of its particular component.

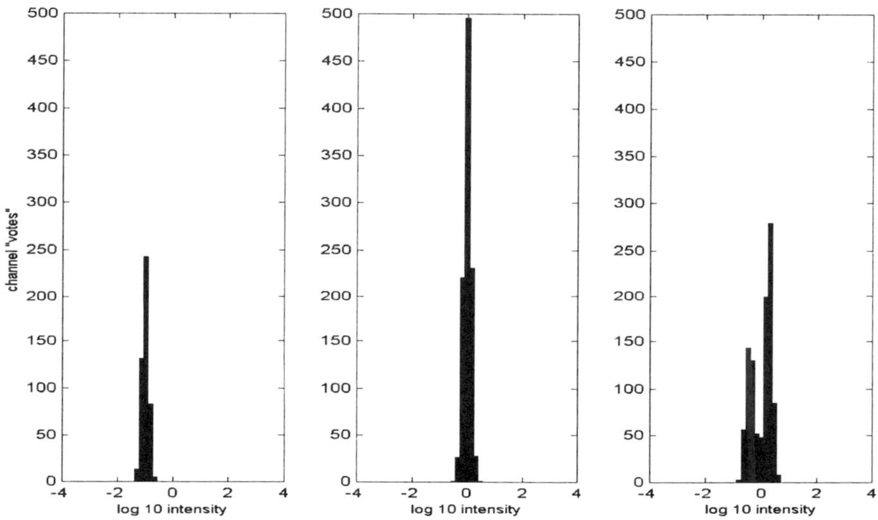

Figure 8. Simulation results for a situation representing a weak 'object' (a=0.02) in the presence of a stronger 'background' (b = 1.0). Graphs are arranged left to right, a-c.

How large should N be?

In rodent biology, the number of cell types N is about 2000. We have intuitive feelings for the case n=3 from color vision. If two light sources are mixed, all channels will usually be effected by both, and there is no reliable way to tell if a 'target' hue is present or not. For example, light of 5100 Å and of 4900 Å wavelengths are quite distinguishable shades of green, but drive chiefly green and blue cones. Because of this, a particular mixture of 5100 Å and 4500 Å light produces an exact visual color match to 4900 Å light. n = 3 is a small number.

In the case of olfaction, if n = 5 and random binding constants picked in the prescribed range, 6% of odors will have significant binding to only one channel. Odors can then be divided into a set which drive more than one channel, and five sets each containing about 1% of all molecules. Within each

1% set all members are utterly indistinguishable. Appropriate levels of five primary odorant chemicals could duplicate any odor. For any primary odorant, unknown backgrounds which drive that unique odorant receptor (and about 50% of backgrounds will do so) completely prevent the reliable detection of the target. About 25% of odors will drive two receptor types. For these the ratio of the bindings can be used, generating some reliability, but there is only one ratio, and about 4% of odors will drive those two receptors in the same ratio. Benefits of large numbers are suggestive, but unreliable. For n=100, many ratios are always available and the statistics already reliable. (For example, the probability at saturation of driving fewer than 25 channels is less than 10^{-6}.) It is not known at present whether the 2000 receptor types of neurobiology fall into a much small number of 'independent' classes. While the system functions well when the binding constant range is spread over a factor of 106, a range of 1000 is already rather useful. To illustrate these points, Fig. 9 shows results analogous to Fig 1-2 for 100 receptor types spread over only a range of 10^3 (rather than 10^6) in binding constants. Though the ability to discern weak odors in the presence of strong unknown backgrounds is lost, it tolerates simple and complex backgrounds which are 10 times stronger than the target, and five components can still be separated.

Neural implementation of computational algorithm

At some level, the elementary odor learning and memory represented by task 1) is an example of associative memory. An elementary neural network description, called LIMAX, of simple odor learning and conditioning made use of the dynamical attractors of simple neural networks to encapsulate much of the simple odor responses of Limax maximus (29). In it, odors were viewed as preprocessed to a standard intensity, and components were then made binary (0-1). This binarized view worked acceptably for simple learning problems and in the absence of background odors, but is entirely inadequate for dealing with the more sophisticated problems of strong backgrounds or multiple sniffs of mixed odors.

For more difficult problems 2) 3), the histogram algorithm has a natural representation on a 'biological' neural network. When use is made of the underlying oscillatory rhythm common to the 'front ends' of all olfactory systems, intensity information can be encoded in terms of the timing of action potentials with respect to this underlying rhythm (72). Since the desired encoding is logarithmic, it is significant that recent studies of receptor cell responses are approximately logarithmic in their responses (73). With this encoding, a decoding based on time-delays can be done such that the time-dependent input to a recognition neuron during one period of the underlying rhythm is exactly the histograms of Figs 6,7,8, with the intensity axis converted to time. The time at which each action potential arrives is its 'vote' in this histogram. If the near-simultaneous arrival of perhaps 50 action potentials is necessary to drive a receiving cell to fire, then that cell is performing a thresholding operation on the histogram.

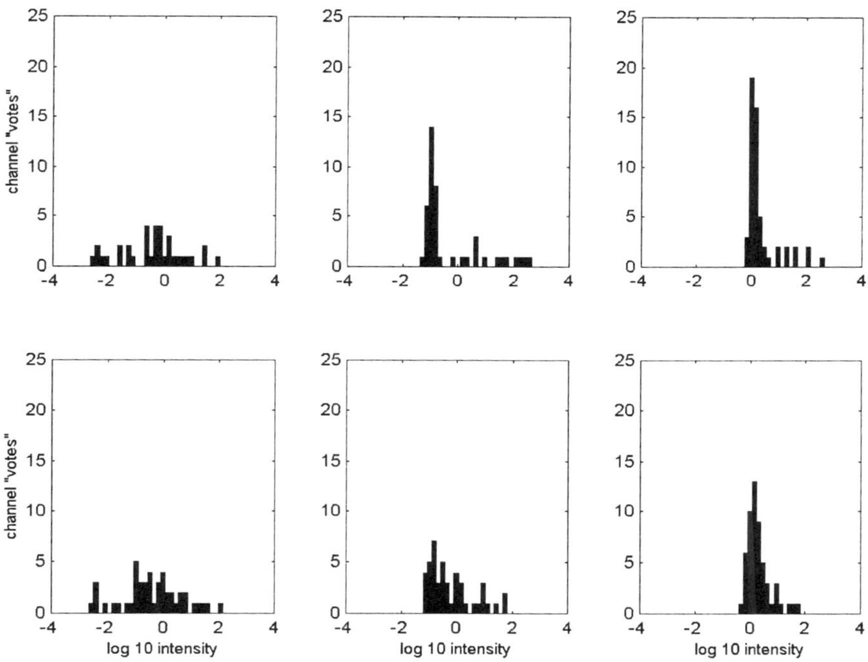

Figure 9. Results analogous to Fig 6-7 for 100 receptor types spread over only a range of 10^3 (rather than 10^6) in binding constants. Though the ability to discern weak odors in the presence of strong unknown backgrounds is lost, it tolerates simple and complex backgrounds which are 10 times stronger than the target, and five components can still be separated. Graphs are arranged left to right, a-c across the top, and d-f across the bottom

Implementation of the 'two sniff' algorithm requires some kind of short-term memory in order to compare a sniff with an immediately subsequent sniff. While in LIMAX and in biological long-term memory, the physical representation of the memory is in synapse change, cellular adaptation appears to be an adequate form of memory for the comparisons between the immediate past and the present. Using a Ca^{+2} based adaptation of threshold for firing (or alternatively, for membrane resting potential), the response of a set of neurons to a second sniff, when properly analyzed, shows all the essential features of the two-sniff histogram of Fig. 9. The essentials for this decomposition include inputs that represent the logarithm of the intensities, at least one neuron for every receptor cell type, and firing rate adaptation. While it is conceptually clear that the information necessary to the two sniff paradigm is now clearly represented in the neurobiology and easily computed from the cellular responses, there is as yet no complete neural system described for carrying out the desired computation. While the original description involved action potential computation, rate coding may in fact be adequate for this problem.

Summary

Studies of biological, computational and electronic olfaction are progressing in a synergistic way to produce both new insights into the information processing algorithms which may be used in biological systems and new functionality in simulated or engineered olfactory systems. The operational principles which allow a biological olfactory system to self assemble during development and retain functionality during the continual addition of new central processing neurons and continual death and replacement of sensory input neurons are very likely to be of general relevance in other sensory processing systems. The evolution of electronic olfactory systems, spurred by incorporation of new design features from biological olfaction, will yield small low power devices useful in process control and medical diagnosis. It can also produce devices that rival or surpass the trained canine in the extremely demanding task of locating buried land mines.

References

1. Sarpeshkar, R. 1998. *Analog versus digital: extrapolating from electronics to neurobiology*. Neural Comput 10:1601-1638.
2. Hopfield, J. J. 1999. *Odor space and olfactory processing: collective algorithms and neural implementation*. Proc Natl Acad Sci USA 96:12506-12511
3. Devos, M., F. Patte, J. Rouault, and P. Laffort. 1990. Standardized Human Olfactory Thresholds. IRL Press, Oxford, UK.

4. Lancet, D. 1992. *Olfactory reception: From transduction to human genetics.* In D. P. Corey and S. D. Roper (eds.), Sensory Transduction, pp. 74-91. The Rockefeller University Press, New York.
5. Kaissling, K.-E. 1986. *Chemo-electrical transduction in insect olfactory receptors.* Ann. Rev. Neurosci. 9:121 - 145.
6. Gelperin, A. 1999. *Oscillatory dynamics and information processing in olfactory systems.* J Exp Biol 202:1855-1864.
7. Hopfield, J. J. 1991. *Olfactory computation and object perception.* Proc. Natl. Acad. Sci., USA 88:6462 - 6466.
8. Gelperin, A., J. L. Dawson, S. M. Cazares, and H. S. Seung. 1999. *Rapid fruit cultivar identification by an artificial olfactory system.* In W. J. Hurst (ed.) Electronic Noses and Sensor Array Based Systems, pp. 263-274. Technomic Pub Co, Lancaster, PA.
9. Thesen, A., J. B. Steen, and K. B. Doving. 1993. *Behaviour of dogs during olfactory tracking.* J Exp Biol 180:247-251.
10. Steen, J. B., I. Mohus, T. Kvesetberg, and L. Walloe. 1996. *Olfaction in bird dogs during hunting.* Acta Physiol Scand 157:115-119.
11. Buck, L. B. 1996. *Information coding in the vertebrate olfactory bulb.* Ann Rev Neurosci 19:517-544.
12. Mombaerts, P., F. Wang, C. Dulac, R. Vassar, S. K. Chao, A. Nemes, M. endelsohn, J. Edmondson, and R. Axel. 1996. *The molecular biology of olfactory perception.* Cold Spring Harbor Symp Quant Biol 61:135-145.
13. Clyne, P. J., C. G. Warr, M. R. Freeman, D. Lessing, J. H. Kim, and J. R. Carlson. 1999. *A novel family of divergent seven-transmembrane proteins: Candidate odorant receptors in Drosophila.* Neuron 22:327-338.
14. Gao, Q. and A. Chess. 1999. *Identification of candidate Drosophila olfactory receptors from genomic DNA sequence.* Genomics 60:31-39.15.
15. Vosshall, L. B., H. Amrein, P. S. Morozov, A. Rzhetsky, and R. Axel. 1999. *A spatial map of olfactory receptor expression in the Drosophila antenna.* Cell 96:725-736.
16. Friedrich, R. W. and S. I. Korsching. 1997. *Combinatorial and chemotopic odorant coding in the zebrafish olfactory bulb visualized by optical imaging.* Neuron 18:737-752.
17. Rubin, B. D. and L. C. Katz. 1999. *Optical imaging of odorant representations in the mammalian olfactory bulb.* Neuron 23:499-511.
18. Hildebrand, J. G. and G. M. Shepherd. 1997. *Mechanisms of olfactory discrimination: Converging evidence for common principles across phyla.* Ann Rev Neurosci 20:595-631.
19. Galizia, C. G., S. Sachse, A. Rappert, and R. Menzel. 1999. *The glomerular code for odor representation is species specific in the honeybee Apis mellifera.* Nature Neurosci 2:473-478.

20. Ressler, K. J., S. L. Sullivan, and L. B. Buck. 1994. *Information coding in the olfactory system: evidence for a stereotyped and highly organized epitope map in the olfactory bulb.* Cell 79:1245-1255.
21. Vassar, R., S. K. Chou, R. Sitcheran, J. M. Nuñez, L. B. Vosshall, and R. Axel. 1994. *Topographic organization of sensory projections to the olfactory bulb.* Cell 79:981-991.
22. Laurent, G. 1999. *A systems perspective on early olfactory coding.* Science 286:723-728.
23. Dodd, G. H., P. N. Bartlett, and J. W. Gardner. 1992. *Odours-the stimulus for an electronic nose.* In J. W. Gardner and P. N. Bartlett (eds.), Sensors and Sensory Systems for an Electronic Nose, pp. 1-11. Kluwer Academic Publishers, Dordrecht, The Netherlands.
24. Löfstedt, C., W. M. Herrebout, and S. B. J. Menken. 1991. *Sex pheromones and their potential role in the evolution of reproductive isolation in small ermine moths* (Ydnomeutidae). Chemoecology 2:20-28.
25. Murlis, J., J. S. Elkinton, and R. T. Carde. 1992. *Odor plumes and how insects use them.* Ann Rev Entomol 37:505-532.
26. Valeur, P. G., B. S. Hansson, and C. Lofstedt. 1999. *Real-time measurement of pheromone release from individual female moths and synthetic dispensers in a wind tunnel by recording of single receptor neurone responses.* Physiol Entomol 24:240-250.
27. Gelperin, A. 1974. *Olfactory basis of homing behavior in the giant garden slug, Limax maximus.* Proc. Natl. Acad. Sci. USA 71:966-970.
28. Sahley, C. L. 1990. *The behavioral analysis of associative learning in the terrestrial mollusc Limax maximus : The importance of interevent relationships.* In S. Hanson and C. Olson (eds.), Connectionist Modeling and Brain Function: The Developing Interface., pp. 36 - 73. MIT Press, Cambridge, MA.
29. Gelperin, A., J. J. Hopfield, and D. W. Tank. 1986. *The logic of Limax learning.* In A. I. Selverston (ed.) Model Neural Networks and Behavior, pp. 237 - 261. Plenum Press, New York.
30. Persaud, K. and G. H. Dodd. 1982. *Analysis of discrimination mechanisms of the mammalian olfactory system using a model nose.* Nature 299:352-355.
31. Gardner, J. W. and P. N. Bartlett. 1999. Electronic Noses: Principles and Applications. Oxford University Press, Oxford.
32. Severin, E. J., B. J. Doleman, and N. S. lewis. 2000. *An investigation of the concentration dependence and response to analyte mixtures of carbon black/insulating organic polymer composite vapor detectors.* Anal Chem 72:658-668.

33. Kress-Rogers, E. 1997. Handbook of biosensors and electronic noses. CRC Press, Boca Raton. FL.
34. Göpel, W., C. Ziegler, H. Breer, D. Schild, R. Apfelbach, J. Joerges, and R. Malaka. 1998. *Bioelectronic noses: A status report Part I.* Biosensors & Bioelectronics 13:479-493.
35. Pearce, T. C. 1997. *Computational parallels between the biological olfactory pathway and its analogue 'The Electronic Nose' 1. Biological olfaction.* Biosystems 41:43-67.
36. Dickinson, T. A., J. White, J. S. Kauer, and D. R. Walt. 1998. *Current trends in 'artificial-nose' technology.* Trends in Biotech 16:250-258.
37. Ziegler, C., W. Gopel, H. Hammerle, H. Hatt, G. Jung, L. Laxhuber, H. L. Schmidt, S. Schutz, F. Vogtle, and A. Zell. 1998. *Bioelectronic noses: A status report.* Biosensors & Bioelectronics 13:539-571.
38. Dodd, G. H. 1996. *The lipid membrane hypothesis of schizophrenia: Implications for possible clinical breath tests.* Prostaglandins Leukotrienes and Essential Fatty Acids 55:95-99.
39. Ping, W., T. Yi, H. B. Xie, and F. R. Shen. 1997. *A novel method for diabetes diagnosis based on electronic nose.* Biosensors & Bioelectronics 12:1031-1036.
40. Schiffman, S. S., B. G. Kermani, and H. T. Nagle. 1997. *Analysis of medication off-odors using an electronic nose.* Chemical Senses 22:119-128.
41. Gardner, J. W., M. Craven, C. Dow, and E. L. Hines. 1998. *The prediction of bacteria type and culture growth phase by an electronic nose with multi-layer perceptron network.* Measurement Science & Technology 9:120-127.
42. Hopfield, J. J. 1995. *Pattern recognition computation using action potential timing for stimulus representation.* Nature 376:33 - 36.
43. Linster, C. and M. Hasselmo. 1997. *Modulation of inhibition in a model of olfactory bulb reduces overlap in the neural representation of olfactory stimuli.* Behavioural Brain Research 84:117-127.
44. Doleman, B. J., E. J. Severin, and N. S. Lewis. 1998. *Trends in odor intensity for human and electronic noses: Relative roles of odorant vapor pressure vs. molecularly specific odorant binding.* Proc Natl Acad Sci USA 95:5442-5447.
45. Hendin, O., D. Horn, and M. V. Tsodyks. 1998. *Associative memory and segmentation in an oscillatory neural model of the olfactory bulb.* J Computational Neurosci 5:157-169.
46. White, J., T. A. Dickinson, D. R. Walt, and J. S. Kauer. 1998. *An olfactory neuronal network for vapor recognition in an artificial nose.* Biol Cybernet 78:245-251.
47. Ache, B. W. 1991. *Phylogeny of smell and taste.* In T. V. Getchell and R. L. Doty and L. M. Bartoshuk and J. B. Snow (eds.), Smell and Taste in Health and Disease, pp. 3-18. Raven Press, New York.

48. Sobel, E. C. and D. W. Tank. 1993. *Timing of odor stimulation does not alter patterning of olfactory bulb unit activity in freely breathing rats.* J Neurophysiol 69:1331-1337.
49. Atema, J. 1995. *Chemical signals in the marine environment- Dispersal, detection and temporal signal analysis.* Proc Natl Acad Sci USA 92:62-66.
50. Gelperin, A., D. Kleinfeld, W. Denk, and I. R. C. Cooke. 1996. *Oscillations and gaseous oxides in invertebrate olfaction.* J Neurobiol 30:110-122.
51. Hopfield, J. J. 1996. *Transforming neural computations and representing time.* Proc Natl Acad Sci USA 93:15440-15444.
52. Christensen, T. A., B. R. Waldrop, and J. G. Hildebrand. 1998. *Multitasking in the olfactory system: Context-dependent responses to odors reveal dual GABA-regulated coding mechanisms in single olfactory projection neurons.* J Neurosci 18:5999-6008.
53. Mattheis, J. P., D. A. Buchanan, and J. K. Fellman. 1998. *Volatile compounds emitted by 'Gala' apples following dynamic atmosphere storage.* J Amer Soc Hort Sci 123:426-432.
54. Plotto, A., J. P. Mattheis, D. S. Lundahl, and M. R. McDaniel. 1998. *Validation of gas chromatography olfactometry results for 'Gala' apples by evaluation of aroma-active compound mixtures.* In C. Mussinan and M. Morello (eds.), Flavor Analysis: Developments in Isolation and Characterization, pp. 290-302. American Chemical Society, Washington, DC.
55. Hinterholzer, A. and P. Schieberle. 1998. *Indentification of the most odour-active volatiles in fresh, hand-extracted juice of Valencia late oranges by odour dilution techniques.* Flavour Fragrance J 13:49-55.
56. Tonder, D., M. A. Petersen, L. Poll, and C. E. Olsen. 1998. *Discrimination between freshly made and stored reconstituted orange juice using GC odour profiling and aroma values.* Food Chem 61:223-229.
57. Bolle, R. M., J. H. Connell, N. Haas, R. Mohan, and G. Taubin. 1996. *Produce Recognition System.* United States Patent No. 5546475, International Business Machines Corp., Armonk, N.Y.
58. Kanali, C., H. Murase, and N. Honami. 1998. *Three-dimensional shape recognition using a charge-simulation method to process primary image features.* J Agricult Eng Res 70:195-208.
59. Recce, M., A. Plebe, J. Taylor, and G. Tropiano. 1998. *Video grading of oranges in real-time.* Artificial Intell Rev 12:117-136.
60. Hansen, K., L. Poll, and M. J. Lewis. 1992. *The influence of picking time on the post-harvest volatile ester production of 'Jonagold' apples.* Lebensw Wiss Technol 25:451-456.
61. Vanoli, M., C. Visai, and A. Rizzolo. 1995. *The influence of harvest date on the volatile composition of Starksspur-Golden apples.* Postharvest Biol Tech 6:225-234.

62. Fan, X. T., J. P. Mattheis, and D. Buchanan. 1998. *Continuous requirement of ethylene for apple fruit volatile synthesis*. J Ag Food Chem 46:1959-1963.
63. Stopfer, M., S. Bhagavan, B. H. Smith, and G. Laurent. 1997. *Impaired odour discrimination on desynchronization of odour-encoding neural assemblies*. Nature 390:70-74.
64. Teyke, T. and A. Gelperin. 1999. *Olfactory oscillations augment odor discrimination not odor identification by Limax CNS*. NeuroReport 10:1061-1068.
65. Katz, H. E. and Z. Bao. 2000. *The physical chemistry of organic field-effect transistors*. J Physical Chem B 104:671-678.
66. Ohmori, Y., H. Takahashi, K. Muro, M. Uchida, T. Kawai, and K. Yoshino. 1991. *Gas-sensitive Schottky gated field effect transistors utilizing poly(3-alkylthiophene) films*. Jap J Appl Physics 30:1247-1249.
67. Torsi, L., A. Dodabalapur, L. Sabbitini, and P. G. Zambonin. 2000. *Multi-parameter gas sensors based on organic thin-film transistors. Sensors and Actuators B*: Chemical 67:312-316.
68. Crone, B., A. Dodabalapur, A. Gelperin, L. Torsi, H. E. Katz, A. J. Lovinger, and Z. Bao. 2001. *Electronic sensing of vapors with organic transistors*. Appl. Physics Lett. 78:2229-2231.
69. Crone, B., A. Dodabalapur, Y. Y. Lin, R. W. Filas, Z. Bao, A. LaDuca, R. Sarpeshkar, H. E. Katz, and W. Li. 2000. *Large-scale complementary integrated circuits based on organic transistors*. Nature 403:521-523.
70. Sicard, G. and A. Holley. 1984. *Receptor cell responses to odorants: similarities and differences among odorants*. Brain Res 292:283-296.
71. Malnic, B., J. Hirono, T. Sato, and L. Buck. 1999. *Combinatorial receptor codes for odors*. Cell 96:713-723.
72. Holley, A. 1991. *Neural coding of olfactory information*. In T. V. Getchell and R. L. Doty and L. M. Bartoshuk and J. B. Snow (eds.), Smell and Taste in Health and Disease, pp. 329-343. Raven press, New York.
73. Duchamp-Viret, P., A. Duchamp, and M. A. Chaput. 2000. *Peripheral odor coding in the rat and frog: Quality and intensity specification*. J Neurosci 20:2383-2390.

Chapter 23

Cross-Reactive Optical Sensing Arrays

Caroline L. Schauer[1], Shannon E. Stitzel[2], and David R. Walt[2]

[1]Consultant, 904 Whitehall Street, Silver Spring, MD 20901
[2]Department of Chemistry, Tufts University, 62 Talbot Avenue, Medford, MA 02155

Abstract

Vertebrate olfaction and gustation, noted for their sensitivity and selectivity, provide intriguing models for creating cross-reactive sensor arrays. To mimic these sensing systems, a sensor array was developed comprising thousands of bead sensors, randomly dispersed across an etched optical fiber array. These bead sensors are impregnated with solvatochromic dyes, which alter their fluorescence emission spectra in response to microenvironmental changes due to the analyte's polarity. Each sensor type is cross-reactive and has unique fluorescence response patterns to different analytes. Such a sensor array can quickly accommodate new bead types and is an improvement over other sensors in terms of size, response and recovery times, ease of preparation, sensitivity, and reproducibility. Possible applications range from pollution monitoring, medical diagnosis, food quality control, and land-mine detection.

Introduction

Vertebrate olfaction and gustation are biological chemical sensing systems that utilize temporal response patterns to distinguish analytes. The vertebrate olfactory system employs approximately 1000 types of cross-reactive sensory neurons, randomly distributed throughout the nose to distinguish analytes (1). The interactions of the analyte with the neuronal receptors create a temporal and spatial pattern that is interpreted by the brain as a specific odor (2). Vertebrate gustation utilizes five cell types of taste receptors: salt, sweet, sour, bitter, and Umami to distinguish different flavors. The taste buds use synaptic contacts between the taste cells and sensory nerve fibers, which respond to stimuli with a non-uniform spatial, temporal response (3). We have developed cross-reactive optical sensors that mimic some of the features of olfaction and gustation.

A typical optical sensor is composed of an optical fiber with an indicator immobilized on the fiber tip (4). By immobilizing different analyte sensitive regions onto the fiber tip, the fiber can be tuned to distinguish between a particular analyte or a range of analytes. By combining multiple fibers into an array format, a sensor array can be designed to detect many specific analytes. Electrolytes, glucose, pH, CO_2, O_2, penicillin, amplified DNA, unlabeled DNA, and cell activity have been detected by employing fiber optic-based sensor arrays (5-12). A cross-reactive array can also be made with fiber-optic bundles. In this approach, the individual sensors are not selective but react with a multitude of analytes. The response of each sensor is generally insufficient to identify a particular analyte; rather it is the *pattern* of response across the array of sensors that provides the identification. Several, recent, general reviews describe these optical cross-reactive array systems (13,14).

Optical Fibers & Instrumentation

An optical fiber is comprised of a plastic, glass, or fused silica core surrounded by a lower refractive index cladding. Due to the differences in the refractive indices, entering light is totally internally reflected at the interface of the two materials. Therefore, light coupled into the optical fiber can be propagated over long distances. Fiber-optic imaging bundles can be created by coherently fusing together individual optical fibers; therefore each fiber carries an independent optical signal from one end of the fiber to the other. Figure 1 shows a typical imaging fiber, which can be comprised of between 3000-

100,000 individually clad optical fibers, each with diameters of 3-10 μm. Image resolution depends on both the size of the individual fiber elements within the high-density optical imaging fiber and on the size of the detector pixels.

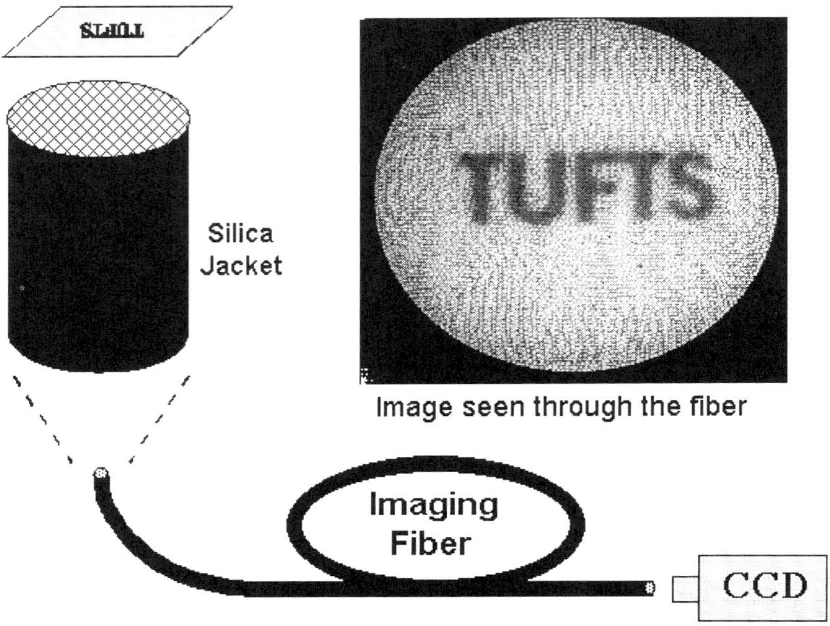

Figure 1: Schematic of an imaging fiber bundle. An imaging fiber consists of thousands of individual optical fibers coherently drawn together. The image is transferred through the fiber and read by a CCD camera.

To create a fiber-optic sensor, a sensing chemistry is attached to the distal tip of an optical fiber. A multitude of transduction mechanisms may be utilized for detecting the analyte including fluorescence intensity and/or lifetime, polarization, spectral shape, absorbance, and reflectance. In our laboratory, fluorescence-based fiber-optic sensors are positioned onto a custom-designed imaging system as seen in Figure 2. The imaging system is computer controlled and consists of an inverted fluorescence microscope, a Xenon arc lamp light source, dichroic mirrors and optical filters, a CCD detector, and imaging processing software.

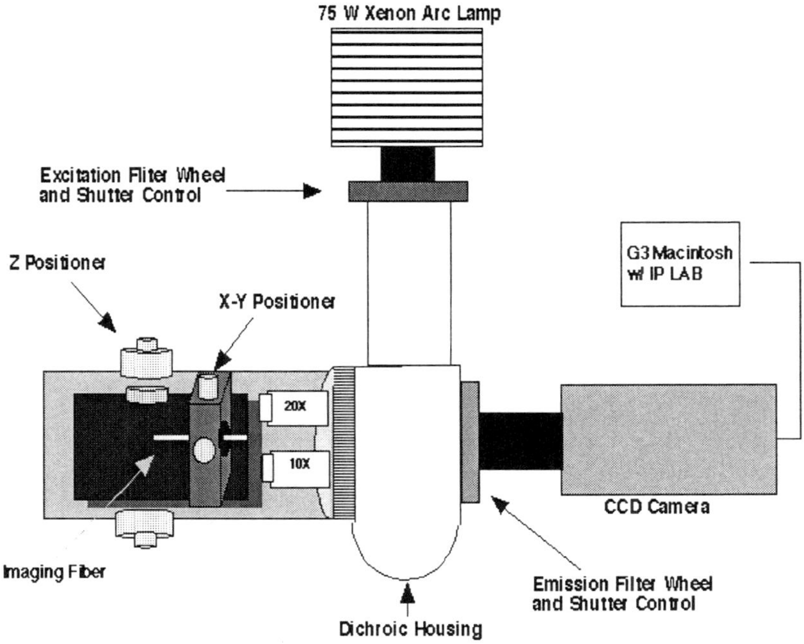

Figure 2: Schematic of an optical imaging system. Components are a Xenon arc lamp excitation source, excitation and emission filter wheels, dichroic mirror, inverted fluorescence microscope, CCD camera and a Macintosh computer to control the system.

The light sources and detectors can be adjusted depending on the sensing application. If an increase in sensitivity is desired, a photomultiplier tube (PMT), avalanche photodiodes, or an intensified charge coupled device (CCD) may be employed for detection. Light excites the sensors, which isotropically fluoresce, and some of this fluorescence is captured back through the fiber and transmitted to the detector. During exposure to target analytes, the fluorescence is monitored and the detector measures the returning light's change in intensity at the detection wavelength.

Randomly Ordered Sensor Arrays

Optical imaging fibers are ideally suited for sensor arrays because many sensor elements can be incorporated onto the distal face of the same imaging fiber. Each of the thousands of fused optical fibers in an imaging array can incorporate one sensing material to produce an enhanced, high-density sensor array capable of multi-analyte sensing on the tip of a 500-1000 μm optical imaging fiber (15). Assembling the high-density sensor arrays consists of two steps: (a) etching microwells on the fiber's distal face and (b) positioning microbead sensors in the wells. The polished tip of an optical imaging fiber bundle can be selectively etched in a buffered hydrofluoric acid solution. Depending on the fiber's composition, either the cladding or core materials will be etched away, leaving either ordered micro*tip* arrays (16) in the former, or ordered micro*well* arrays (17) in the latter. Each well in the array of microwells etched onto the fiber's distal face can be individually interrogated because it is optically 'wired' to an individual fiber in the bundle. Figure 3A shows an Atomic Force Micrograph (AFM) scan of a portion of the etched face of an optical imaging fiber.

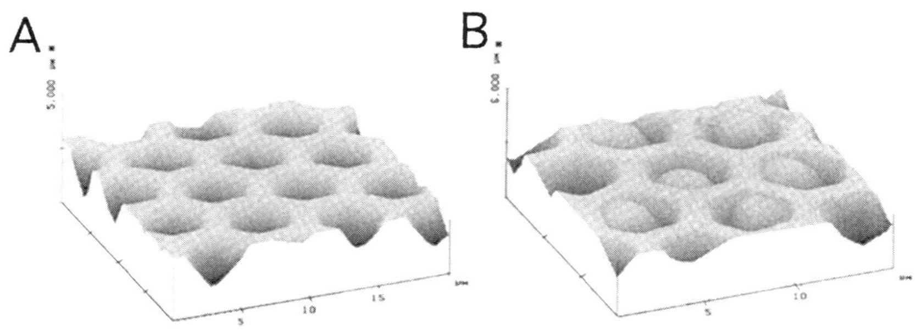

Figure 3: A) Atomic force micrograph of ~3.6-μm diameter wells etched into the surface of an imaging fiber bundle. The wells are ~3-μm deep. B) Atomic force micrograph of ~3.6-μm diameter wells with ~3.1-μm bead sensors in the wells. Reprinted with permission from Anal. Chem, ***1998****, 70, 1242-1248. Copyright 1998 Am. Chem. Soc.*

The ordered microwells serve as a high-density array platform for the distribution of microsensors. Many different sensing tasks have been solved with this array platform by incorporating complementary-sized bead sensors such that one bead is contained in each microwell. Microbead dispersion into the microwells occurs either in solution or dry. A drop of a bead suspension can be placed onto the tip of an etched fiber. As the solution evaporates, the microbead sensors fall into the etched wells with one sensor per well, and the excess beads are removed. Alternatively, beads can be placed in wells by simply pressing the etched fiber face into a stock of dry beads and then removing excess beads. Figure 3B shows the etched microwells containing bead sensors. The microwell arrays provide a new architecture for optical sensing.

Bead Sensor Fabrication and Encoding

Microbead sensors are made on 3 or 5 μm diameter beads of polymer, silica or silica with polymer coatings. Depending on the specific application of a bead sensor, the details of the fabrication process can vary, but all beads have a fluorescent indicator either covalently attached or absorbed to the beads. Each sensor type is made in a batch process, creating billions of microsensors; one milliliter of a bead suspension contains ~6×10^9 identical sensors. Sensor stocks can be reproducibly fabricated from one batch to the next and, in some instances, sensor shelf life has been shown to be at least ten months (18). The ability to prepare billions of identical beads in a single batch allows identical sensor arrays to be made from the bead stocks. Different bead sensor types can be mixed together and distributed into the wells, creating random arrays of sensors. Due to the randomized nature of the sensor platform, new sensors can be readily incorporated into the array simply, by adding a new bead type to the mixture. The array takes minutes to fabricate and some arrays have been stored for many months.

Although the fabrication of a random array is much easier than that of an ordered array, the random array sensors must be positionally registered after the array is prepared. Therefore each sensor type requires an encoding scheme so that different sensors in the random array can be later identified. In one approach, we employ optical bar codes, by encoding bead sensors with different dyes (Figure 4) (9,15,19). To independently identify each bead in an array, the optical bar code consists of a mixture of fluorescent dyes with different excitations, emissions, and intensities. Encoding simplifies the fabrication of sensor arrays by allowing sensors to be randomly dispersed instead of specifically positioned in an array.

An alternative to using specific encoding dyes to label sensor types, the bead sensors can be 'self-encoded'. Self-encoding means that each sensor type can be discriminated from other sensors based on a unique, intrinsic sensor

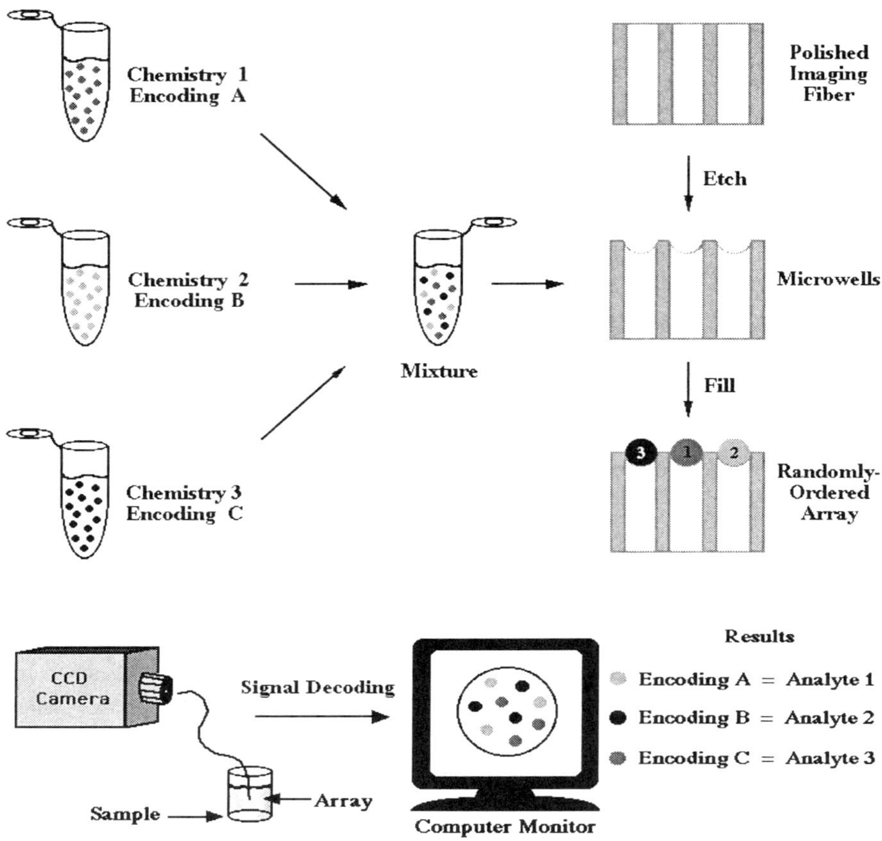

Figure 4: Schematic concept of randomly distributed microbead sensors in a microwell array. The top shows three sensors types with different optical encoding signals on each type. The sensors are mixed together and randomly distributed throughout the array. The array is then exposed to a sample solution and the different sensor types are simultaneously decoded and positionally registered with a CCD detector. Reprinted with permission from Anal. Chem, ***1998****, 70, 1242-1248. Copyright 1998 Am. Chem. Soc.*

property. Such properties can include spectral properties of the sensor material such as the maximum emission wavelength, or the unique sensor response to a known specific analyte. An example of self-encoding is seen in our cross-reactive fiber-optic vapor detection system. Each sensor type has its own unique response to a given vapor. Therefore, the beads can easily be located after placement into the image guide wells by simply exposing the array to a known test vapor. The resulting response plots are matched to standard plots obtained beforehand for each bead type. In this way, the bead sensors are 'self-encoding' (19).

Data processing and Detection Limit Enhancements

There are several data processing steps involved in the analysis of randomly distributed optical arrays. The first step is registering or locating the sensors in the array as discussed above. The second step involves calibrating or training the array by exposing it to known analytes. If the array is cross-reactive, the response from the sensors is used to train a pattern recognition program to recognize specific response patterns as specific analytes. Currently, several different pattern recognition techniques are being explored including principal component analysis, learning vector quantization, neural networks, and high-dimensional data processing. Once the array is calibrated or trained for the analytes of interest, the array can be used to collect data.

Another processing step that can help improve analyte detection limits involves bead summing. By summing the low-level responses of a large number of individual sensing elements (polymer beads), significant enhancements in the overall sensitivity are obtained for the array. In the optical format, each element is individually addressable making it possible to combine the signals from large numbers of like beads randomly dispersed throughout the array. Summing the responses from several beads leads to a substantial improvement in the signal-to-noise (S/N) ratio of a measurement. The S/N ratio increases by $\sqrt{n}$ where n is the number of sensors analyzed. This signal enhancement removes noise from the individual sensors by averaging responses over many sensors. If each sensor within the array responds with a small signal to a low-level target, then their identical response profiles can be combined to greatly improve the overall response. Sensor redundancy and signal summing have led to the detection of low vapor concentrations (18,19) and zeptomole levels of fluorescent labeled DNA targets.

Applications

Nose

One application for fiber-optic sensors is the optical nose. Bead sensors can be modified with a solvatochromic dye that exhibits a fluorescence change with a shift in the polarity of the dye's environment. Unique temporal response patterns are obtained for each vapor/sensor interaction, by monitoring the change in fluorescence as a vapor pulse is delivered to the sensors. In Figure 5, data are plotted for a three-component nose system. The beads are self-encoded as discussed earlier, and methanol is used to decode the positions of the beads, as it provides a distinct response for the three bead types. Once the beads have been registered, the array is exposed to three other vapors: dichloromethane, toluene, and acetone. Each vapor produces a distinct response for each sensor and the combined use of the three sensors allows for the differentiation between the four saturated vapors. The resulting data are normalized to clarify the display and assist in the response-shape comparisons.

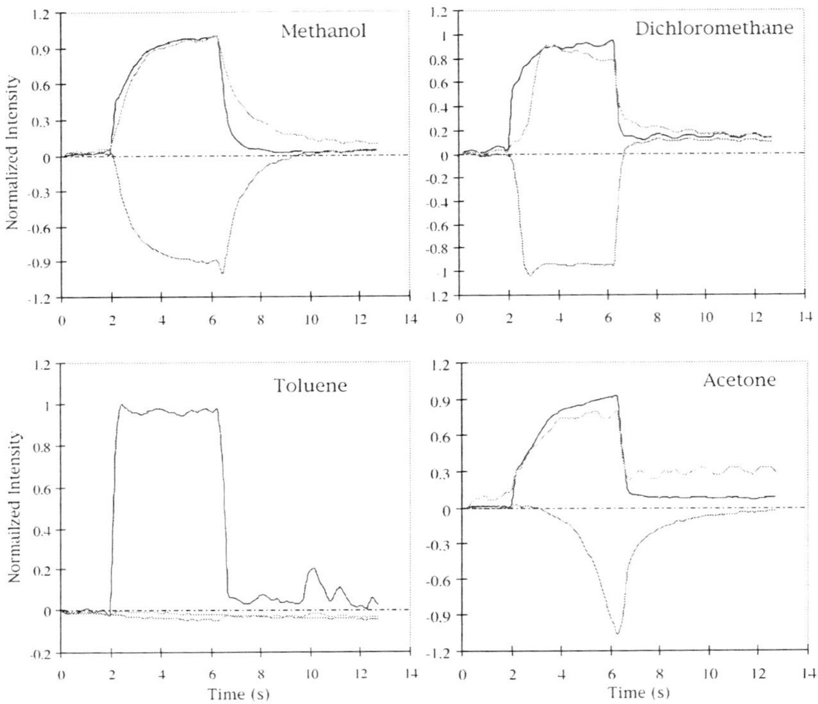

Figure 5: This figure demonstrates normalized temporal responses of three different bead sensors to saturated methanol, dichloromethane, toluene and acetone. Example of the diversity of sensor responses to a vapor pulse. Reprinted with permission from Anal. Chem, ***1999****, 71, 2192-2198. Copyright 1999 Am. Chem. Soc.*

The nose offers several advantages over other vapor-sensing arrays. Using a high-density bead array allows for bead summing to enhance S/N as discussed above. The cross-reactivity of the sensors allows more analytes to be identified than there are sensor types in the array. Also, as seen in Figure 5, the bead sensors respond very quickly to the presence of vapors with response times on the order of a few hundred milliseconds. A 5-second experiment is ample time for most beads to fully respond to and recover from a pulse of vapor at any given concentration. All of these advantages combined make the artificial nose an attractive chemical detection system.

Solution

Toka *et.al.* created a multichannel taste sensor, or "electronic tongue" by employing lipid/polymer membranes, which gave different outputs for chemical substances with different qualities, such as saltiness or sourness (20). McDevitt, Shear, and Neikirk investigated an electronic tongue formed from a bead sensor immobilized into pits on a silicon wafer. The bead sensor version of the electronic tongue distinguishes between different taste analytes, however the sensors are not cross-reactive (21). By combining the approach of the cross-reactive optical nose with the solution based DNA sensors (9) we designed a cross-reactive solution sensor.

The cross-reactive solution sensor had some initial design difficulties to overcome. The solvatochromic dyes used to make cross-reactive sensors had typically been adsorbed onto a bead surface. This became a problem when the sensors were put into solution because the dye would leach off the bead surface in certain solvents. Therefore, the dye is either covalently attached to the beads, or the beads are used in a buffer that does not cause dye leaching. The dyed beads can then react with analytes and the resulting temporal fluorescence data are used to characterize specific analytes. Work is presently underway to generate a cross-reactive sensor array for distinguishing between different sugar or protein solutions. Another application of the optical tongue is to determine the components of different liquid organic solvent systems.

The cross-reactivity of the optical tongue does not have to arise from different bead types. Many enzymes are considered selective; for example, L-glutamate oxidase reacts primarily with L-glutamate. Other enzymes are considered cross-reactive, such as, L-amino acid oxidase, which reacts with a range of L-amino acids at different rates. The incorporation of these non-specific cross-reactive enzymes into a sensor array creates a cross-reactive solution sensor. Our initial efforts have focused on esters. The designed enzymatic sensor array uses the 96-well microtiter plate format to combine esterases, esters, and a pH sensitive dye. As the esterase reacts with the ester, a carboxylic acid product is formed, which alters the pH of the solution. A microtiter plate reader monitors the changes in fluorescence of the pH indicator due to ester hydrolysis. The rates of reaction were used to group the analytes through principal component analysis, and, so far nine esterases can cluster over twenty esters.

A distinguishing feature of the fiber optic system is the need for very small sample volumes. The fluorescence-based array platform enhances observed detection limits to an unprecedented level. The system is inherently sensitive for measuring small sample volumes (<10 μL) of targets and sensor redundancy within the array provides detection limit enhancements. Incorporating small features with extremely large spatial resolution allows chemical concentrations and gradients to be measured on scales smaller than individual cells. Reduced measurement volumes also provide highly localized target concentrations. Smaller sensors enable the interrogation of even smaller absolute numbers of molecules without sacrificing sensitivity.

Conclusions

Imaging fibers are versatile optical substrates for multi-analyte sensing applications. The sensor technology has advanced from employing single-core optical fibers and fiber bundles to high-density microsensor arrays incorporating thousands of sensors for simultaneous measurements. The need for small, fast responding detection systems is expanding and the fiber optic platform offers a different approach to multi-analyte sensing through small sensor design. These arrays have the ability to detect analytes in very small volumes with detection limit enhancements built into their design. Cross-reactive fiber-optic arrays have distinct advantages over other systems presently in use such as size, response and recovery times, ease of fabrication, sensitivity, and reproducibility. The self-encoded bead sensor array is an attractive approach because it can quickly accommodate new sensor bead types. Microspheres have an additional advantage as individual sensing elements because they are smaller in size than any other currently employed techniques. The high-density sensor arrays (15) have been employed to detect low-level vapors (18,19), develop immunoassays (22,23), detect fluorescently labeled DNA targets in the zeptomole range, and unlabeled DNA targets in a complex environment (9). The sensors also have fast response times, which allows for the possibility of real-time monitoring in a variety of applications, such as pollution monitoring, medical diagnosis, food quality control, and land mine detection. All of these features make optical bead sensing arrays a versatile chemical sensing technique.

Acknowledgement - The authors wish to thank Keith Albert for helpful discussions and DARPA, Office of Naval Research, and National Institutes of Health for financial support.

References

1. Buck, L.; Axel, R. *Cell* **1991**, *65*, 175-187.
2. Kauer, J. S. *Trends Neurosci.* **1991**, *14*, 79-85.
3. Scott, T. R. In *Neurobiology of Taste and Smell*; Finger, T. E., Silver, W. L., Eds.; John Wiley and Sons, Inc.: Malabar, 1991.
4, Seitz, W. R. *Crit. Rev. Anal. Chem.* **1988**, *19*, 135-173.
5. Healey, B. G.; Li, L.; Walt, D. R. *Biosens. Bioelectron.* **1997**, *12*, 521-529.
6. Li, L.; Walt, D. R. *Anal. Chem.* **1995**, *67*, 3746-3752.
7. Ferguson, J. A.; Boles, T. C.; Adams, C. P.; Walt, D. R. *Nat. Biotechnol.* **1996**, *14*, 1681-1684.
8. Healey, B. G.; Matson, R. S.; Walt, D. R. *Anal. Biochem.* **1997**, *251*, 270-279.
9. Steemers, F.; Ferguson, J.; Walt, D. *Nat. Biotech.* **2000**, *18*, 91-94.
10. Ferguson, J. A.; Healey, B. G.; Bronk, K. S.; Barnard, S. M.; Walt, D. R. *Anal. Chim. Acta* **1997**, *340*, 123-131.
11. Goyet, C.; Walt, D. R.; Brewer, P. G. *Deep-Sea Research Part A* **1992**, *39*, 1015-1026.
12. Walt, D. R.; Tabacco, M. B.; Uttamial, M.; Schanzle, J. A. *Abstr. Pap. Am. Chem. Soc.* **1998**, *216*, U885-U885.
13. Steemers, F. J.; Walt, D. R. *Mikrochimica Acta* **1999**, *131*, 99-105.
14. Dickinson, T. A.; White, J.; Kauer, J. S.; Walt, D. R. *Trends Biotechnol.* **1998**, *16*, 250-258.
15. Michael, K. L.; Taylor, L. C.; Schultz, S. L.; Walt, D. R. *Anal. Chem.* **1998**, *70*, 1242-1248.
16. Pantano, P.; Walt, D. R. *Rev. Sci. Instrum.* **1997**, *68*, 1357-1359.
17. Pantano, P.; Walt, D. R. *Chem. Mat.* **1996**, *8*, 2832-2835.
18. Albert, K. A.; Walt, D. R. *Anal. Chem.* **2000**, *In Press*.
19. Dickinson, T.; Michael, K.; Kauer, J.; Walt, D. *Anal. Chem.* **1999**, *71*, 2192-2198.
20. Toko, K. *Biosens. Bioelectron.* **1998**, *13*, 701-709.
21. Lavigne, J. J.; Savoy, S.; Clevenger, M. B.; Ritchie, J. E.; McDoniel, B.; Yoo, S.-J.; Anslyn, E. V.; McDevitt, J. T.; Shear, J. B.; Neikirk, D. *JACS* **1998**, *120*, 6429-6430.
22. Szurdoki, F.; Michael, K. L.; Agrawal, D.; Taylor, L. C.; Schultz, S. L.; Walt, D. R. Pathogen Detection and Remediation for Safe Eating, Bellingham, WA, 1999; p 52-62.
23. Szurdoki, F.; Michael, K. L.; Taylor, L. C.; Schultz, S. L.; Walt, D. R. 216th ACS National Meeting, Boston, MA, 1998; p ANYL-034.

INDEXES

Author Index

Anslyn, Eric V., 276
Bachmanov, Alexander A., 40
Beauchamp, Gary K., 40
Blake, Anthony, 158
Breslin, Paul A. S., 65
Bryant, Bruce, 202
Buletic, Zivjena, 65
Cayeux, Isabelle, 140
Davidson, Jim M., 166
DeGroot, Lucy, 166
Delwiche, Jeannine F., 65
Doty, Richard L., 123
Drewnowski, Adam, 52
Ennis, Daniel M., 32
Firestein, Stuart, 110
Gelperin, Alan, 289
Gilbertson, Timothy A., 180
Graham, Brevick G., 94
Guinard, Jean-Xavier, 227
Hollowood, Tracey A., 166
Hopfield, J. J., 289
Kim, Insook, 180
Krizhanovsky, V., 18
Lavigne, John J., 276
Linforth, Rob S. T., 166
McDevitt, John T., 262, 276
Metzger, Axel, 276
Mezine, Igor, 202
Moskowitz, Howard R., 214
Naim, M., 2, 18
Neikirk, Dean, 276
Nir, S., 2
Noble, A. C., 2, 192
Peri, I., 2
Plata-Salamán, Carlos R., 78
Reed, Danielle R., 40
Rodin, S., 2
Rosenzweig, S., 18
Roy, Glenn, 262
Samuelov-Zubare, M., 2
Schauer, Caroline L., 318
Schiffman, Susan S., 94
Schneider, Stephen E., 276
Scott, Thomas R., 78
Shear, Jason B., 276
Spielman, A. I., 2, 18
Stitzel, Shannon E., 318
Taylor, Andrew J., 166
Tordoff, Michael G., 40
Velazco, Maria Inés, 140
Vuilleumier, Christine, 140
Walt, David R., 318
Warrenburg, Stephen, 243
Westall, Holly L., 94
Wiskur, Sheryl L., 276
Yamamoto, T., 18
Yan, W., 18
Zervakis, Jennifer, 94

Subject Index

A

Absolute threshold, definition, 125
Acesulfame potassium
  discrepancy, 90
  relationship among sweet stimuli, 88
  stimulus in macaque study, 84*t*
Acetic acid, physiological measurement, 246
Acetone, normalized temporal responses of three bead sensors, 326*f*
Adenosine triphosphate (ATP), support bound single analyte chemosensor, 283–284
Affinity, receptor binding, 35
AFFIRM® technology
AFFIRM® measurement of acetone, ethyl butyrate, and ethanol in breath of volunteer, 162*f*
  analysis of different constructions of peanut butter and grape jelly sandwich (PBGJS), 164*f*
  atmospheric pressure chemical ionization mass spectrometry, 160–161
  food composition influencing flavor release and perception, 163
  gas phase concentrations, 143–144, 155
  improving comprehension of flavor perception, 163
  peanut butter and grape jelly analysis, 164*f*
  physical makeup of food, 165
  schematic, 161*f*
  See also Dose-response curves of odor and taste stimuli
Aftertaste
  lingering, of non-sugar sweeteners, 5, 14
  taste-peripheral phenomena, 3
Age-related chemosensory losses
  analgesic/anti-inflammatory drugs, 98*t*, 99*t*
  antihistamines, 99*t*
  antimicrobial drugs, 99*t*, 100*t*
  average percent suppression of bitterness for each blocker, 106*t*
  cardiovascular drugs, 97*t*, 98*t*
  chemosensory disorders from medications, 95
  drugs interfering with taste system, 97*t*, 98*t*, 99*t*, 100*t*, 101*t*, 102*t*
  effect of bitter inhibitors on bitter taste perception of urea, quinine HCl, magnesium chloride, and caffeine, 105–107
  effect of topical application of drugs to lingual surface, 96
  elevation of norepinephrine from bitter taste of drug, 103–105
  HIV drugs, 101*t*, 102*t*
  levels of norepinephrine in blood after single taste of stimuli, 104*f*
  medications altering quality and intensity of tastants, 107
  medications impacting taste perception, 95–96
  miscellaneous drugs, 100*t*, 101*t*
  psychotropic drugs, 97*t*
  sedatives and antiemetics, 99*t*
  thresholds for elderly vs. younger subjects, 96

Amino acids, quality control, 269
interaction with liposomal membranes and translocation into liposomes, 5–6
ionic charge during experiments, 8*t*
multilamellar vesicles (MLVs) preparation, 6
permeating taste cells, 8–10
support for receptor-independent mechanisms, 9–10
Analgesic drugs, interference with taste system, 98*t*, 99*t*
Analgesics, quality control, 269
Animals
models of human biology, 32–33
Sac locus and taste receptor 1 (T1R1) family of putative taste receptors, 47–48
sweetness studies, 45–48
sweet perception and preference in mice, 45–46
See also Molecular mixture models
Antiemetics, interference with taste system, 99*t*
Antihistamines, interference with taste system, 99*t*
Anti-inflammatory drugs, interference with taste system, 98*t*, 99*t*
Antimicrobial drugs, interference with taste system, 99*t*, 100*t*
Antithyroid compounds, bitterness, 66
Apples
data clustering from three varieties, 294*f*
electronic olfaction, 291–292, 294
pattern classification algorithms, 295
patterns of sensor responses, 293*f*
Aromatherapy, effects on blood pressure, 245–246
Arrays, optical sensing. *See* Cross-reactive optical sensing arrays
Amphipathic tastants
Ascending method of limits procedure (AML), odor detection threshold measurement, 126–127
Aspartame
average time intensity curves for sweetness, 4*f*
macaque neural response vs. human perception, 89
orthonasal dose-response curves of lemon flavor by olfactometry, 151*f*
relationship among sweet stimuli, 88
sensory and analytical thresholds, 173*t*
stimulus in macaque study, 84*t*
structure, 143
temporal properties, 3
See also Dose-response curves of odor and taste stimuli
Astringency
average intensity of, of black tea, 197*f*
average intensity of, over time of red wine, 197*f*
average maximum intensity of, for black tea vs. salivary flow rate, 198*f*
average time-intensity curves for, of skin and seed tannin fractions, 196*f*
effect of salivary flow rate, 196, 198
effect of sensory methodology, 195–196
effect of tannin composition, 194–195
tactile sensation, 193
See also Flavonoid phenols
Atomic force microscopy (AFM), optical imaging fibers, 322*f*

B

Background elimination, computational olfaction, 300
Bead sensor, fabrication and encoding, 323–324
Bead summing, data processing, 325
Beer, quality control, 269
Beverages
 models for mixture effects, 37
 quality control, 263, 269
 See also Flavonoid phenols
Biological olfactory systems, molecular recognition, 290
Bitterness
 average percent suppression by blockers, 106*t*
 correlation coefficients of bitter ratings and rankings, 68*t*
 definition, 193
 effect of tannin composition, 194–195
 food processing, 19
 idiosyncratic patterns of perceived bitter intensity, 67–70
 intensity scaling, 55, 57
 oscillation mode, 268–269
 physiological transduction mechanisms, 66
 technology to reduce, 19
 See also Clustering bitter compounds; Flavonoid phenols; *n*-6-Propylthiouracil (PROP)
Bitter tastants
 amphipathic, 3
 amphipathic, interacting with liposomal membranes and translocation into liposomes, 5–6
 amphipathic, permeating taste cells, 8–10
 average time intensity curves for bitterness of caffeine and denatonium, 4*f*
 bitter-taste inhibitor (BMI-60), 11, 14
 cyclo(Leu-Trp) and quinine translocation, 6, 7*f*
 inhibition of permeation of quinine and cycle(Leu-Trp) into rat circumvallate (CV) taste bud cells in presence of BMI-60, 12*f*
 interaction with phospholipid-based bitter taste inhibitors, 10–11, 14
 mechanism of permeation or translocation through liposomal membranes, 6
 signal transduction, 3
 slow taste onset and lingering aftertaste, 5, 14
 temporal properties, 3–5
 translocation of quinine and cyclo(Leu-Trp) into multilamellar vesicles (MLVs) and bitter-taste inhibitor BMI-60, 12*f*
 See also Clustering bitter compounds
Bitter taste
 aversive response to potentially toxic compounds, 18
 categorizing, 65–66
 correlations among papillae density and, 74–75
 debittering processes, 62
 effect of bitter inhibitors on perception of urea, quinine HCl, magnesium chloride, and caffeine, 105–107
 phytochemicals and, 60–62
 polymorphism, 44
Bitter taste inhibitors
 amphipathic bitter tastants interaction with, 10–11, 14
 average percent suppression of bitterness, 106*t*
 BMI-60, 11, 13*f*, 14
 effect on bitter taste perception of urea, quinine HCl, magnesium chloride, and caffeine, 105–107
Bitter taste transduction
 behavioral and electrophysiological results, 22

behavioral experiments, 19–20
changes in cAMP level in intact circumvallate (CV) taste cells and in intact non-sensory epithelial cells, 27, 29*f*
electrophysiological experiments, 20
electrophysiological taste responses of glossopharyngeal nerve of rats to taste stimulation by bitter stimuli, 23*f*
fast pippeting system (FPS) experiments, 20–22
lack of effect on cGMP, 30
materials and methods, 19–22
monitoring release of signal molecules in intact taste cells by FPS, 21–22
QFM and FPS complementary, 27, 30
QFM experiments, 20
reduction in cAMP by limonin and catechin, 30
signal transduction experiments, 20–22
signal transduction results, 22, 27
summary of taste transduction pathways by selected bitter tastants, 27*t*
technology to reduce bitterness, 19
time-course of catechin-stimulated IP_3 and cAMP in rat taste tissue, 24*f*
time-course of cyclo(Leu-Trp)-induced cAMP in rat taste tissue, 26*f*
time-course of limonin-stimulated IP_3 and cAMP production in rat taste tissue, 28*f*
time-course of naringin-stimulated IP_3 and cAMP in rat taste tissue, 25*f*
two-choice short-term preference tests between water and solution with bitter taste stimulus, 23*f*
Black tea
average intensity of astringency of two concentrations, 197*f*
average maximum intensity of astringency for successive sips of black tea vs. salivary flow rate, 198*f*
effect of salivary flow rate on astringency, 196, 198
See also Flavonoid phenols
Bombyx mori, pheromone, 290
Broccoli, *n*-6-propylthiouracil (PROP) tasting, 61*f*
Brussel sprouts
n-6-propylthiouracil (PROP) responsiveness and preferences, 59*t*
PROP tasting, 60, 61*f*
Butyric acid, physiological measurement, 246
Buzzing sensation. *See* Trigeminal chemical sensitivity

C

Cabbage
n-6-propylthiouracil (PROP) responsiveness and preferences, 59*t*
PROP tasting, 61*f*
Cabernet franc
average time-intensity curves for astringency of skin and seed tannin fractions, 196*f*
skin vs. seed tannins, 195
Cabernet Sauvignon
average time-intensity curves for astringency of skin and seed tannin fractions, 196*f*
skin vs. seed tannins, 195
Caffeine
average time intensity curves for bitterness, 4*f*
bitterness, 66–67
cluster analyses, 69*f*

correlation coefficients of bitter ratings and rankings, 68*t*
efficacy of bitter blockers on, 105–107
subclass grouping, 72
temporal properties, 5
Camphor, physiological measurement, 246
Cancer prevention, fruit and vegetable consumption, 62–63
Capsaicin
plasma norepinephrine levels, 105
taste stimulus testing neuroendocrine levels, 103–105
Carbohydrates
macaque neural response vs. human perception, 89
relationship in macaque study, 88
Carbonated water, taste stimulus testing neuroendocrine levels, 103–105
Cardiovascular drugs, interference with taste system, 97*t*, 98*t*
Catechin
bitterness and astringency, 194–195
electrophysiological taste responses of glossopharyngeal nerve of rats, 22, 23*f*
n-6-propylthiouracil (PROP) tasting, 60
reduction in cAMP, 30
taste transduction pathways by bitter tastants, 27*t*
time-course for release of IP_3 and cAMP by bitter taste stimulation, 24*f*
See also Flavonoid phenols
Categorizing compounds. *See* Clustering bitter compounds
Cauliflower, *n*-6-propylthiouracil (PROP) tasting, 61*f*
Cheese, taste transduction pathways by bitter tastants, 27*t*
Chemical sensitivity. *See* Trigeminal chemical sensitivity
Chemoreception of fat
cholecystokinin (CCK)-secreting duodenal enterocytes, 182
cloned delayed rectifying K^+ (DRK) channels, 184
contributions from gustatory mechanisms, 181
correlation of fatty acid responsiveness and dietary fat preference, 187
digestion of triglycerides, 182
effect of fatty acids on DRK currents in CCK-secreting duodenal enterocyte cell line STC-1, 189*f*
effect of fatty acids on DRK currents in insulin-secreting pancreative beta cell line HIT-T15, 189*f*
fatty acids activating taste cells via inhibition of DRK channels, 183–186
fatty acids as primary signaling molecules, 182–183
fatty acids directly activating fat sensitive cell types, 188
fatty acid sensitive Shaker Kv1.5 K^+ channels, 182
implications of fatty acids altering activity of taste receptor cells, 186
inhibition of DRK channels by fatty acids in vallate taste receptor cell, 185*f*
Kv1.5-like protein in peripheral tissues, 188*f*
nutrient detection, 181–182
pancreatic beta cell line, 182
roles of peripheral gustatory system, 181
sensory cue for dietary fat, 181
taste system and nutrient detection, 181
von Ebner's gland proteins (VEG-P), 182–183
Chemosensory losses. *See* Age-related chemosensory losses
Chemosensory signals, foods, 95

Chemo-sensory stimuli
understanding reaction to, 224
See also Drivers of liking
Chewing gum, flavor release and perception, 171–172
Circumvallate (CV) taste-bud sheets
estimated accumulation of amphiphathic tastants, 9*t*
preparation, 8
Citral
air-liquid phase partition coefficients, 148*t*
gaseous concentrations in finalized olfactometric tests, 146*t*
gas phase-liquid phase partition coefficient in nose, 148*t*
headspace concentration of lemon, 147*t*
mapping of olfactometric data, 152*f*
olfactometric and tasting mappings, 154*f*
orthonasal dose-response curves by olfactometry, 151*f*
sniffing mapping, 153*f*
structure, 142
test concentrations, 147*t*
See also Dose-response curves of odor and taste stimuli
Citrate
binding studies, 279–280
concentration in beverages by NMR and competition assay, 280*t*
ingredient in citrus beverages, 278
See also Single analyte sensors
Citrus, taste transduction pathways by bitter tastants, 27*t*
Clustering bitter compounds
categorizing, 65–66
correlation coefficients of bitter ratings and rankings, 68*t*
correlations among papillae density and bitter taste, 74–75
differences in bitterness, 66
groupings into subclasses, 72
human psychophysical studies, 73–74
hypothesis of clustering, 72
idiosyncratic patterns of perceived bitter intensity, 67–70
impact of *n*-6-propylthiouracil (PROP) sensitivity on idiosyncratic patterns, 70–72
individuals of groups differing only quantitatively, 72
molecular studies suggesting multiple bitter transduction sequences, 73
physiological transduction mechanisms, 66
PROP status influence on bitter intensities and rankings, 71*f*
sensitivity variations, 66–67
underlying factors for groupings, 70
unrotated principal component analyses (PCA) and cluster analyses, 69*f*
Coffee
complex mixtures, 290
distribution of three sensory segments as function of country, 219*t*
n-6-propylthiouracil (PROP) responsiveness and preferences, 59*t*
quality control, 269
relation between bitter taste from expert panel and liking of consumer panel, 218*f*
See also Drivers of liking
Common sense, bipolar dimensions of mood states, 250
Communication
signaling between living systems, 158–159
See also AFFIRM® technology
Component separation, computational olfaction, 300
Computational olfaction

approach through large numbers of receptor types, 304–307
available information, 301
background elimination, 300
basic olfaction, 299
component separation, 300
constructing target odor on basis of first sniff, 308
distinguishing between target and non-target odors, 305–306
fractional coverage of binding sites, 303
histograms of number of receptor cell types (N) voting for concentrations of target odor, 305*f*
identifying odor in presence of unknown background, 306
levels of analysis, 299
LIMAX network, 310–312
modeling odors in absence of additional information, 302
neural implementation of computational algorithm, 310–312
odorant modeling, 301–303
odor separation, 300
olfactory receptor types of rat, 300
olfactory tasks, 299–300
olfactory tasks in least-squared error algorithm, 303
separating unknown odors using fluctuations, covariation, and large N, 307–309
simulation results for weak object in stronger background, 308, 309*f*
size for N, 309–310
statistics of large numbers and chance occurrence of non-interference, 306–307
See also Electronic olfaction
Consumer Fragrance Thesaurus (CFT)
best combination page, 256*f*
beyond Mood Mapping®, 257
See also Emotion in olfactory research
Consumer products
development, 225
link between model systems and real products, 223–224
successful optimization, 228
See also Drivers of liking; Pasta sauce
Consumers
fitting to models, 235, 236*f*, 237*f*, 238*f*
ice cream case study, 229
techniques examining preferences, 228–229
uses and attitudes measures, 241
vocabulary describing sensory perceptions, 228
See also Ice cream; Preference mapping
Contingent negative variation (CNV), stimulating vs. relaxing fragrances, 246–247
Cortex. *See* Neural representation of sweet taste in monkey
Cortisol, levels from bitter taste, 103
Cranraspberry juice
relationship among sweet stimuli, 88
stimulus in macaque study, 84*t*
Cross-reactive optical sensing arrays
advantages, 328
AFM (atomic force micrograph) of etched face of optical imaging fiber, 322*f*
AFM of etched microwells containing bead sensors, 322*f*
applications, 325–328
bead sensor fabrication and encoding, 323–324
bead summing, 325
cross-reactive solution sensor, 327
cross-reactivity of optical tongue, 327
data processing and detection limit enhancements, 325
electronic tongue, 327
microbead sensors, 323
multichannel taste sensor, 327

need for small sample volumes, 327–328
normalized temporal responses of three bead sensors to methanol, dichloromethane, toluene, and acetone, 326*f*
optical bar codes, 323
optical fibers and instrumentation, 319–321
optical nose, 325–326
randomly ordered sensor arrays, 322–323
schematic of imaging fiber bundle, 320*f*
schematic of optical imaging system, 321*f*
schematic of randomly distributed microbead sensors in microwell array, 324*f*
solution sensor, 327–328
typical optical sensor, 319
Cruciferous vegetables, *n*-6-propylthiouracil (PROP) tasting, 60, 62
N-(4-Cyanophenyl)-*N'*-[(sodiosulfo)methyl]urea, bitter blockers, 106*t*, 107
Cyclamate
direct activators of G-proteins in vitro, 10
macaque neural response vs. human perception, 89
relationship among sweet stimuli, 88
sensory and analytical thresholds, 173*t*
stimulus in macaque study, 84*t*
Cyclic nucleotides
monitoring levels of cAMP and cGMP in response to bitter taste stimulation, 20–22
See also Bitter taste transduction
Cyclodextrins, bitter blockers, 106*t*, 107
Cyclo(Leu-Trp)
changes in cAMP level in intact circumvallate (CV) taste cells and intact non-sensory epithelial cells, 27, 29*f*
electrophysiological taste responses of glossopharyngeal nerve of rats, 22, 23*f*
estimated accumulation in CV taste cells, 9*t*
inhibition of permeation into rat CV taste-bud cells, 11, 13*f*
ionic charge, 8*t*
permeation through liposomal membranes, 6
taste transduction pathways by bitter tastants, 27*t*
time-course of cAMP in rat taste tissue, 26*f*
translocation into liposomes, 6, 7*f*
translocation into multilamellar vesicles (MLVs), 11, 12*f*
two-choice short-term preference tests, 22, 23*f*

D

Debittering, processes, 62
Delayed rectifying K^+ (DRK) channels
cloning, 184
fatty acids activating taste cells by inhibiting, 183–186
fatty acid transporter (FAT) in vallate taste cells, 186
implications of fatty acids altering activity of taste receptor cells, 186
inhibition of DRK channels in vallate taste receptor cell, 185*f*
See also Chemoreception of fat
Denatonium
average time intensity curves for bitterness, 4*f*
temporal properties, 5
Denatonium benzoate
cluster analyses, 69*f*

correlation coefficients of bitter ratings and rankings, 68*t*
subclass grouping, 72
Detection threshold, definition, 125
Dichloromethane, normalized temporal responses of three bead sensors, 326*f*
Differential threshold, definition, 126
Dioctyl phenylphosphonate (DOPP), plasticizer, 267
Dose-response curves of odor and taste stimuli
Affirm® technology for gas phase concentrations, 143–144, 155
air-liquid phase partition coefficients, 148*t*
analysis of variance (ANOVA), 149
calculation of dose-response curve parameters, 149–150
curves by sniffing and tasting solutions, 147–148
data treatment, 149–150
determination of parameters for dose-response curve, 150*f*
determining gas phase concentrations of chemicals, 143–144
experimental device to measure liquid-gas partition coefficients, 146*f*
experimental protocol, 145–148
experimental results and interpretation, 150–154
flavor ingredients for lemon flavor, 142
gaseous concentration in finalized olfactometric tests, 146*t*
gas phase-liquid phase partition coefficient in nose, 148*t*
general approach, 142–144
headspace concentration of lemon, 147*t*
intensity curves vs. concentrations, 149
mapping of olfactometric data, 152*f*
media containing sucrose and aspartame, 143
methods for panelists evaluating flavor intensity, 143
olfactometric and sniffing mapping, 153*f*
olfactometric and tasting mapping, 154*f*
orthonasal, curves using olfactometers, 145–147
orthonasal dose-response curves by olfactometry, 151*f*
orthonasal dose-response curves for lemon flavor by olfactometry, 151*f*
panel and sensory evaluation, 144
reproducibility, 151*f*
scaling procedure, 144
scheme of olfactometer, 145*f*
sniffing mapping, 153*f*
test concentrations for chemicals and lemon flavor tested, 147*t*
test sessions, 144
Drivers of liking
approach to developing sensory-liking curves, 216
change in, as physical stimulus magnitude changes, 215, 216*f*
chemo-sensory stimulus changing degree, 215
complex product exciting different senses, 219
determining sensory liking curve, 216
distribution of sensory segments for coffee as function of country, 219*t*
equation relating liking to attributes, 215
importance of analysis for consumer product development, 225
important sensory inputs, 220
individual differences, 217
link between model systems and real products, 223–224
quantifying magnitude, 222

reaction to chemo-sensory stimuli, 224
relation between bitter taste of coffee from expert panel and overall liking by consumer panel, 218*f*
relation between sensory level and liking, 216*f*
relative importance of attributes as, 223*t*
research protocol for pasta sauce, 220
sensory attributes driving overall liking for three sensory segments, 221*f*
sensory attributes driving overall liking for two user groups, 221*f*
sensory liking curves for full panel and usage subgroups, 221–222
sensory segmentation as organizing principle, 224–225
slope of relation between attribute liking and overall liking, 220*t*
using sensory-liking optimum to create operationally defined segments, 217–218
See also Pasta sauce; Preference mapping

Drugs
analgesic/anti-inflammatory, 98*t*, 99*t*
antihistamines, 99*t*
antimicrobial, 99*t*, 100*t*
cardiovascular, 97*t*, 98*t*
effect of topical application to lingual surface, 96
HIV, 101*t*, 102*t*
interference with taste system, 97–102
miscellaneous, 100*t*, 101*t*
psychotropic, 97*t*
quality control, 269
sedatives and antiemetics, 99*t*
See also Age-related chemosensory losses; Medications

E

Efficacy, association for second reaction, 35
Elderly persons
demographics of world's population, 94
taste and smell losses, 95
See also Age-related chemosensory losses
Electromyogram (EMG), facial, emotional reaction to odors, 247
Electronic olfaction
accomplishing produce recognition, 292
apples, 291–292, 294
artificial systems mimicking biological systems, 291
background odors and variability, 296
data clustering from three apple varieties, 294*f*
differentiation between closely related odor objects, 292
identifying odor objects by sniffing, 291
odor sensors for produce, 295–296
odor spectra for apples, 292, 293*f*
oranges, 291–292
pattern classification algorithms, 295
patterns of sensor responses for apples, 293*f*
test objects for complex odor sources, 291–292
See also Computational olfaction
Electronic tongue, fiber optic sensors, 327
Elliptical model, fitting consumers to, 235, 237*f*
Emotion in olfactory research
Aroma-Chology term, 244–245
beyond Mood Mapping®, 257
common sense, 250
Consumer Fragrance Thesaurus (CFT), 256*f*, 257

contingent negative variation (CNV), 246–247
directions for future research, 257
distinguishing pleasant from unpleasant odorants, 246
effects of nutmeg on blood pressure in stress conditions, 245–246
facial electromyogram (EMG), 247
factor analysis (FA), 248, 249–250
mood adjective checklist (MACL), 248–249
mood adjectives, 248
mood categories by MDS of 44 mood adjectives, 252*f*
Mood Mapping®, 253–257
Mood Mapping® of flavors, 255, 257
Mood Map® of fragrance materials, 255*f*
Mood Map® of Living® vs. picked mint, 254*f*
mood measurement of fragrance effects, 251–252
mood profiling, version 2, 253
multidimensional scaling (MDS), 250–251
perceptual and cognitive process of smelling, 244
physiological effects, 245–247
physiological self-report methods, 248–251
plot of 28 terms in two dimensional map, 251*f*
profile of mood states (POMS), 249
theoretical approaches to emotion, 247–248

Epicatechin
bitterness and astringency, 194–195
cluster analyses, 69*f*
correlation coefficients of bitter ratings and rankings, 68*t*
subclass grouping, 72

Epinephrine, levels from bitter taste, 103

Ermine moths, pheromone mixtures, 290

Ethyl acetoacetate, physiological measurement, 246

Ethyl maltol, bitter blockers, 106*t*, 107

External preference map. *See* Ice cream; Preference mapping

F

Facial electromyogram (EMG), emotional reaction to odors, 247

Factor analysis (FA)
advantage of multidimensional scaling (MDS), 250–251
application to mood changes, 249
problems with FA research, 249–250
relationships among mood terms, 248
See also Emotion in olfactory research

Family
preferences for individual sweet substances, 43
sweet taste sensitivity and preference, 44
See also Genetics

Fast pippeting system (FPS)
monitoring cellular events, 19
See also Bitter taste transduction

Fat. *See* Chemoreception of fat

Fatty acids
activating fat sensitive cell types, 188, 189*f*
activating taste cells via inhibition of delayed rectifying K^+ (DRK) channels, 183–186
correlation of responsiveness and dietary fat preference, 187
inhibition of DRK channels in vallate taste receptor cell, 185*f*
primary signaling molecules, 182–183
sensitivity of cloned DRK channels, 184

See also Chemoreception of fat
Flavonoid phenols
astringency, 193
average intensity of astringency of two concentrations of black tea, 197*f*
average intensity of astringency over time of red wine, 197*f*
average maximum intensity of astringency for successive sips of black tea vs. salivary flow rate, 198*f*
average time-intensity curves for astringency of skin and seed tannin fractions, 196*f*
bitterness, 193
effect of salivary flow rate, 196, 198
effect of sensory methodology, 195–196
effect of tannin composition, 194–195
maximum intensity of bitterness and astringency of monomers, dimers, and trimers, 194*f*
presence in beverages, 193
skin vs. seed tannins in wine industry, 195
Flavor industry
atmospheric pressure chemical ionization mass spectrometry, 160–161
emphasis changes, 159–160
improving use in food systems, 160
problems as issues of communication and cognitive science, 159
See also AFFIRM® technology
Flavor perception
measuring stimuli, 167–168
relation to volatile release in vivo, 169
release and, in viscous solutions, 171
release of flavor compounds from foods, 167
study of mechanisms and processes, 166–167
See also Taste release
Flavor release. *See* Taste release
Flavors
Mood Mapping®, 255, 257
See also Emotion in olfactory research
Fluorescence resonance energy transfer (FRET)t binding, 283
Food
models for mixture effects, 37
quality control, 263, 269
Food choices, *n*-6-propylthiouracil (PROP) tasting, 60, 62
Fragrance
mood measurement of effects, 251–252
mood profiling, 253–255
See also Emotion in olfactory research
Fragrances, distinguishing between stimulating vs. relaxing, 246–247
Fructose, stimulus in macaque study, 84*t*
Fructose/glucose, models for mixtures, 36–37
Fruit
electronic olfaction, 291–295
quality control, 269
See also Apples; Oranges
Fruit consumption, genetic taste markers, 62–63

G

Genetics
animal model studies of sweetness, 45–48
bitter taste polymorphism, 44
family and twin studies of preferences for individual sweet food items, 43
family and twin studies of sweet taste sensitivity and preference, 44
prospectus, 48

Sac locus and taste receptor 1 (T1R1) family of putative taste receptors, 47–48
studies of sweetness, 42–44
sweet food preferences, 44
sweet perception and preference in mice, 45–46
twin studies of preferences for individual sweet substances, 43*t*
understanding sweet perception, 41
See also Sweetness
Glucose
preferences of mice, 45–46
stimulus in macaque study, 84*t*
taste-responsive neurons of macaque, 82, 83*f*
Glucose/fructose, models for mixtures, 36–37
Glucosinolates, *n*-6-propylthiouracil (PROP) tasting, 60
G-protein-coupled receptors (GPCRs), signal transduction of tastants, 3
Grape jelly, peanut butter and. *See* AFFIRM® technology
Grape seeds and skins, bitterness and astringency, 194–195
Green tea, *n*-6-propylthiouracil (PROP) tasting, 59
Gustation
biological chemical sensing, 319
description, 277

H

Headspace analyses
flavor perception, 168
volatile flavors, 167*f*
See also Taste release
Healthful eating, barrier of taste, 54
Hedonics
aspect of taste and smell, 215
See also Drivers of liking
HIV drugs, interference with taste system, 101*t*, 102*t*
HO-α-sanshool (HOαS)
activation of trigeminal neurons, 206, 207*f*
effect on intraneuronal calcium, 209–210
nerve responses to cooling and warming before and after application to tongue, 206, 208*f*
sensation to tongue, 204
See also Trigeminal chemical sensitivity
Humans
animals as models, 32–33
individual variation in sweet taste perception and preference, 41–42
macaque sharing sweet taste sensitivity of, 79, 90
neural responses of macaque and relationship to human perception of sweetness, 89–90
psychophysical studies, 73–74
See also Genetics; Molecular mixture models; Neural representation of sweet taste in monkey
Human sensory tests, unsaturated alkyl amides, 210
Hydrocolloid concentration, relationship of, to sensory properties of solution, 173–174
Hydrogen chloride
stimulus in macaque study, 84*t*
taste-responsive neurons of macaque, 82, 83*f*
Hydroxymethylcellulose (HPMC)
analysis of solutions, 174
effect of concentration on sweetness perception, 175*f*

I

Ice cream
case study, 229
composition of samples, 229*t*

external preference map examples, 239*f*, 240*f*
external preference mapping, 233, 235, 241
fitting consumer to models, 235, 236*f*, 237*f*, 238*f*
internal preference clustering, 230, 233
internal preference mapping, 230, 231*f*, 232*f*
measuring sensory properties, 229
principle component analysis (PCA) of matrix of descriptive ratings, 234*f*
See also Preference mapping
Idiosyncratic patterns
impact of PROP sensitivity, 70–72
perceived bitter intensity, 67–70
Imaging fiber, schematic of bundle, 320*f*
Impedance measurement, taste sensing, 266
In-nose analysis
proximal signal, 168
sensory perception of strawberry flavor vs. maximum volatile concentration, 174, 175*f*
volatile flavors, 167*f*
See also Taste release
Internal preference map. *See* Ice cream; Preference mapping
Inverase
plasma norepinephrine levels, 105
taste stimulus testing neuroendocrine levels, 103–105
In vitro taste sensors
adsorption of quinine ions onto membrane surfaces, 266
applications for food and beverage quality control (QC), 269
astringent and pungent substances using multichannel taste sensor, 267
basic taste substances, 266
charged membranes, 266
commercial QC device, 263
differential measurement method for light addressable potentiometric (LAP) sensor, 268
dioctyl phenylphosphonate (DOPP) as plasticizer, 267
electrical selectivity and order of detection threshold of membrane, 267
future perspective, 270
impedance measurement based sensing, 266
integration of artificial lipid membranes on semiconductor surface, 268
materials of sensor composition, 263–264
membrane infiltration with dioleyl phosphate (DOPH) and dioleoyl phosphatidylethanolamine, 266
miniaturization of LAP system, 268
optical fiber sensors, 268
optical method using arachidic acid Langmuir–Blodgett (LB) doped film, 267–268
oscillation modes, 268–269
quality control, 263
role of transducer, 264
taste modality differentiation by principle component analysis (PCA) of electronic responses from lipid/polymer membrane, 265*f*
taste recognition and signal processing, 264
technology, 263–269
treating taste modalities with multivariate analysis and pattern recognition, 264
voltammetric techniques, 267
washing membranes, 266
Isoboles
glucose/fructose mixtures, 36*f*
implications of mixture models, 37
linear and non-linear, 34*f*

synergy and, 33–34
See also Molecular mixture models

J

Japanese green tea, *n*-6-propylthiouracil (PROP) tasting, 59
Jelly, peanut butter and grape. *See* AFFIRM® technology

K

Kale, *n*-6-propylthiouracil (PROP) tasting, 61*f*

L

Langmuir–Blodgett (LB) films. *See* In vitro taste sensors
Lavender, physiological measurement, 246
Law of mass action, models, 34–35
Lecithin, inhibiting bitterness, 10
Lemon flavor
headspace concentration, 147*t*
ingredients, 142
methods for panelists evaluating flavor intensity, 143
olfactometric and tasting mappings, 154*f*
orthonasal dose-response curves of, by olfactometry, 151*f*
sniffing mapping, 153*f*
test concentrations, 147*t*
See also Dose-response curves of odor and taste stimuli
Light addressing potentiometers (LAP)
differential measurement method, 268
miniaturization, 268
sensing technique, 264
Liking. *See* Drivers of liking
LIMAX, neural network description, 310–312
Limax maximus, odor cues, 291
Limonene
air-liquid phase partition coefficients, 148*t*
gaseous concentrations in finalized olfactometric tests, 146*t*
gas phase-liquid phase partition coefficient in nose, 148*t*
headspace concentration of lemon, 147*t*
mapping of olfactometric data, 152*f*
olfactometric and tasting mappings, 154*f*
orthonasal dose-response curves by olfactometry, 151*f*
sniffing mapping, 153*f*
structure, 142
test concentrations, 147*t*
(+)-Limonene. *See* Dose-response curves of odor and taste stimuli
Limonin
bitterness in citrus, 19
changes in cAMP level in intact circumvallate (CV) taste cells and intact non-sensory epithelial cells, 27, 29*f*
electrophysiological taste responses of glossopharyngeal nerve of rats, 22, 23*f*
reduction in cAMP, 30
taste transduction pathways by bitter tastants, 27*t*
time-course for IP_3 and cAMP production in rat taste tissue, 28*f*
two-choice short-term preference tests, 22, 23*f*
Linalool
air-liquid phase partition coefficients, 148*t*
gaseous concentrations in finalized olfactometric tests, 146*t*

gas phase-liquid phase partition coefficient in nose, 148*t*
headspace concentration of lemon, 147*t*
mapping of olfactometric data, 152*f*
olfactometric and tasting mappings, 154*f*
orthonasal dose-response curves by olfactometry, 151*f*
sniffing mapping, 153*f*
structure, 142
test concentrations, 147*t*
(+)-Linalool. *See* Dose-response curves of odor and taste stimuli
Lingual lipase, role in oral cavity, 182
Lipoprotein, bitter-taste inhibitor, 10
Liposomes, amphipathic bitter and sweet tastants, 5–6

M

Macaque
afferent limb of central taste system, 79, 80*f*, 81*f*, 82
electrophysiological experiments at each level of taste system, 82
primary taste cortex, 82
sweet taste sensitivity of humans and, 79, 90
See also Neural representation of sweet taste in monkey
Magnesium chloride, efficacy of bitter blockers on, 105–107
Magnesium sulfate
cluster analyses, 69*f*
correlation coefficients of bitter ratings and rankings, 68*t*
subclass grouping, 72
Magnitude estimation
scales, 129–130
tasks, 127, 129
Malate
calibration curves, 282*f*
concentrations in grape derived beverages, 282*t*
testing sensor, 281–282
See also Single analyte sensors
Maltol, bitter blockers, 106*t*, 107
Maltose, stimulus in macaque study, 84*t*
Mammalian senses, 276
Mapping. *See* Preference mapping
Mapping, preference. *See* Preference mapping
Market segmentation. *See* Preference mapping
Mass transport, effect of viscosity, 176
Maximum stimulus intensity (SImax), method, 169
Medications
chemosensory disorders, 95
impacting taste perception, 95–96
See also Age-related chemosensory losses; Drugs
Membranes. *See* In vitro taste sensors
Methanol, normalized temporal responses of three bead sensors, 326*f*
2-(4-Methoxyphenoxy)propionic acid, bitter blockers, 106*t*, 107
Mice
Sac locus, 46
sweet perception and preference, 45–46
Microbead sensors, fabrication and encoding, 323–324
Mint
Mood Map® of Living® vs. picked, 254*f*
mood profiling, 253–254
See also Emotion in olfactory research
Mint flavor
mean time intensity curve from panelists, 173*f*
perception, 171
perception in chewing gum, 171–172
Miso, quality control, 269

Mixtures
 exposure of senses, 33
 occurrence in environment, 33
 See also Molecular mixture models
Models
 fitting consumers to elliptical, 235, 237*f*
 fitting consumers to quadratic, 235, 238*f*
 fitting consumers to vectorial, 235, 236*f*
 linking volatile stimuli and odor perception, 168–169
 odorant modeling, 301–303
 See also Molecular mixture models
Molecular mixture models
 affinity, 35
 applications to food and beverage products, 37
 application to glucose/fructose mixtures, 36–37
 efficacy, 35
 implications, 37–38
 improvements of new models, 35
 isobole data and two model fits, 36*f*
 linear and non-linear isoboles, 34*f*
 models based on law of mass action, 34–35
 points of subjective equality, 33–34
 synergy and isoboles, 33–34
Molecular recognition
 biological olfactory systems, 290
 single analyte, 277–282
 See also Single analyte sensors
Monellin
 discrepancy, 90
 relationship among sweet stimuli, 88–89
 stimulus in macaque study, 84*t*
Monkey. *See* Neural representation of sweet taste in monkey
Monoammonium glycyrrhizinate, bitter blockers, 106*t*, 107
Monosodium glutamate, sensory and analytical thresholds, 173*t*
Mood adjective checklist (MACL), describing mood changes, 248–249
Mood Mapping®
 Consumer Fragrance Thesaurus (CFT), 256*f*, 257
 flavors, 255, 257
 fragrance materials, 255*f*
 Living(R) vs. picked mint, 254*f*
 mood profiling fragrances, 253–257
 See also Emotion in olfactory research
Mood profiling
 alternative technique, 253
 See also Emotion in olfactory research
Mouse. *See* Mice
Multichannel taste sensor
 astringent and pungent substances, 267
 See also In vitro taste sensors
Multi-component analyte sensors
 bar graphs showing color attenuation by charge-coupled device (CCD), 286*f*
 detection of Ca^{2+} at various pHs, 285–286
 mimicking sense of taste, 285
 rat fungiform papilla by scanning electron microscopy (SEM), 285*f*
 SEM image of micro-machined silicon well, 285*f*
 use of charge-coupled device (CCD), 285
Multidimensional scaling (MDS)
 advantages over factor analysis (FA), 250–251
 MDS solution for 28 affect words, 251*f*
 See also Emotion in olfactory research
Myoinositol

relationship among sweet stimuli, 88
stimulus in macaque study, 84*t*

N

Nanospray-mass spectroscopy (MS), non-sugar sweeteners, 172, 173*t*
Naringin
bitterness in citrus, 19
changes in cAMP level in intact CV taste cells and intact non-sensory epithelial cells, 27, 29*f*
electrophysiological taste responses of glossopharyngeal nerve of rats, 22, 23*f*
n-6-propylthiouracil (PROP) tasting, 59, 60
taste transduction pathways by bitter tastants, 27*t*
time-course for release of IP_3 and cAMP by bitter taste stimulation, 25*f*
two-choice short-term preference tests, 22, 23*f*
Neohesperidin dihydrochalcone
direct activators of G-proteins in vitro, 10
influence of *n*-6-propylthiouracil (PROP)-tasting, 53
relationship among sweet stimuli, 88
stimulus in macaque study, 84*t*
Neural representation of sweet taste in monkey
activity in macaque cortex, 79
afferent limb of central taste system in macaque, 80*f*, 81*f*
dendrogram showing degree of similarity among tastants plus water, 87*f*
electrophysiological experiments at each level of macaque taste system, 82
neural responses and relationship to human sweetness perception, 89–90
primary taste cortex, 82
recording activity of isolated cells, 85
relationship among sweet stimuli, 88–89
relationship between sweet and non-sweet stimuli, 85, 88
representation of taste quality, 85
response characteristics, 85
stimuli abbreviations, concentrations, molecular weights, and chemical groups, 84*t*
tracing of transverse section through left hemisphere at anteroposterior level of anterior clinoid process of sphenoid bone, 83*f*
two-dimensional space representing relative similarity among stimulus profiles, 86*f*
value of primate model for human taste, 79
Neural studies, unsaturated alkyl amides, 210
Non-sugar sweeteners
amphipathic, 3
chewing gum, 172
sampling analysis, 170–171
slow taste onset and lingering aftertaste, 5, 14
taste thresholds, 172, 173*t*
See also Taste release
Non-volatile compounds, interactions of volatile and, 170
Norepinephrine
elevation from bitter taste of drug, 103–105
levels in blood after single taste of stimuli, 104*f*
Nose, optical, application for fiber-optic sensors, 325–326

Nucleotides, cyclic. *See* Bitter taste transduction
Nutmeg, effect on blood pressure in stress, 245–246

O

Odor. *See* Dose-response curves of odor and taste stimuli
Odor identification tests
 reliability, 135
 suprathreshold procedures, 131, 133
 University of Pennsylvania Smell Identification Test (UPSIT), 131, 132*f*
Odor memory tests, suprathreshold procedures, 133, 134*f*
Odor receptors, cloning, 113–114
Odor sensors, organic transistors as, 296–299
Odor separation, computational olfaction, 300
Olfaction
 biological olfactory systems, 290
 complex mixtures, 290–291
 electronic, 291–296
 electronic olfactory systems, 290
 sense in animals, 119, 122
 See also Computational olfaction; Electronic olfaction
Olfactometers
 orthonasal dose-response curves using, 145–147
 schematic, 145*f*
 See also Dose-response curves of odor and taste stimuli
Olfactory discrimination
 genetic basis, 117–118
 receptive field of receptors, 118–119
Olfactory psychophysics
 ascending method of limits procedure (AML), 126–127
 definitions of threshold, 125–126
 detection or absolute threshold, 125
 differential threshold, 126
 forced-choice procedures, 125
 formal development of threshold methodology, 124–125
 modern odor detection threshold measurement procedures, 126–127
 odor identification tests, 131, 133
 odor memory tests, 133, 134*f*
 rating and magnitude estimation scales, 129–130
 recognition threshold, 125–126
 reliability of psychophysical olfactory tests, 133, 135–136
 scaling and magnitude estimation tasks, 127, 129
 single staircase procedure (SS), 126–127
 Smell Threshold Test™ (STT), 127, 128*f*
 suprathreshold procedures, 127, 129–133
 threshold procedures, 124–127
 University of Pennsylvania Smell Identification Text (UPSIT), 131, 132*f*
Olfactory receptor neurons
 adaptation, 116–117
 cloning of odor receptors, 113–114
 elements of second messenger transduction pathway, 115*f*
 expression of large family of odorant receptors, 113–114
 genetic basis of olfactory discrimination, 117–118
 location and morphology, 111–113
 molecular models of compounds with different levels of activity at I7 receptor, 121*f*
 odor discrimination, 117–119
 patterns of receptor expression, 114
 receptive field of olfactory receptors, 118–119
 responses of cells with I7 receptor to series of saturated and unsaturated aliphatic aldehydes, 120*f*

schematic of mammalian olfactory epithelium, 112*f*
second messenger pathway, 114, 116
signaling, 114–117
system in vertebrates, 119, 122
Olfactory receptors, receptive field, 118–119
Olfactory research. *See* Emotion in olfactory research
Optical bar codes, description, 323
Optical fibers
description, 319–320
schematic of imaging fiber bundle, 320*f*
See also In vitro taste sensors
Optical imaging
schematic of system, 321*f*
sensor arrays, 322
Optical nose, application for fiber-optic sensors, 325–326
Optical sensing arrays
typical optical sensor, 319
See also Cross-reactive optical sensing arrays
Optical tongue, cross-reactivity, 327
Oranges, electronic olfaction, 291–292
Organic transistors
odor sensors, 296–299
organic semiconductors and odor sensitivity, 297
possible mechanism for odor-induced changes, 299
responses of two materials to set of straight chain alcohols, 298*f*
typical geometry of device for odor stimulation, 297*f*
typical response to application of odor, 298*f*
Oscillation mode, taste sensing, 268–269

P

Papillae density, correlations among, and bitter taste, 74–75
Pasta sauce
important sensory inputs, 220
product exciting different senses, 219
quantifying magnitude of driver, 222, 223*t*
relative importance of attributes as drivers of liking, 223*t*
research protocol, 220
sensory liking curves, 221–222
slope of relation between attribute and overall liking, 220*t*
See also Drivers of liking
Peanut butter and grape jelly. *See* AFFIRM® technology
Perception
consumer vocabulary describing sensory, 228
exploring sensory, 224–225
individual variation in sweet taste, in people, 41–42
sweet, in mice, 45–46
See also Sweetness
Phenols. *See* Flavonoid phenols
Phenylalanine
cluster analyses, 69*f*
correlation coefficients of bitter ratings and rankings, 68*t*
subclass grouping, 72
Phenylthiocarbamide (PTC)
bitterness, 66
bitter or tasteless, 52–53
bitter taste polymorphism, 44
human psychophysical studies, 73–74
See also *n*-6-Propylthiouracil (PROP)
Pheromones, moths, 290
Phospholipids, mechanisms for inhibiting bitterness, 10–11
Physiology, effects of odorants, 245–247
Phytochemicals, bitter taste, 60–62

Polycose
relationship among sweet stimuli, 88
stimulus in macaque study, 84*t*
Polydextrose, bitter blockers, 106*t*, 107
Polymorphism, bitter taste, 44
Preference
family and twin studies, 44
family and twin studies of, for individual sweet foods, 43
individual variation in sweet taste, in people, 41–42
sweet, in mice, 45–46
sweet food, 44
See also Sweetness
Preference mapping
case study with ice cream, 229
composition of ice cream samples, 229*t*
example of consumer fitted by elliptical model, 237*f*
example of consumer fitted by quadratic model, 238*f*
example of consumer fitted by vectorial model, 236*f*
example of external map, 239*f*
example of external map showing response of satisfied consumers, 240*f*
external, 233, 235, 241
internal, 230
internal map matrix of hedonic ratings for ice cream samples showing consumers, 231*f*
internal map matrix of hedonic ratings for ice cream samples showing products, 232*f*
internal preference clustering, 230, 233
limitation of external, 241
measurement of sensory properties of ice creams, 229
principle component analysis (PCA) of descriptive analysis data for ice cream samples, 234*f*
product optimization, 228
relating consumer and sensory data, 228
techniques examining consumer preferences, 228–229
uses and attitude measures of consumers, 241
Principle component analysis (PCA). *See* Preference mapping
Produce
electronic olfaction, 291–295
odor sensors, 295–296
n-6-Propylthiouracil (PROP)
bitterness, 66
bitterness intensity and hedonic ratings for 5 PROP solutions by PROP taster status for 538 women, 58*f*
bitterness intensity scaling, 55, 57
bitter or tasteless, 52–53
bitter taste polymorphism, 44
cluster analyses, 69*f*
correlation coefficients of bitter ratings and rankings, 68*t*
debittering processes, 62
detection threshold procedure, 54–55
distribution of PROP taste detection thresholds for 538 female respondents, 56*f*
distribution of summed bitterness ratings for 5 PROP solutions for 538 female respondents, 56*f*
human psychophysical studies, 73–74
identification of PROP supertasters, 57
impact of PROP sensitivity on idiosyncratic patterns, 70–72
influencing perceived sweetness, 53
PROP status influence on bitter intensities and rankings, 71*f*
PROP-tasting and enhanced sensitivity to bitter taste, 53
PROP tasting and other bitter compounds, 59–60

relationship between measures of PROP responsiveness and self-reported preferences for selected foods, 59*t*
response to other bitter compounds, 54
scatterplot of mean hedonic ratings for PROP against mean preferences for cruciferous vegetables, 61*f*
sensory response to dietary fats, 53–54
subclass grouping, 72
summed hedonic ratings for PROP solutions as function of summed bitterness, 57*f*
tasters and nontasters, 54–57
See also Bitter taste
Psychology
self-report methods of mood, 248–251
See also Emotion in olfactory research
Psychophysical studies, human, 73–74
Psychophysics
categories, 124
description, 123
See also Olfactory psychophysics
Psychotropic drugs, interference with taste system, 97*t*
Pungency. *See* Trigeminal chemical sensitivity

Q

Quadratic model, fitting consumers to, 235, 238*f*
Quality control (QC)
applications in food and beverages, 269
commercial device, 263
food and beverage industry, 263
future perspective, 270
See also In vitro taste sensors
Quench flow module (QFM)
monitoring cellular events, 19
See also Bitter taste transduction
Quinine
adsorption onto membrane surfaces, 266
direct activators of G-proteins in vitro, 10
electrophysiological taste responses of glossopharyngeal nerve of rats, 22, 23*f*
estimated accumulation in circumvallate (CV) taste cells, 9*t*
inhibition of permeation into rat CV taste-bud cells, 11, 13*f*
ionic charge, 8*t*
oscillation mode, 268
permeation through liposomal membranes, 6
sensitivity, 66–67
sensory and analytical thresholds, 173*t*
subclass grouping, 72
translocation into liposomes, 6, 7*f*
translocation into multilamellar vesicles (MLVs), 11, 12*f*
Quinine hydrochloride
cluster analyses, 69*f*
correlation coefficients of bitter ratings and rankings, 68*t*
efficacy of bitter blockers on, 105–107
stimulus in macaque study, 84*t*
taste-responsive neurons of macaque, 82, 83*f*
Quinine sensitivity, human psychophysical studies, 74

R

Radishes, *n*-6-propylthiouracil (PROP) tasting, 61*f*
Rating scales, suprathreshold procedures, 129–130
Rats

animal experimental model to study bitter taste, 22
olfactory receptor types, 300, 309–310
See also Bitter taste transduction
Receptor expression, patterns, 114
Recognition threshold, definition, 125–126
Red wine. *See* Wine
Reliability, psychophysical olfactory tests, 133, 135–136
Rice, quality control, 269

S

Saccharin
control for two-choice short-term preference tests, 22, 23*f*
direct activators of G-proteins in vitro, 10
estimated accumulation in circumvallate (CV) taste cells, 9*t*
ionic charge, 8*t*
macaque neural response vs. human perception, 89
permeation through liposomal membranes, 6
relationship among sweet stimuli, 88
sensory and analytical thresholds, 173*t*
stimulus in macaque study, 84*t*
translocation into liposomes, 6, 7*f*
Sac locus
mouse, 46
possibly coding for sweet receptor in animals, 47–48
Sake, quality control, 269
Saltiness, oscillation mode, 268–269
Scaling, suprathreshold procedures, 127, 129
Second messenger pathway
elements of transduction, 115*f*
olfactory receptor neuron signaling, 114, 116
Sedatives, interference with taste system, 99*t*
Segmentation. *See* Preference mapping
Sensation, measuring stimuli, 167–168
Senses, mammalian, 276
Sensing arrays
atomic force micrographs, 322*f*
randomly ordered, 322–323
See also Cross-reactive optical sensing arrays
Sensitivity
family and twin studies of sweet taste, 44
organic semiconductors and odor, 297
See also Clustering bitter compounds; Trigeminal chemical sensitivity
Sensors. *See* In vitro taste sensors; Multi-component analyte sensors; Organic transistors; Single analyte sensors
Sensory cell. *See* Olfactory receptor neurons
Sensory drivers. *See* Drivers of liking
Sensory liking curves
approach to developing, 216
creating operationally defined segments, 217–218
determination, 216
pasta sauce, 221–222
typical, 216*f*
See also Drivers of liking
Sensory segmentation
exploring sensory perception, 224–225
organizing principle, 224–225
See also Drivers of liking
Signaling, olfactory receptor neurons, 114–117
Signal transduction. *See* Bitter taste transduction
Silk moth, pheromone, 290

Single analyte sensors
analysis of citrate concentration in beverages by NMR and competition assay, 280*t*
binding isotherms for Ser-Tyr-Ser with ATP, AMP, and GTP, 284*f*
binding studies using ^{1}H NMR and UV/vis spectroscopy, 279–280
citrate, 278–280
receptor selective for tartaric acid, 280–281
sensor employing competition assay using 5-carboxyfluorescein, 278
sequencing results for peptide chains, 284*t*
single analyte molecular recognition, 277–282
support bound single analyte sensing, 283–284
tartaric acid, 280–282
tartrate and malate concentrations in grape derived beverages by NMR and colorimetric assay, 282*t*
UV/vis and fluorescence calibration curves, 279*f*
UV/vis calibration curves for tartrate, malate and others for tartrate chemosensor, 282*f*
UV/vis spectra for alizarin complexone, 281*f*
UV/vis spectra of citrate binding to host and fluorophore 5-carboxyfluorescein, 279*f*
See also Multi-component analyte sensors
Single staircase procedure (SS), odor detection threshold measurement, 126–127
Slugs, terrestrial, odor cues, 291
Smell
chemical senses in foods, liquids, and air, 94–95
losses in elderly, 95
Smell Threshold Test™ (STT), commercial, 127, 128*f*
Sodium chloride
stimulus in macaque study, 84*t*
taste-responsive neurons of macaque, 82, 83*f*
Sodium saccharin. *See* Saccharin
Sodium salts, bitter blockers, 106*t*, 107
Solution sensor, fiber optic, 327–328
Sorbitol, stimulus in macaque study, 84*t*
Sorbose
relationship among sweet stimuli, 88
stimulus in macaque study, 84*t*
Sourness, oscillation mode, 268–269
Soybean seeds, quality control, 269
Soy sauce, quality control, 269
Spearman–Brown Prophesy Formula, reliability of odor identification tests, 135
Spinach, *n*-6-propylthiouracil (PROP) responsiveness and preferences, 59*t*
Stevens Law, relationship between stimulus and perception, 168
Stevioside
macaque neural response vs. human perception, 89
relationship among sweet stimuli, 88
stimulus in macaque study, 84*t*
Stimulus intensity, maximum (SImax), method, 169
Strawberry flavor, sensory perception vs. maximum volatile concentration in-nose, 175*f*
Structure-activity relationships, unsaturated alkyl amides, 210
Sucrose
average time intensity curves for sweetness, 4*f*
bitter blockers, 106*t*, 107
combinations with mint, 172, 173*f*
gelatin-sucrose gel systems, 173–174
influence of PROP-tasting, 53
orthonasal dose-response curves of lemon flavor by olfactometry, 151*f*
perception of mint flavor with and without, 171

preferences of mice, 45–46
stimulus in macaque study, 84*t*
structure, 143
temporal properties, 3
See also Dose-response curves of odor and taste stimuli
Sucrose octaacetate (SOA)
bitterness, 66–67
cluster analyses, 69*f*
correlation coefficients of bitter ratings and rankings, 68*t*
subclass grouping, 72
Support bound, single analyte sensing, 283–284
Suprathreshold procedures
odor identification tests, 131, 133
odor memory tests, 133, 134*f*
rating and magnitude estimation scales, 129–130
reliability, 135–136
scaling and magnitude estimation tasks, 127, 129
University of Pennsylvania Smell Identification Test (UPSIT), 131, 132*f*
See also Olfactory psychophysics
Sweetening agents
factors affecting volatile components of soft drink, 141
influence on flavor release and perception, 141–142
psychological effects, 141–142
quality control, 269
See also Dose-response curves of odor and taste stimuli; Non-sugar sweeteners
Sweetness
animal model studies, 45–48
bitter taste polymorphism, 44
effect of hydroxymethylcellulose concentration on, perception, 175*f*
family and twin studies of preferences for individual sweet food items, 43
family and twin studies of sweet taste sensitivity and preference, 44
genetic studies, 42–44
human work summary, 45
individual variation in perception and preference in people, 41–42
oscillation mode, 268–269
Sac locus and taste receptor 1 (T1R1) family of putative taste receptors, 47–48
summary and prospectus, 48
sweet food preferences, 44
sweet perception and preference in mice, 45–46
twin studies of preferences for individual sweet substances, 43*t*
See also Neural representation of sweet taste in monkey; *n*-6-Propylthiouracil (PROP)
Sweet tastants
amphipathic, interacting with liposomal membranes and translocation into liposomes, 5–6
amphipathic, permeating taste cells, 8–10
average time intensity curves for sweetness of aspartame and sucrose, 4*f*
signal transduction, 3
temporal properties, 3–5
See also Neural representation of sweet taste in monkey
Synergy
implications of mixture models, 37
isoboles and, 33–34
See also Molecular mixture models

T

Tannins
effect of composition on astringency and bitterness, 194–195
red wines, 195
See also Wine

Tartaric acid
calibration curves, 282*f*
ingredient in wines, 280
receptor selective for, 280–281
tartrate concentration in grape derived beverages by NMR and colorimetric assay, 282*t*
See also Single analyte sensors
Taste
barrier against healthful eating habits, 54
chemical senses in foods, liquids, and air, 94–95
description, 277
losses in elderly, 95
primary tastes, 276
See also Dose-response curves of odor and taste stimuli
Taste buds, concept, 277
Taste receptor 1 (T1R1), putative sweet taste receptor, 47–48
Taste receptor cells
concept, 277
See also Chemoreception of fat
Taste release
aim of current work, 170
effect of hydroxymethylcellulose (HPMC) concentration on sweetness perception, 175*f*
experimental, 170–171
flavor compounds from foods, 167
flavor release and perception in chewing gum, 171–172
flavor release and perception in viscous solutions, 171
gelatin-sucrose gel systems, 173–174
headspace analyses, 168
HPMC solutions, 174
in-nose sampling, 168
interactions of volatile and non-volatile compounds, 170
maximum stimulus intensity (SI) method, 169
mean time intensity curve from panelists given mint/sucrose solution, 173*f*
mint flavor perception, 172
models linking volatile stimuli and perception of odor, 168–169
non-sweeteners in gum, 172
perception of mint flavor with and without sucrose, 171
plot of sensory perception of strawberry flavor against maximum volatile concentration in-nose, 175*f*
relating volatile flavor perception to volatile release in vivo, 169
relationship between analyses for volatile flavors and distal and proximal perceptual stimuli, 167*f*
relationship between hydrocolloid concentration and sensory properties of hydrocolloid, sugar, and flavor solutions, 174*f*
sampling flavor compounds in vivo and their analysis, 170–171
sensory and analytical thresholds for non-sugar sweeteners, 173*t*
sucrose and mint combinations, 172, 173*f*
taste thresholds of non-sugar sweeteners, 172, 173*t*
time intensity (TI) analysis, 169
time release (TR) curves, 169
viscosity affecting mass transport, 176
Taste sensors. *See* In vitro taste sensors
Tea, quality control, 269
Terrestrial slugs, odor cues, 291
Tetralone®
cluster analyses, 69*f*
correlation coefficients of bitter ratings and rankings, 68*t*
subclass grouping, 72
Threshold
definitions, 125–126

sensory and analytical, of non-sugar sweeteners, 173*t*
Time intensity (TI), sensory analysis, 169
Time release (TR), aroma intensity over time, 169
Tingling. *See* Trigeminal chemical sensitivity
Toluene, normalized temporal responses of three bead sensors, 326*f*
Tongue. *See* Trigeminal chemical sensitivity
Transduction
molecular studies suggesting multiple bitter sequences, 73
See also Bitter taste transduction
Transistors. *See* Organic transistors
Trigeminal chemical sensitivity
activation of subpopulation of somatosensory neurons, 211
compounds producing tingling or buzzing sensation in mouth, 203
discharge rate of single acid/cool sensitive nerve fiber, 208*f*
effects of HO-α-sanshool (HOαS) on intraneuronal calcium, 209–210
human sensory tests, 210
naturally occurring unsaturated alkyl amides (UAAs) and synthetic analogs, 205*f*
nerve responses to cooling and warming before and after application of HOαS to tongue, 208*f*
neural responses to UAAs, 204, 206, 209
neural studies, 210
oral sensations of unsaturated fatty acid amides, 204
sensory mechanisms underlying sensation of UAAs, 211
spontaneous discharge rates of four single trigeminal fibers, 207*f*
structure-activity relationships, 210
structures of UAAs, 205*f*
trigeminal nerve fiber discharges, 207*f*
UAA compounds, 203–204
Triglycerides, digestion, 182
Tryptophan
cluster analyses, 69*f*
correlation coefficients of bitter ratings and rankings, 68*t*
macaque neural response vs. human perception, 89
relationship among sweet stimuli, 88
stimulus in macaque study, 84*t*
subclass grouping, 72
Twin studies
preferences for individual sweet substances, 43
sweet taste sensitivity and preference, 44
See also Genetics

U

Umami
oscillation mode, 268–269
signal transduction, 3
taste, 266
University of Pennsylvania Smell Identification Test (UPSIT), odor identification, 131, 132*f*
Unsaturated alkyl amides (UAAs)
activation of subpopulation of somatosensory neurons, 211
compounds, 203–204
human sensory tests, 210
neural responses, 204, 206, 209
neural studies, 210
sensory mechanisms underlying sensation, 211
structure-activity relationships, 210
structures, 205*f*
tingling or buzzing sensation in mouth, 203
See also Trigeminal chemical sensitivity